T0213012

Lecture Notes in Computer Science 10432

Commenced Publication in 1973
Founding and Former Series Editors:
Gerhard Goos, Juris Hartmanis, and Jan van Leeuwen

More information about this series at http://www.springer.com/series/7407

Srečko Brlek · Francesco Dolce
Christophe Reutenauer · Élise Vandomme (Eds.)

Combinatorics on Words

11th International Conference, WORDS 2017
Montréal, QC, Canada, September 11–15, 2017
Proceedings

 Springer

Editors
Srečko Brlek
Université du Québec à Montréal
Montreal, QC
Canada

Francesco Dolce
Université du Québec à Montréal
Montreal, QC
Canada

Christophe Reutenauer
Université du Québec à Montréal
Montreal, QC
Canada

Élise Vandomme
Université du Québec à Montréal
Montreal, QC
Canada

ISSN 0302-9743   ISSN 1611-3349   (electronic)
Lecture Notes in Computer Science
ISBN 978-3-319-66395-1   ISBN 978-3-319-66396-8   (eBook)
DOI 10.1007/978-3-319-66396-8

Library of Congress Control Number: 2017949524

LNCS Sublibrary: SL1 – Theoretical Computer Science and General Issues

Printed on acid-free paper

This Springer imprint is published by Springer Nature
The registered company is Springer International Publishing AG
The registered company address is: Gewerbestrasse 11, 6330 Cham, Switzerland

Preface

This volume of Lecture Notes in Computer Science contains the proceedings of the 11th International Conference WORDS 2017 which was organized by the Laboratoire de Combinatoire et d'Informatique Mathématique (LaCIM) and held during September 11–15, 2017, in Montreal, Canada. WORDS is the main conference series devoted to the mathematical theory of words. In particular, the combinatorial, algebraic and algorithmic aspects of words are emphasized. Input may also come from other domains such as theoretical computer science, bioinformatics, digital geometry, symbolic dynamics, numeration systems, text processing, number theory, etc.

The conference WORDS takes place every two years. The first conference of the series was held in Rouen, France in 1997. Since then, the locations of WORDS conferences have been: Rouen, France (1999), Palermo, Italy (2001), Turku, Finland (2003 and 2013), Montreal, Canada (2005), Marseille, France (2007), Salerno, Italy (2009), Prague, Czech Republic (2011), and Kiel, Germany (2015).

For the third time in the history of WORDS, a refereed proceedings volume has been published in the Lecture Notes in Computer Science series of Springer. There were 26 submissions, from 17 countries, and each of them was reviewed by at least two reviewers. The selection process was undertaken by the Program Committee with the help of generous reviewers. From these submissions, 21 papers were selected to be published and presented at WORDS. In addition to the contributed papers, the present volume also includes the abstracts of the lectures given by the five invited speakers:

- Štěpàn Holub (Charles University in Prague, Czech Republic): "Commutation and Beyond".
- Lila Kari (University of Waterloo, Canada): "DNA Words and Languages",
- Anna Frid (Aix-Marseille University, France): "An Unsolved Problem on Palindromes and Sturmian Words",
- Volker Diekert (University of Stuttgart, Germany): "Church-Rosser Systems, Codes with Bounded Synchronization Delay and Local Rees Extensions" (with Lukas Fleischer),
- David Clampitt (Ohio State University, USA): "The Role of Combinatorics on Words in Mathematical Music Theory",

We take this opportunity to warmly thank all the invited speakers and all the authors for their contributions. We are also grateful to all Program Committee members and the additional reviewers for their hard work that led to the selection of papers published in this volume. The reviewing process was facilitated by the EasyChair conference system, created by Andrej Voronkov. Special thanks are due to Alfred Hofmann and Elke Werner and the Lecture Notes in Computer Science team at Springer for having granted us the opportunity to publish this special issue devoted to WORDS 2017 and for their help during the final stage. We are also grateful to the Centre de Recherches Mathématiques (CRM) for its support in the organization of WORDS 2017. Finally, we are much obliged to a number of collaborators who contributed to the success of the conference: the secretary, Johanne Patoine, and our students, Herman Goulet-Ouellet, Nadia Lafrenière, Mélodie Lapointe and Émile Nadeau. Our warmest thanks for their assistance in the organization of the event.

July 2017

Srečko Brlek
Francesco Dolce
Christophe Reutenauer
Élise Vandomme

Organization

WORDS 2017 was hosted by the Laboratoire de Combinatoire et d'Informatique Mathématique (LaCIM) of the Université du Québec à Montréal, Canada.

Steering Committee

Valérie Berthé	Université Paris Diderot, France
Srečko Brlek	Université du Québec à Montréal, Canada
Julien Cassaigne	Aix-Marseille Université, France
Maxime Crochemore	King's College London, UK
Aldo de Luca	Università degli Studi di Napoli Federico II, Italy
Anna Frid	Aix-Marseille Université, France
Juhani Karumäki (Chair)	Turun yliopisto, Finland
Jean Néraud	Université de Rouen, France
Dirk Nowotka	Christian-Albrechts-Universität zu Kiel, Germany
Edita Pelantová	České vysoké učenì technické v Praze, Czech Republic
Dominique Perrin	Université Paris-Est Marne-la-Vallée, France
Antonio Restivo	Università degli Studi di Palermo, Italy
Christophe Reutenauer	Université du Québec à Montréal, Canada
Jeffrey Shallit	University of Waterloo, Canada
Mikhail Volkov	Ural Federal University, Russia

Program Committee

Elena Barcucci	Università degli Studi di Firenze, Italy
Valérie Berthé	Université Paris Diderot, France
Srečko Brlek (Chair)	Université du Québec à Montréal, Canada
Arturo Carpi	Università degli Studi di Perugia, Italy
Émilie Charlier	Université de Liège, Belgium
Sylvie Hamel	Université de Montréal, Canada
Juhani Karhumäki	Turun yliopisto, Finland
Xavier Provençal	Université Savoie Mont Blanc, France
Michaël Rao	Université de Lyon, France
Christophe Reutenauer (Chair)	Université du Québec à Montréal, Canada

Organizing Committee

Srečko Brlek (Chair)	Université du Québec à Montréal, Canada
Francesco Dolce (Co-chair)	Université du Québec à Montréal, Canada
Johanne Patoine (Secretary)	Université du Québec à Montréal, Canada
Élise Vandomme (Co-chair)	Université du Québec à Montréal, Canada

Additional Reviewers

Marilena Barnabei	Università di Bologna, Italy
Alexandre Blondin Massé	Université du Québec à Montréal, Canada
Alexander Burstein	Howard University, USA
Maxime Crochemore	King's College London, UK
Alessandro De Luca	Università degli Studi di Napoli Federico II, Italy
Sergi Elizalde	Dartmouth College, USA
Luca Ferrari	Università degli Studi di Firenze, Italy
Anna Frid	Aix-Marseille Université, France
Amy Glen	Murdoch University, Australia
Gábor Hetyei	University of North Carolina Charlotte, USA
Matthieu Josuat-Verges	Université Paris-Est Marne-la-Vallée, France
Jean-Philippe Labbé	Freie Universität Berlin, Germany
Sébastien Labbé	Université de Bordeaux, France
Marion Le Gonidec	Université de la Réunion, France
Julien Leroy	Université de Liège, Belgium
Florin Manea	Christian-Albrechts-Universität zu Kiel, Germany
Victor Marsault	Université de Liège, Belgium
Jay Pantone	Dartmouth College, USA
Edita Pelantová	České vysoké učenì technické v Praze, Czech Republic
Svetlana Puzynina	Sobolev Institute of Mathematics, Russia
Narad Rampersad	University of Winnipeg, Canada
Gwenaël Richomme	Université Paul-Valéry Montpellier 3, France
Michel Rigo	Université de Liège, Belgium
Aleksi Saarela	Turun yliopisto, Finland
Joe Sawada	University of Guelph, Canada
Patrice Séébold	Université Paul-Valéry Montpellier 3, France
Jeffrey Shallit	University of Waterloo, Canada
Wolgang Steiner	Université Paris Diderot, France
Manon Stipulanti	Université de Liège, Belgium
Élise Vandomme	Université du Québec à Montréal, Canada
Laurent Vuillon	Université Savoie Mont Blanc, France
Luca Zamboni	Université de Lyon, France

Invited Talks

Commutation and Beyond
(Extended Abstract)

Štěpán Holub

Department of Algebra, Charles University, Prague, Czech Republic
holub@karlin.mff.cuni.cz

Abstract. We survey some properties of simple relations between words.

Keywords: Periodicity forcing · Word equations

Supported by the Czech Science Foundation grant number 13-01832S.

DNA Words and Languages

Lila Kari

School of Computer Science, University of Waterloo, Canada
lila.kari@uwo.ca

The practical possibility of encoding symbolic information as "DNA words" and "DNA languages", and the fact that biochemical processes (Watson-Crick complementarity, cut-and-paste of DNA strands, etc.) can be used to perform arithmetic and logic operations, led to the development of the fields of theoretical and experimental DNA computing.

I will describe our work on defining and investigating new concepts in combinatorcs on words and formal language theory that capture the biological reality of DNA- and RNA-encoded information, as well as various bio-operations and their relationships to traditional models of information and computation. Besides its potential significance for the design of programmable DNA-based computational devices, the impact of this research is that it creates a mutually enriching link between theoretical computer science and molecular biology.

I will also describe our research into the mathematical properties of naturally-occurring bioinformation by exploring the connection between the syntactical structure of genomic sequences and species classification. In particular, I will present our ongoing investigation into Chaos Game Representations of DNA sequences as genomic signatures, and its applications to comparative genomics. The potential impact of such an alignment-free universal classification method could be significant, given that 86% of existing species on Earth and 91% of species in the oceans still await classification.

An Unsolved Problem on Palindromes and Sturmian Words

Anna Frid

Aix Marseille Univ, CNRS, Centrale Marseille, I2M, Marseille, France
anna.frid@univ-amu.fr

I will center the talk around the following unsolved problem: Is it true that for any aperiodic infinite word u and for any $K > 0$, there exists a factor of u which cannot be represented as a concatenation of at most K palindromes? In other terms, is the *palindromic length* of factors of any aperiodic infinite word unbounded?

We know that the answer is positive for all infinite words avoiding some power and in fact for a wider class including in particular all morphic words [2]. We also know that the answer is positive if we consider left or right *greedy* palindromic length instead of the usual one [1]. Note that a greedy palindromic length can be much greater than the usual one: for an easy example, we can remark that the usual palindromic length of *aaba* is two, *a aba*, whereas its left greedy palindromic length is three: *aa b a*.

The existing proof for words avoiding a power is not constructive, and in fact, in this talk I am going to discuss how a general constructive proof could look. It does not exist even for the case of Sturmian words. Strictly speaking, I do not know even if the answer to the problem is positive for a Sturmian word with unbounded elements of the continued fraction expansion of the slope!

So, I am going to discuss why standard techniques for Sturmian words do not work in this case (unless somebody finds a working technique before the conference!) and what is missing to develop a constructive proof of the conjecture that for any aperiodic infinite word u and for any $K > 0$, there exists a factor of u of palindromic length greater than K.

References

1. Bucci, M., Richomme, G.: Greedy palindromic lengths, arXiv preprint, https://arxiv.org/abs/1606.05660
2. Frid, A.E., Puzynina, S., Zamboni, L.Q.: On palindromic factorization of words. Adv. Appl. Math. **50**(5), 737–748 (2013)

Church-Rosser Systems, Codes with Bounded Synchronization Delay and Local Rees Extensions

Volker Diekert and Lukas Fleischer

FMI, University of Stuttgart, Stuttgart, Germany
{diekert,fleischer}@fmi.uni-stuttgart.de

In memoriam: Zoltàn Ésik (1951–2016)

Abstract. What is the common link, if there is any, between Church-Rosser systems, prefix codes with bounded synchronization delay, and local Rees extensions? The first obvious answer is that each of these notions relates to topics of interest for WORDS: Church-Rosser systems are certain rewriting systems over words, codes are given by sets of words which form a basis of a free submonoid in the free monoid of all words (over a given alphabet) and local Rees extensions provide structural insight into regular languages over words. So, it seems to be a legitimate title for an extended abstract presented at the conference WORDS 2017. However, this work is more ambitious, it outlines some less obvious but much more interesting link between these topics. This link is based on a structure theory of finite monoids with varieties of groups and the concept of *local divisors* playing a prominent role. Parts of this work appeared in a similar form in conference proceedings [6, 10] where proofs and further material can be found.

Lukas Fleischer—Supported by the German Research Foundation (DFG) under grant DI 435/6-1.

The Role of Combinatorics on Words in Mathematical Music Theory

David Clampitt

The Ohio State University, Columbus, Ohio, USA
clampitt.4@osu.edu

Abstract. Algebraic combinatorics on words plays an important role in mathematical music theory of the past three decades.

Keywords: Sturmian morphisms · Sturmian involution · Christoffel words · Standard words · Central words · Three distance theorem · Diatonic scales · Pentatonic scales · Modes · Guido of Arezzo · Zarlino

A subfield of mathematical music theory studies the properties of basic musical materials, such as diatonic scales (the familiar do re mi scale is an example). The musical octave, an interval between two pitches whose fundamental frequencies are in the ratio 2:1, is effectively a musical equivalence relation: humans learn to recognize pitches separated by octaves as functionally equivalent in most music (men and women singing the same tune will sing in octaves, for example), and in Western music we give the notes referring to such pitches the same name; music theorists speak of pitch classes. Diatonic scales, as periodic patterns modulo the octave, may be represented as circular words. Berstel and Perrin [1], in their essay "The origins of combinatorics on words," mention music as a field (along with astronomy) in which periodic discrete phenomena occur, giving rise to circular words or necklaces. We might represent the diatonic scale as the circular word $\langle TTTSTTS \rangle$, where T stands for the interval tone and S for the interval semitone. As a circular word it is equivalent to $\langle TTSTTTS \rangle$, a more familiar representation for musicians.

In 1989, Carey and Clampitt [3] defined well-formed scales, generalizing the diatonic scale. We departed from the fact that the diatonic scale admits a generating interval, in the sense that every note of a diatonic scale do re mi fa sol la ti is reachable by extending multiples of intervals of a perfect fifth (P5) from fa: fa $\rightarrow$ do $\rightarrow$ sol $\rightarrow$ re $\rightarrow$ la $\rightarrow$ mi $\rightarrow$ ti. "Reachable" here means that some representative of the respective diatonic pitch classes is met as we stack up perfect fifths. What is the definition of a perfect fifth? That depends upon musical and historical context: in equal temperament, to tune a piano, for example, we would theoretically use the frequency ratio $2^{\frac{7}{12}}$; but a more natural definition would be the frequency ratio 3/2. Since we are considering scales as periodic modulo the octave, and combining perfect fifths means taking powers of frequency ratios, we consider base-2 logarithms, and take multiples modulo 1 of rational 7/12 or irrational $log_2 \frac{3}{2}$. Mapping the diatonic pitch classes in generation order to the integers mod 7 by fa $\leftrightarrow$ 0, do $\leftrightarrow$ 1, sol $\leftrightarrow$ 2, re $\leftrightarrow$ 3, la $\leftrightarrow$ 4, mi $\leftrightarrow$ 5, ti $\leftrightarrow$ 6, we note that taking multiples of perfect fifths and reducing them modulo 1 we arrange the pitch class representatives in scale order, $\langle$fa sol la ti do re mi$\rangle$, and the mapping is linear (an automorphism of the cyclic group of order 7), via multiplication by 2 modulo 7: $0 \rightarrow 0$, $1 \rightarrow 2$, $2 \rightarrow 4$, $3 \rightarrow 6$, $4 \rightarrow 1$, $5 \rightarrow 3$, $6 \rightarrow 3$. Generalizing, we define generated scales where this mapping, from generation order to scale order, is linear modulo the cardinality of the scale, to be *well-formed scales*. We show that what characterizes the cardinalities of well-formed scales for a given generating interval size is that they are all and only those integers that are denominators of continued fraction convergents or semi-convergents to the generating interval log-frequency ratio. Thus, whether we take as generating interval $\frac{7}{12}$ or $log_2 \frac{3}{2}$, $\frac{4}{7}$ is a continued fraction approximation to the generating interval, and the 7-note scale generated by the perfect fifth is well-formed. Assuming $log_2 \frac{3}{2}$ as the generating interval yields an infinite hierarchy of well-formed scales, with cardinalities $2, 3, 5, 7, 12, 17, 29, 41, 53, \ldots$, including among them *tetraktys*, pentatonic, and chromatic scales, of cardinalities 3, 5, and 12. The characterizing theorem of course holds irrespective of the size of the generating interval [5]. This conception touches on the Three Distances Theorem, due to V. Sós [8] and others.

Cutting the circular word corresponding to the diatonic scale, the finite word *TTTSTTS* is a Christoffel word, while *TTSTTTS* is a standard word. Indeed, applying the special Sturmian morphisms $GG\tilde{D}$ and GGD of the monoid A^* for $A = \{a, b\}$ to the root word ab yields $aaabaab$ and $aabaaab$, respectively. The application of word theory in this framework brings about a treatment of the modes of the scales, represented by the conjugacy class of the Christoffel word [7]. Moreover, the standard factorization often considered, reflecting the respective images of a and b, has a musical meaning. Thus, the factorization of the Christoffel word $GG\tilde{D}(a, b) = aaab, aab$ is a representation of what is known in music theory as the authentic division of the Lydian mode of the diatonic scale, while the factorization of the standard word $GGD(a, b) = aaba, aab$ represents the authentic division of the Ionian (or major) mode. Similarly for the other morphisms in the conjugation class. The one member of the conjugacy class that is not the image of a special Sturmian morphism in the conjugation class of GGD is the bad conjugate, $baabaaa$, which is similarly rejected by the music theorists (Glarean in 1547 called this mode *hyperaeolius reiectus* [rejected]).

Standard words of length n have prefixes of length $n - 2$ which are central words. Guido of Arezzo in the eleventh century conferred privileged status on the hexachord ut re mi fa sol la, corresponding to the central word *aabaa*. Central words appear frequently in medieval music treatises, as discussed in [4] and in [6]. In particular, Guido presents the word *aabaaabaa*, but he emphasizes its dual under Sturmian involution, discussed below, *abaabaaba*.

There is a duality introduced by Sturmian involution, the anti-automorphism of the monoid of special Sturmian morphisms St_0 that sends f to f^* by fixing G and $\tilde{G}$ and exchanging D and $\tilde{D}$, reversing the order of the constituents [2]. This dual perspective also has a musical interpretation, in terms of what Clampitt and Noll [7] call the folding, a word formed by the pattern of falling perfect fifths and rising perfect fourths that reaches the pitch class representatives of the scale within the modal octave. For example, the word $ab, abababb$ may be taken to represent the folding for the C major scale, where the highest pitch B (ti) of the scale is followed by falling perfect 5th (a) to E (mi), a rising perfect 4th (b) to A (la), falling perfect 5th (a) to D (re), rising 4th (b) to

G (sol), falling 5th (*a*) to the lowest note of the scale C (do), rising 4ths (*b*) to F (fa) and to the excluded boundary note B-flat (just as the scale C D E F G A B (*C'*) needs an excluded boundary pitch *C'* to complete it). The standard morphism *GGD* that results in the scale pattern for the major scale, under Sturmian involution yields the Christoffel morphism *D̄GG*, which applied to the root word *ab* results in the folding for the major scale. The duality leads to music theoretical insights that would not be accessible without the mathematical framework.

Finally, we will consider the musical interpretation of results in the contributed paper to this conference by Clampitt and Noll, "Matching Lexicographic and Conjugation Orders of the Conjugation Class of a Special Sturmian Morphism." This is reflected in the reordering of the diatonic modes effected by Zarlino in 1571, which brought the modes into conformity with the conjugation order of the morphisms, which, as we show, is also the lexicographic order, properly defined, of the morphisms of the conjugation class.

References

1. Berstel, J., Perrin, D.: The origins of combinatorics on words. Eur. J. Comb. **28**(3), 996–1022 (2007)
2. Berthé, V., de Luca, A., Reutenauer, C.: On an involution of Christoffel words and sturmian morphisms. Eur. J. Comb. **29**(2), 535–553 (2008)
3. Carey, N., Clampitt, D.: Aspects of well-formed scales. Music Theory Spectr. **11**(2), 187–206 (1989)
4. Carey, N., Clampitt, D.: Regions: A theory of tonal spaces in early medieval treatises. J. Music Theory. **40**(1), 113–147 (1996)
5. Carey, N., Clampitt, D.: Two theorems concerning rational approximations. J. Math. Music **6** (1), 61–66 (2012)
6. Clampitt, D., Noll, T.: Regions and standard modes. In: Chew, E., Childs, A., Chuan, C.H. (eds.) MCM 2009. CCIS, vol. 38, pp. 81–92. Springer, Heidelberg (2009)
7. Clampitt, D., Noll, T.: Modes, the height-width duality, and handschin's tone character. Music Theory Online **17**(1) (2011)
8. Sós, V.T.: On the distribution mod 1 of the sequence nα. Ann. Univ. Sci. Budap. Rolando Eötvös, Sect. Math. **1**, 127–134 (1958)

Contents

Commutation and Beyond
Extended Abstract

Štěpán Holub[⊠]

Department of Algebra, Charles University, Prague, Czech Republic
holub@karlin.mff.cuni.cz

Abstract. We survey some properties of simple relations between words.

Keywords: Periodicity forcing · Word equations

1 Commutation Forcing

Arguably, the core of the combinatorics on words folklore is the fact that two words commute if and only if they are powers of the same word. In fact, the claim that x and y commute would be written, by a combinatorist on words, most probably as $x, y \in t^*$ (for some t), instead of $xy = yx$.

This fact has a fairly strong and well known generalization: any nontrivial relation of two words forces commutation. This follows easily from the following lemma.

Lemma 1. *Let x and y do not commute and let z be the longest common prefix of xy and yx. Then z is the longest common prefix of any pair of words $u \in x\{x,y\}^*$ and $v \in y\{x,y\}^*$ that are both at least as long as z.*

Another folklore property of words is that every word has a unique *primitive root*: for each word w there is a unique word r such that w is a power of r, and r is a power of itself only. (Therefore, commutation is equivalent to having the same primitive root.) This uniqueness can be also expressed by saying that if

$$s^n = t^m \, ,$$

then both s and t are powers of some r. Also this fact has several generalizations. The first one is the theorem of Lyndon and Schützenberger [20] (which happens to hold in free groups as well):

Theorem 2. *If $x^n y^m = z^p$ with $n, m, p \geq 2$, then all words x, y and z commute.*

A clever short proof of this classical result was given by Harju and Nowotka [14].

This theorem naturally raises a question: for which words $w \in \{x, y\}^+$ the equality $w = z^p$, $p > 2$ implies that x and y commute? The answer is formulated in the following result.

Š. Holub—Supported by the Czech Science Foundation grant number 13-01832S.

S. Brlek et al. (Eds.): WORDS 2017, LNCS 10432, pp. 1–5, 2017.
DOI: 10.1007/978-3-319-66396-8_1

Theorem 3. *Suppose that* $x, y \in A^*$ *do not commute and let* $X = \{x, y\}$*. Let* $\mathcal{C}$ *be the set of all* X*-primitive words from* $X^* \backslash X$ *that are not primitive. Then either* $\mathcal{C}$ *is empty or there is* $k \geq 1$ *such that*

$$\mathcal{C} = \{x^i y x^{k-i} \mid 0 \leq i \leq k\} \quad or \quad \mathcal{C} = \{y^i x y^{k-i} \mid 0 \leq i \leq k\}.$$

A word is X-primitive if it is primitive when understood as a word over the alphabet X.

A weaker version of the theorem was obtained in a paper by Lentin and Schützenberger [19]. The full claim was proved in a paper by Barin-Le Rest and Le Rest [2], and in the dissertation of Spehner [22]. Although very natural and important, Theorem 3 seems to be not very well known.

Another generalization of Theorem 2 was given by Appel and Djorup [1] who proved that words satisfying

$$x_1^k x_2^k \cdots x_n^k = y^k \tag{1}$$

commute if $n \leq k$. This was further generalized by Harju and Nowotka [15]. For $n > k$, a natural question is whether the equality (1) can be simultaneously satisfied by non-commuting words for several different exponents k. It turns out that this is possible for two different exponents but not for three. For exponents $1 < k_1 < k_2 < k_3$, this was proved by Holub [16]. For $1 = k_1 < k_2 < k_3$, there was only a limited knowledge [12], until the recent complete and elegant proof by Saarela [21].

2   Periodicity Forcing

The equality (1) is related to the equality

$$x_1^k x_2^k \cdots x_n^k = y_1^k y_2^k \cdots y_n^k.$$

Due to the symmetry, this equality is better seen not as a relation between words but rather as a property of two morphisms $g : a_i \mapsto x_i$ and $h : a_i \mapsto y_i$, *viz.* that they agree on the word $a_1^k a_2^k \cdots a_n^k$. This idea leads to a concept of periodicity forcing words and/or languages. A set of words L is said to be *periodicity forcing* if the equality $g(w) = h(w)$ for all $w \in L$ implies (with $g \neq h$) that both morphisms g and h are periodic, that is, all their images commute. For two morphisms g and h, $g \neq h$, we can define their *equality set* as

$$\mathsf{Eq}(g, h) = \{w \mid w \text{ nonempty}, g(w) = h(w)\}.$$

Elements of equality sets of non-periodic morphisms are called *equality words*, which is a complementary property to being periodicity forcing. The question whether $\mathsf{Eq}(g, h)$ is nonempty, for given morphisms g and h, is known as the (in general undecidable) *Post Correspondence Problem*. That is why the question whether a given word or language is periodicity forcing is also known as the *Dual Post Correspondence Problem*. The problem was explicitly or implicitly studied in many particular cases, for example in relation with test sets [18]. Recently, the question was investigated also on a general level [5,6].

The most simple nontrivial version of the problem is the question on equality words over the binary alphabet. Although the binary Post Correspondence Problem is decidable in polynomial time [13], classification of binary equality words turns out to be surprisingly difficult. For instance, it is presently not known whether *abbaaaa* is an equality word or not. Binary equality languages were first studied by Čulík II and Karhumäki [3] almost forty years ago. Soon it was shown that binary equality languages are generated either by one word, or by two words, or are of a special form $\alpha\gamma^*\beta$ [7]. The latter possibility, however, was conjectured to be impossible. The conjecture was confirmed in 2003 [16]. Moreover, it was shown that the only possible form of a two-generated binary equality language is (up to the symmetry of letters) $\{a^i b, ba^i\}$ [17].

There is a series of papers studying words that can be generators of binary equality sets [4,8,10,11]. The current state of the art is captured in the Ph.D. thesis of Jana Hadravová [9] by the Fig. 1 below. Coordinates represent the number of a's and b's in the potential equality word. Black dots indicate that there is a known equality set generator with this numbers of letters, and the grey area delimits cases where all equality words are known. The asymmetry is

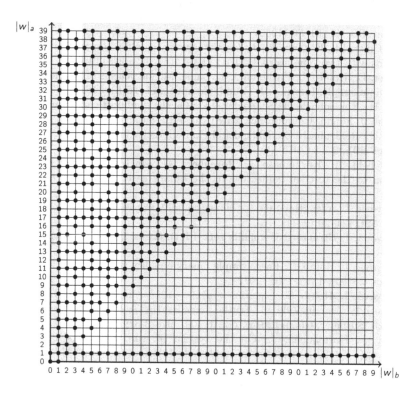

Fig. 1. Binary equality words

given by the assumption that $h(b)$ is not shorter than the other three images $g(a)$, $g(b)$ and $h(a)$.

Example 4. We give an example illustrating how to read the figure. No generator with ten b's and twelve a's is recorded in the figure. The corresponding empty place is within the gray area which means that we can prove that no such generator exists.

On the other hand, there is a dot for five b's and six a's. In fact, we know three such generators: a^6b^5, b^5a^6 and $(ab)^5a$. The coordinate $(5,6)$ is not in the gray area, which means that we are not presently sure that there is no other word (although we believe so).

Note that also $(b^5a^6)^2$ is a binary equality word, which may seem to be at odds with our first example. But that word is not a generator of an equality language since any two morphisms agreeing on $(b^5a^6)^2$ agree also on b^5a^6 as one can easily see.

References

1. Appel, K.I., Djorup, F.M.: On the equation $z_1^n z_2^n z_k^n = y_n$ in a free semigroup. Trans. Am. Math. Soc. **134**(3), 461–470 (1968)
2. Barbin-Le Rest, E., Le Rest, M.: Sur la combinatoire des codes à deux mots. Theor. Comput. Sci. **41**, 61–80 (1985)
3. Karel Culik, I.I., Karhumäki, J.: On the equality sets for homomorphisms on free monoids with two generators. RAIRO ITA **14**(4), 349–369 (1980)
4. Czeizler, E., Holub, Š., Karhumäki, J., Laine, M.: Intricacies of simple word equations: an example. Internat. J. Found. Comput. Sci. **18**(6), 1167–1175 (2007)
5. Day, J.D., Reidenbach, D., Schneider, J.C.: On the dual post correspondence problem. Int. J. Found. Comput. Sci. **25**(8), 1033–1048 (2014)
6. Day, J.D., Reidenbach, D., Schneider, J.C.: Periodicity forcing words. Theor. Comput. Sci. **601**, 2–14 (2015)
7. Ehrenfeucht, A., Karhumäki, J., Rozenberg, G.: On binary equality sets and a solution to the test set conjecture in the binary case. J. Algebra **85**(1), 76–85 (1983)
8. Hadravová, J.: A length bound for binary equality words. Commentat. Math. Univ. Carol. **52**(1), 1–20 (2011)
9. Hadravová, J.: Structure of equality sets. Ph.D. thesis, Charles University, Prague (2015)
10. Hadravová, J., Holub, Š.: Large simple binary equality words. Int. J. Found. Comput. Sci. **23**(06), 1385–1403 (2012)
11. Hadravová, J., Holub, Š.: Equation $x^i y^j x^k = u^i v^j u^k$ in Words. In: Dediu, A.-H., Formenti, E., Martín-Vide, C., Truthe, B. (eds.) LATA 2015. LNCS, vol. 8977, pp. 414–423. Springer, Cham (2015). doi:10.1007/978-3-319-15579-1_32
12. Hakala, I., Kortelainen, J.: On the system of word equations $x_1^i x_2^i \ldots x_m^i = y_1^i y_2^i \ldots y_n^i (i = 1, 2, \ldots)$ in a free monoid. Acta Informatica **34**(3), 217–230 (1997)
13. Halava, V., Holub, Š.: Binary (generalized) post correspondence problem is in P. Technical report 785, Turku Centre for Computer Science (2006)
14. Harju, T., Nowotka, D.: The equation $x^i = y^j z^k$ in a free semigroup. Semigroup Forum **68**(3), 488–490 (2004)

15. Harju, T., Nowotka, D.: On the equation $x^k = z_1^{k_1} z_2^{k_2} \cdots z_n^{k_n}$ in a free semigroup. Theor. Comput. Sci. **330**(1), 117–121 (2005)

16. Holub, Š.: Local and global cyclicity in free semigroups. Theor. Comput. Sci. **262**(1), 25–36 (2001)

17. Holub, Š.: A unique structure of two-generated binary equality sets. In: Ito, M., Toyama, M. (eds.) DLT 2002. LNCS, vol. 2450, pp. 245–257. Springer, Heidelberg (2003). doi:10.1007/3-540-45005-X_21

18. Holub, Š., Kortelainen, J.: Linear size test sets for certain commutative languages. Theor. Inform. Appl. **35**(5), 453–475 (2001)

19. Lentin, A., Schützenberger, M.-P.: A combinatorial problem in the theory of free monoids. In: Algebraic Theory of Semigroups. North-Holland Pub. Co., Amsterdam, New York (1979)

20. Lyndon, R.C., Schützenberger, M.-P.: The equation $a^m = b^n c^p$ in a free group. Michigan Math. J. **9**(4), 289–298 (1962)

21. Saarela, A.: Word equations where a power equals a product of powers. In: STACS: Leibniz International Proceedings in Informatics, vol. 66, pp. 55:1–55:9. Dagstuhl, Germany (2017)

22. Spehner, J.-P.: Quelques problèmes d'extension, de conjugaison et de presentation des sous-monoïdes d'un monoïde libre. PhD thesis, Université Paris VII, Paris (1976)

Church-Rosser Systems, Codes with Bounded Synchronization Delay and Local Rees Extensions

Volker Diekert and Lukas Fleischer[✉]

FMI, University of Stuttgart, Stuttgart, Germany
{diekert,fleischer}@fmi.uni-stuttgart.de

In memoriam: Zoltàn Ésik (1951–2016)

Abstract. What is the common link, if there is any, between Church-Rosser systems, prefix codes with bounded synchronization delay, and local Rees extensions? The first obvious answer is that each of these notions relates to topics of interest for WORDS: Church-Rosser systems are certain rewriting systems over words, codes are given by sets of words which form a basis of a free submonoid in the free monoid of all words (over a given alphabet) and local Rees extensions provide structural insight into regular languages over words. So, it seems to be a legitimate title for an extended abstract presented at the conference WORDS 2017. However, this work is more ambitious, it outlines some less obvious but much more interesting link between these topics. This link is based on a structure theory of finite monoids with varieties of groups and the concept of *local divisors* playing a prominent role. Parts of this work appeared in a similar form in conference proceedings [6,10] where proofs and further material can be found.

1 Introduction

Ceci n'est pas une introduction.[1] The present paper does not claim to provide any new results. Its purpose is to give an overview on a theory developed over the past twenty years, having its origins in a construction derived from the *Habilitationsschrift* of Thomas Wilke [27] which the first author was refereeing in 1997. Inspired by this construction (which also appears in [27]), he distilled the concept of a local divisor of a finite monoid without knowing that this concept existed long before in commutative algebra [14] (denoted by Kurt Meyberg as *local algebra*) and without giving any special name to it. The term *local divisor* was coined 2012 in [9] only.[2]

L. Fleischer—Supported by the German Research Foundation (DFG) under grant DI 435/6-1.
[1] Following "La trahison des images" by René Magritte.
[2] The pointer to [14] is due to Benjamin Steinberg and that a "local divisor" is a monoid divisor in the usual sense was observed by Daniel Kirsten, first. Thanks!

© Springer International Publishing AG 2017
S. Brlek et al. (Eds.): WORDS 2017, LNCS 10432, pp. 6–16, 2017.
DOI: 10.1007/978-3-319-66396-8_2

Originally, the concept was solely used as a tool to simplify existing proofs. Still, this was particularly helpful in [3] which introduced this proof technique to the semigroup community. However, over the last decade, it gave rise to new results. Amazingly, it was powerful enough to solve long-standing open problems. It is hard to formally pinpoint where the power of method comes from or why, on the other hand, it has clear limitations. Let us conclude with an *étale* statement: There are not enough local submonoids, so the role of local submonoids transfers to local divisors and there are plenty of them. This seems to be useful.

2 Preliminaries

Throughout the paper A denotes a finite alphabet and M, N denote monoids. If not stated otherwise, M and N will be finite. A *divisor* of a monoid M is a monoid N which is a homomorphic image of a subsemigroup of M. A *variety* of finite monoids is a nonempty family of finite monoids $\mathbf{V}$ which is closed under taking divisors and finite direct products. A variety of finite groups is a variety of finite monoids where each of the monoids is a group.

The largest group variety is $\mathbf{G}$, the variety of all finite groups. If $\mathbf{H}$ is a variety of finite groups, $\overline{\mathbf{H}}$ denotes the class of finite monoids where all subgroups are members of $\mathbf{H}$. It turns out that for every group variety $\mathbf{H}$, the class $\overline{\mathbf{H}}$ is a variety, see [11]. Actually, it is the greatest variety of finite monoids such that $\overline{\mathbf{H}} \cap \mathbf{G} = \mathbf{H}$. Clearly, $\overline{\mathbf{G}}$ is the class of all finite monoids which we denote by **Mon**. The most prominent subclass is $\overline{\mathbf{1}}$, the variety of aperiodic monoids **Ap**. Here, $\mathbf{1}$ denotes the smallest group variety, containing the trivial group $\{1\}$ only.

Given a variety $\mathbf{V}$, we denote by $\mathbf{V}(A^*)$ the set of languages $L \subseteq A^*$ such that $L = \varphi^{-1}(\varphi(L))$ for some homomorphism $\varphi : A^* \to M$ where $M \in \mathbf{V}$. From formal language theory, we know that $\mathbf{Mon}(A^*)$ is the set of all regular languages in A^*.

3 Church-Rosser Thue Systems

A *semi-Thue system* is a set of rewriting rules $S \subseteq A^* \times A^*$ over some alphabet A. (For simplicity, throughout this paper, semi-Thue systems are assumed to be finite.) A system S defines a finitely presented quotient monoid

$$A^*/S = A^*/\{\ell = r \mid (\ell, r) \in S\},$$

and the system is called *Church-Rosser* (with respect to the length function) if S is confluent and length-reducing. The interest in Church-Rosser systems stems from the fact that we can compute irreducible normal forms in linear time (as the system is finite and length-reducing) and that the irreducible normal forms of two words u, v are identical if and only if u and v represent the same word in A^*/S (as the system is confluent). Thus, if a monoid M has a presentation as $M = A^*/S$, then the word problem of M is solvable in linear time. The notion of a *Church-Rosser language* is an offspring of that observation and appeared first in Narendran's PhD thesis [15], followed by a systematic study of that concept in [13]. As a result, [13] defines a language class strictly larger than the class of deterministic context-free languages for which the word problem is solvable in linear time. The authors of this work also define a restricted class which is incomparable with the class of (deterministic) context-free languages.

A language $L \subseteq A^*$ is called *Church-Rosser congruential*, if there exists a finite, confluent, and length-reducing semi-Thue system $S \subseteq A^* \times A^*$ such that L is a finite union of congruence classes modulo S. If, in addition, the index of S is finite (i.e., the monoid A^*/S of all congruence classes is finite) then L is called *strongly* Church-Rosser congruential. Strongly Church-Rosser congruential languages are necessarily regular. It was conjectured (but open for more than 25 years until 2012) that all regular languages are (strongly) Church-Rosser congruential. Some partial results were known before 2012 but commutativity in the syntactic monoid seemed to be a major obstacle. For example, it is easy to verify the conjecture provided the syntactic monoid is a finite non-Abelian simple group like $\mathcal{A}_5$. On the other hand, it is surprisingly hard to prove the result for the Klein group $\mathbb{Z}/2\mathbb{Z} \times \mathbb{Z}/2\mathbb{Z}$. Nevertheless, [7] proved a stronger result. Given a regular language $L \subseteq A^*$ and any weight function $\gamma : A \to \mathbb{N}\backslash\{0\}$, there exists a finite confluent and weight-reducing semi-Thue system S such the quotient monoid is A^*/S is finite and such that L is a (necessarily finite) union of congruence classes. This result is indeed stronger because the mapping $w \mapsto |w|$ is just one particular weight function.

4   Star-Freeness and Bounded Synchronization Delay

The class of *star-free languages* over some alphabet A, denoted by $\mathrm{SF}(A^*)$, is the least class of languages which contains all finite languages over A and which is closed under both Boolean operations (finite union and complementation) and concatenation. As the name suggests, we do not allow the Kleene star. Nevertheless, B^* is star-free for all $B \subseteq A$. A fundamental result of Schützenberger characterizes the class of star-free languages by aperiodic monoids [21]. That is, a regular language belongs to $\mathrm{SF}(A^*)$ if and only if all subgroups in its syntactic monoid are trivial. By slight abuse of notation, one usually abbreviates this result by $\mathrm{SF} = \mathbf{Ap}$ as a short version of $\mathrm{SF}(A^*) = \mathbf{Ap}(A^*)$. Schützenberger found another, but less prominent characterization of SF: the star-free languages are exactly the class of languages which can be defined inductively by finite languages and closure under finite union, concatenation, and the Kleene star restricted to prefix codes of bounded synchronization delay [23]. This result is abbreviated by $\mathrm{SD} = \mathbf{Ap}$.

A language $K \subseteq A^+$ is called *prefix code* if it is *prefix-free*, i.e., $u \in K$ and $uv \in K$ implies $u = uv$. A prefix-free language K is a code since every word $u \in K^*$ admits a unique factorization $u = u_1 \cdots u_k$ with $k \geq 0$ and $u_i \in K$. A prefix code K has *bounded synchronization delay* if for some $d \in \mathbb{N}$ and for all $u, v, w \in A^*$ with $uvw \in K^*$ and $v \in K^d$, we have $uv \in K^*$. Note that the condition implies that for all $uvw \in K^*$ with $v \in K^d$, we have $w \in K^*$, too.

The idea is as follows: assume that a transmission of a code message is interrupted and we receive a fragment of the form $u''vw$ where $v \in K^d$ and $w \in K^*$. Then, we know that the original message was of the form $u'u''vw$ with $u'u''v \in K^*$ and $w \in K^*$. Hence, we can decode w as part of the original message. With a delay of d code words the decoding can be synchronized. For $B \subseteq A$ and $c \in A \setminus B$, the star-free language B^*c is a prefix code of delay 1 and $(B^*c)^+ = (B \cup \{c\})^*c$ is star-free. The block code A^2 is finite, but not of bounded synchronization delay. Moreover, $(A^2)^*$ is not star-free as its syntactic monoid is the cyclic group of order two.

Schützenberger's result $\mathbf{Ap} \subseteq \mathrm{SD}$ is actually stronger than the well-known $\mathrm{SF} = \mathbf{Ap}$ because proving the inclusions $\mathrm{SD} \subseteq \mathrm{SF} \subseteq \mathbf{Ap}$ is relatively easy, see [17, Chapter VIII], so $\mathrm{SF} = \mathbf{Ap}$ follows from $\mathbf{Ap} \subseteq \mathrm{SD}$. A simple proof for $\mathbf{Ap} = \mathrm{SD}$ including an extension to infinite words (which was not known before) was obtained much later in [5]. It could be achieved thanks to the same algebraic decomposition into submonoids and local divisors.

5 Local Divisors

In this section $e \in M$ denotes an idempotent, that is $e^2 = e$. For such an idempotent, the set $M_e = eMe$ forms a monoid with e as the identity element. It is called the *local monoid* at e. A *local divisor* generalizes this concept by considering any element $c \in M$ and the set $M_c = cM \cap Mc$. Note that $eMe = eM \cap Me$, so local monoids are indeed a special case of local divisors. The next step is to define a multiplication $\circ$ on $cM \cap Mc$ by letting

$$xc \circ cy = xcy$$

for all $x, y \in M$. A straightforward calculation shows that the structure $(M_c, \circ, c)$ defines a monoid with this operation where the neutral element of M_c is c. This works for every $c \in M$. If c is a unit, then M_c is isomorphic to M. If c is idempotent, then M_c is the local monoid at the idempotent c. However, in general M_c does not appear as a subsemigroup in M.

The important fact is that M_c is always a divisor of M. Indeed, the mapping $\lambda_c : \{x \in M \mid cx \in Mc\} \to M_c$ given by $\lambda_c(x) = cx$ is a surjective homomorphism. Moreover, if c is not a unit, then $1 \notin cM \cap Mc$, hence $M_c \subsetneq M$. This makes the construction suitable for induction.

6 Rees Extensions

Let N, L be monoids and $\rho : N \to L$ be any mapping. The *Rees extension* over N, L, ρ is a classical construction for monoids [18,19], frequently described in terms of matrices. It was used in the synthesis theory of Rhodes and Allen [20] which says that we can represent every finite monoid as a divisor of iterated Rees extensions, starting with groups. The "advantage" is that starting with a variety of groups $\mathbf{H}$, the construction produces monoids in $\overline{\mathbf{H}}$, only. This is not true for taking wreath products which are used in Krohn-Rhodes theory. For example, the symmetric group over three elements is not nilpotent, but appears as a subgroup of the wreath product of $\mathbb{Z}/3\mathbb{Z}$ and $\mathbb{Z}/2\mathbb{Z}$. Our definition of a Rees extension is similar, but not the same as the classical one. It avoids matrices and it is exactly as in [16]. The carrier set is

$$\mathrm{Rees}(N, L, \rho) = N \cup (N \times L \times N).$$

Let $n_1, n_1', n_2, n_2' \in N$ and $m, m' \in L$. Then the multiplication $\cdot$ on $\mathrm{Rees}(N, L, \rho)$ is given by

$$n \cdot n' = nn',$$
$$n \cdot (n_1, m, n_2) \cdot n' = (nn_1, m, n_2 n'),$$
$$(n_1, m, n_2) \cdot (n_1', m', n_2') = (n_1, m\rho(n_2 n_1')m', n_2').$$

For a variety $\mathbf{V}$ of finite monoids let $\mathrm{Rees}(\mathbf{V})$ be the least variety which contains $\mathbf{V}$ and which is closed under Rees extensions $\mathrm{Rees}(N, L, \rho)$. Almeida and Klíma called a variety $\mathbf{V}$ *bullet-idempotent* if $\mathbf{V} = \mathrm{Rees}(\mathbf{V})$, see [1]. They showed $\mathrm{Rees}(\mathbf{V}) \subseteq \overline{\mathbf{H}}$ where $\mathbf{H} = \mathbf{V} \cap \mathbf{G}$ and asked whether all bullet-idempotent varieties are of that form. The answer is "yes" [10] and can be proved by showing the stronger result that so-called *local Rees extensions* suffice to capture all of $\overline{\mathbf{H}}$. To define these objects, consider a finite monoid M (which is not a group), an element $c \in M$, and a smaller submonoid N of M such that N and c generate M. Then, let M_c be the local divisor at c and let ρ_c be the mapping $\rho_c : N \to M_c$ with $\rho_c(x) = cxc$. The *local Rees extension* $\mathrm{LocRees}(N, M_c)$ is defined as the Rees extension $\mathrm{Rees}(N, M_c, \rho_c)$. Thus, a local Rees extension is a special case of a Rees extension. Still the result in [10] shows that $\overline{\mathbf{H}} = \mathrm{LocRees}(\mathbf{H})$. Here, $\mathrm{LocRees}(\mathbf{H})$ denotes the least variety which contains the group variety $\mathbf{H}$ and which is closed under local Rees extensions. Since $\mathrm{Rees}(\mathbf{V}) \subseteq \overline{\mathbf{H}}$ for $\mathbf{H} = \mathbf{V} \cap \mathbf{G}$ we obtain

$$\mathrm{Rees}(\mathbf{V}) \subseteq \overline{\mathbf{H}} = \mathrm{LocRees}(\mathbf{H}) \subseteq \mathrm{Rees}(\mathbf{V}).$$

Hence, all varieties appearing in the line above coincide.

7   The Local Divisor Technique and Green's Lemma

For a survey on the local divisor technique we refer to [4]. In general, there are more local divisors than local monoids, so having information about the structure in all local divisors tells us more about the structure of M than just looking at the local monoids. Before we continue let us revisit Green's Lemma as sort of a "commercial break" for the local divisor technique in semigroup theory.

The following section is based on [2,8] and closely follows the presentation in [8, Corollary 7.45] where full proofs are given. Green's relations are classical. There are three basic equivalence relations $\mathcal{L}$, $\mathcal{R}$, and $\mathcal{J}$ which relate elements in a monoid M generating the same left- (resp. right-, resp. two-sided-) ideal.

$$x\mathcal{L}y \iff Mx = My, \quad x\mathcal{R}y \iff xM = yM, \quad x\mathcal{J}y \iff MxM = MyM.$$

The other two relations are defined by $\mathcal{H} = \mathcal{L} \cap \mathcal{R}$ and $\mathcal{D} = \mathcal{L} \circ \mathcal{R}$. In particular,

$$x \mathcal{D} y \iff \exists z : x \mathcal{L} z \wedge z \mathcal{R} y.$$

A standard exercise shows that $\mathcal{J} = \mathcal{D}$ for finite monoids. (For infinite monoids this false, in general.) As $\mathcal{J}$ is symmetric, $\mathcal{J} = \mathcal{D}$ implies $\mathcal{D} = \mathcal{L} \circ \mathcal{R} = \mathcal{R} \circ \mathcal{L}$. The latter assertion is independent of that: $\mathcal{L} \circ \mathcal{R} = \mathcal{R} \circ \mathcal{L}$ holds in infinite monoids, too. Therefore, all relations above are equivalence relations. If $\mathcal{G}$ is any of them and $s \in M$, then we write $\mathcal{G}(s) = \{t \in M \mid s\mathcal{G}t\}$ for the equivalence class of s.

In the following, we assume that M is finite. If G is a subgroup of M with neutral element e, then G is a subgroup in $\mathcal{H}(e)$; and $\mathcal{H}(e)$ itself is a group. Now, Green's Lemma says that the groups $\mathcal{H}(e)$ and $\mathcal{H}(f)$ are isomorphic if e and f are idempotents belonging to the same $\mathcal{D}$-class. The classical proof uses $\mathcal{D} = \mathcal{L} \circ \mathcal{R}$. Hence, $e\mathcal{R}z\mathcal{L}f$ for some $z \in M$. Then one shows that the right multiplication $\cdot v$, mapping x to xv, induces a bijection $Me \to Mz$, $x \mapsto xv$. By symmetry, we obtain a bijection between $\mathcal{H}(e)$ and $\mathcal{H}(f)$ which turns out to be an isomorphism of groups.

The proof is somewhat "mysterious" because the isomorphism passes through $\mathcal{H}(z)$ which is not subgroup of M, in general. Using local divisors however, the proof becomes fully transparent and reveals a more general fact. For this, consider any two $\mathcal{R}$-equivalent (or symmetrically $\mathcal{L}$-equivalent) elements s and t. Whether or not s or t are idempotent, we can define the local divisors M_s and M_t. For $s\mathcal{R}t$ we can write $t = sv$ and now, the right multiplication $\cdot v$ defines an isomorphism $M_s \to M_t$. Moreover, as a set, $\mathcal{H}(s)$ is the group of units in M_s. In the case that $s = e$ is an idempotent $M_s = M_e$ is a local monoid and $\mathcal{H}(s) = \mathcal{H}(e)$ is a subgroup of M. Thus, as in the scenario of $e\mathcal{R}z\mathcal{L}f$ with idempotents e and f we see that three groups are isomorphic: $\mathcal{H}(e)$, $\mathcal{H}(z)$ as the group of units in $(zM \cap Mz, \circ, z)$, and $\mathcal{H}(f)$. There is no mystery in Green's Lemma if we view it from a more general perspective.

8   The Common Theme: Local Divisor Proofs

Let us now discuss the common theme in Church-Rosser systems, bounded synchronization delay, and Rees extensions. From an abstract viewpoint these deal with properties $\mathcal{P}$ which can be defined for regular languages. Assume we know that a property $\mathcal{P}$ of regular languages is true for all languages where the syntactic monoid belongs to some variety of groups $\mathbf{H}$. Then $\mathcal{P}$ holds for all languages where the syntactic monoid belongs to $\overline{\mathbf{H}}$ if and only if and we can show the following implication for local Rees extensions $\mathrm{LocRees}(N, M_c)$:

$$\mathcal{P}(N) \wedge \mathcal{P}(M_c) \implies \mathcal{P}(\mathrm{LocRees}(N, M_c)). \tag{1}$$

Actually, it is enough show an implication without mentioning $\mathrm{LocRees}(N, M_c)$:

$$\mathcal{P}(N) \wedge \mathcal{P}(M_c) \implies \mathcal{P}(M). \tag{2}$$

The reason that we mention the "complicated" implication (1) is that the power of the method lies in the underlying algebraic connection between N, M_c and M which is best reflected by the local Rees extension. For simplicity of notation we just focus on the equivalent condition (2). This implication is particularly appealing for aperiodic monoids. Indeed, any nontrivial property which is closed under taking submonoids must also hold for the trivial group $\{1\}$. So, the base for the induction is trivial for the variety $\mathbf{1}$. In order to prove that $\mathcal{P}$ holds for all aperiodic languages, one only needs to show (2). Sometimes this is very easy. Remember SF = $\mathbf{Ap}$, the probably most cited result of Schützenberger. The inclusion SF $\subseteq \mathbf{Ap}$ is rather straightforward and the assertion $\mathbf{1}(A^*) \subseteq \mathrm{SF}(A^*)$ is trivial since $\emptyset$ and its complement A^* are star-free. Now proving, (2) is possible within less than a page, see [12]. Almost the same holds for the less famous but more general result $\mathbf{Ap} = \mathrm{SD}$, see [5].

What about Krohn-Rhodes theory? It goes beyond $\mathbf{Ap}$, but the group case is built-in! The theory says that every monoid can be constructed by iterated wreath products, starting from finite simple groups and the so-called *reset monoid* U_2. According to [19, page 241] the monoid U_2 is "essentially junk" whereas the "groups are gems". Showing (2) for the Krohn-Rhodes property was done in [9] and led to a surprisingly easy proof of the Krohn-Rhodes decomposition theorem.

Returning to prefix codes of bounded synchronization delay, it is worth mentioning that Schützenberger did not stop this line of research by showing that $\mathbf{Ap} = \mathrm{SD}$. In [22] he was able to prove an analogue of $\mathbf{Ap} = \mathrm{SD}$ for languages where syntactic monoids have Abelian subgroups, only. For several years, no such characterization was known beyond $\overline{\mathbf{Ab}}$.

8.1   Schützenberger's SD Classes

Let $\mathbf{H}$ be a variety of finite groups. Consider a prefix code K with bounded synchronization delay which can be written as a disjoint union $K = \bigcup \{K_g \mid g \in G\}$ where $G \in \mathbf{H}$ and each K_g is regular in A^*. The $\mathbf{H}$-*controlled star* (more precisely, the G-*controlled star*) associates with such a disjoint union the following language:

$$\{u_{g_1} \cdots u_{g_k} \in K^* \mid u_{g_i} \in K_{g_i} \wedge g_1 \cdots g_k = 1 \in G\}.$$

Another view of the G-controlled star of K is the following: Let $\gamma_K : K \to G$ be a mapping such that $K_g = \gamma_K^{-1}(g)$ and let $\gamma : K^* \to G$ denote the canonical extension of γ_K to a homomorphism from the free submonoid $K^* \subseteq A^*$ to G, then the G-controlled star of K is exactly the set $\gamma^{-1}(1)$. Let $\mathcal{C}$ be any class of languages. We say that $\mathcal{C}$ is closed under $\mathbf{H}$-controlled star if for all K and for every group $G \in \mathbf{H}$, the following closure property holds: if $K = \bigcup \{K_g \mid g \in G\}$ is a prefix code with bounded synchronization delay such that $K_g \in \mathcal{C}$ for all $g \in G$, then the G-controlled star $\gamma^{-1}(1)$ is in $\mathcal{C}$ as well. By $\mathrm{SD}_{\mathbf{H}}(A^*)$ we denote the smallest class of regular languages containing all finite subsets of A^* and being closed under finite union, concatenation, and $\mathbf{H}$-controlled star.

Note that the definition of $\mathrm{SD}_\mathbf{H}(A^*)$ does not use any complementation. Using different notation, Schützenberger showed that $\mathrm{SD}_\mathbf{H}(A^*) \subseteq \overline{\mathbf{H}}(A^*)$ in [22], but he proved the converse inclusion only for $\mathbf{H} \subseteq \mathbf{Ab}$. The main result in [10] states that $\mathrm{SD}_\mathbf{H}(A^*) = \overline{\mathbf{H}}(A^*)$ for all $\mathbf{H}$. In retrospective, it is hard to say why Schützenberger did not prove this general result. Perhaps he was not interested in that, but we believe that this is unlikely because he proved half of it. More likely, he tried to use the Krohn-Rhodes decomposition as in [22] which involves wreath products and they may take you outside $\overline{\mathbf{H}}$. Perhaps, Krohn-Rhodes theory was simply the wrong tool for this result. Local Rees extensions, on the other hand, are perfectly suitable for this kind of applications.

8.2 Church-Rosser Thue Systems Revisited

In the following M, denotes a finite monoid. The results in Sect. 3 have their origins in formal language theory and led to the notion of Church-Rosser congruential languages. As mentioned before, for more than 25 years it was open whether or not all regular languages are Church-Rosser congruential. A positive answer was given in [7], and the corresponding theorem has a purely algebraic formulation. It says that for each homomorphism φ from A^* to M factorizes through A^*/S where S is a finite confluent and length-reducing semi-Thue system of finite index. Thus, $\varphi(\ell) = \varphi(r)$ and $|\ell| > |r|$ for all $(\ell, r) \in S$. Moreover, A^*/S is a finite monoid.

For the inductive argument, one crucial idea is to consider weight functions $\gamma : A \to \mathbb{N} \setminus \{0\}$. The statement then becomes "for every weight function and every homomorphism $\varphi : A^* \to M$ there exists a finite confluent semi-Thue system S of finite index such that $\varphi(\ell) = \varphi(r)$ and $\gamma(\ell) > \gamma(r)$ for all $(\ell, r) \in S$". Instead of weight-reducing systems we can also define the notions of Parikh-reducing and subword-reducing systems. For a letter a and a word $w \in A^*$ we let $|w|_a$ be the number of a's which occur in w. This defines a canonical homomorphism $\pi : A^* \to \mathbb{N}^A$ by $\pi(w) = (a \mapsto |w|_a)$. The vector $\pi(w)$ is usually called the *Parikh-image* of w. We say that S is *Parikh-reducing* if $(\ell, r) \in S$ implies $|\ell|_a \geq |r|_a$ for all $a \in A$ and $|\ell|_a > |r|_a$ for at least one $a \in A$. Clearly, a *Parikh-reducing* system is weight-reducing for every weight function. In the following, when using the term "subword" we mean "scattered subword". More precisely, a word u is called a *subword* of w if there exists a factorization $u = a_1 \cdots a_k$ such that $w \in A^* a_1 A^* \cdots a_k A^*$. We say that S is *subword-reducing* if $(\ell, r) \in S$ implies $\ell \neq r$ and that r is a subword of ℓ. Clearly, a subword-reducing system is Parikh-reducing. The induction scheme (2) introduced in the beginning of Sect. 8 works for all variants, but the group case is quite different. The trivial group leads to the subword-reducing system $\{(a, 1) \mid a \in A\}$. Consequently, the result in [16] speaks about subword-reducing systems and this is the strongest result. The PhD thesis of Tobias Walter [26] shows that for all homomorphisms to Abelian groups there exists a Parikh-reducing Church-Rosser system as desired, thereby allowing him to construct such systems for all languages in $\overline{\mathbf{Ab}}$. Additionally, he proves that for all regular languages L over a two letter alphabet there exists a Parikh-reducing Church-Rosser system S of finite index such that L is recognized by A^*/S. This shows that the existence of Parikh-reducing presentations is not limited to the variety $\overline{\mathbf{Ab}}$.

9   Conclusion and Open Problems

This extended abstract deals with the recurring theme of proving results for varieties of finite monoids and their associated language classes. The most prominent example is the variety $\mathbf{Ap}$ of aperiodic monoids, but our methods go beyond. We have seen deep connections between apparently quite different objects where the technique allows to transfer results from a group variety $\mathbf{H}$ to its closure $\overline{\mathbf{H}}$.

Let us conclude with some open problems, starting with the new perspective on Church-Rosser systems given in the previous subsection.

For subword-reducing and Parikh-reducing Church-Rosser systems, only partial results are known. To date, it is still open whether subword-reducing (resp. Parikh-reducing) Church-Rosser systems exist for every regular language. It is tempting to believe that Parikh-reducing systems exist for all regular languages, but we refrain from any conjecture in this case.

The notion of local Rees extensions gives rise to various interesting combinatorial problems concerning the complexity of Rees decompositions. For a finite monoid M, a *Rees decomposition tree* of M is a rooted node-labeled tree such that the following conditions are satisfied.

- The root has label M.
- Every inner node with label M' has two children labeled by N, M'_c such that M' is a divisor of the local Rees extension LocRees(N, M'_c).
- Every leaf is labeled by a group which divides M.

In [26], it was shown that if M is a monoid having n elements which are not units, then there exists a decomposition tree of M having at most $\mathcal{O}(3^{n/3})$ nodes. However, it is not clear whether this bound optimal. Actually, it is not even known whether the size of the tree be bounded by a polynomial function. Regardless of whether tight bounds can be obtained in the general case, it would also be interesting to analyze subclasses of $\mathbf{Mon}$. For example, it is easy to see that for commutative monoids with a fixed number of generators, there indeed is a polynomial bound. What happens if the number of generators is not fixed?

The starting point of our journey was the characterization of SD and SF by aperiodic monoids. Having this theme in mind, another interesting question about the limits of the method arises. In [25], Straubing showed that the so-called Mal'cev product of $\mathbf{Ap}$ and a group variety $\mathbf{H}$, denoted by $\mathbf{Ap} \circledM \mathbf{H}$, corresponds to the closure of $\mathbf{H}(A^*)$ under concatenation product. Following the proof of SD $= \mathbf{Ap}$ using local divisors, it is tempting to ask whether the local divisor technique can also be applied to obtain a new, possibly more general proof of Straubing's result. In particular, it would be interesting to see whether there is a natural language characterization of $\mathbf{Ap} \circledM \mathbf{H}$ that relies on prefix codes with bounded synchronization delay. A major obstacle to initial attempts is a result of Steinberg [24] that $\mathbf{Ap} \circledM \mathbf{H}$ is strictly contained in $\overline{\mathbf{H}}$ for all non-trivial group varieties $\overline{\mathbf{H}}$.

References

1. Almeida, J.M., Klíma, O.: On the irreducibility of pseudovarieties of semigroups. J. Pure Appl. Algebra **220**, 1517–1524 (2016)
2. Costa, A., Steinberg, B.: The schützenberger category of a semigroup. Semigroup Forum **91**(3), 543–559 (2015)
3. Diekert, V., Gastin, P.: Pure future local temporal logics are expressively complete for Mazurkiewicz traces, pp. 232–241. Springer, Heidelberg (2004)
4. Diekert, V., Kufleitner, M.: A survey on the local divisor technique. CoRR, abs/1410.6026 (2014)
5. Diekert, V., Kufleitner, M.: Omega-rational expressions with bounded synchronization delay. Theory Comput. Syst. **56**(4), 686–696 (2015)
6. Diekert, V., Kufleitner, M., Reinhardt, K., Walter, T.: Regular languages are Church-Rosser congruential. In: Czumaj, A., Mehlhorn, K., Pitts, A., Wattenhofer, R. (eds.) ICALP 2012. LNCS, vol. 7392, pp. 177–188. Springer, Heidelberg (2012). doi:10.1007/978-3-642-31585-5_19
7. Diekert, V., Kufleitner, M., Reinhardt, K., Walter, T.: Regular languages are Church-Rosser congruential. J. ACM **62**(5), 39:1–39:20 (2015)
8. Diekert, V., Kufleitner, M., Rosenberger, G., Hertrampf, U.: Discrete Algebraic Methods. Arithmetic, Cryptography, Automata and Groups. De Gruyter, Berlin (2016)
9. Diekert, V., Kufleitner, M., Steinberg, B.: The Krohn-Rhodes theorem and local divisors. Fundam. Inform. **116**(1–4), 65–77 (2012)
10. Diekert, V., Walter, T.: Characterizing classes of regular languages using prefix codes of bounded synchronization delay. In: Chatzigiannakis, I., Mitzenmacher, M., Rabani, Y., Sangiorgi, D. (eds.) ICALP, LIPIcs, vol. 55, pp. 129:1–129:14. Schloss Dagstuhl - Leibniz-Zentrum fuer Informatik (2016)
11. Eilenberg, S.: Automata, Languages, and Machines. Academic Press Inc., Orlando (1974)
12. Kufleitner, M.: Star-Free Languages and Local Divisors, pp. 23–28. Springer International Publishing, Cham (2014)
13. McNaughton, R., Narendran, P., Otto, F.: Church-Rosser Thue systems and formal languages. J. ACM **35**(2), 324–344 (1988)
14. Meyberg, K.: Lectures on Algebras and Triple Systems. University of Virginia, Charlottesville (1972)
15. Narendran, P.: Church-Rosser and related Thue systems. Ph.D. thesis, Department of Mathematical Sciences, Rensselaer Polytechnic Institute, Troy (1984)
16. Niemann, G., Otto, F.: The Church-Rosser languages are the deterministic variants of the growing context-sensitive languages. Inf. Comput. **197**(1–2), 1–21 (2005)
17. Perrin, D., Pin, J., Words, I.: Automata, Semigroups, Logic and Games, Pure and Applied Mathematics, vol. 141. Elsevier Science, Amsterdam (2004)
18. Pin, J.-É.: Varieties of Formal Languages. Plenum Publishing Co., New York (1986)
19. Rhodes, J., Steinberg, B.: The q-theory of finite semigroups. Springer Monographs in Mathematics, New York (2009)
20. Rhodes, J.L., Allen, D.: Synthesis of the classical and modern theory of finite semigroups. Adv. Math. **11**(2), 238–266 (1973)
21. Schützenberger, M.P.: On finite monoids having only trivial subgroups. Inf. Control **8**, 190–194 (1965)
22. Schützenberger. M.P.: Sur les monoides finis dont les groupes sont commutatifs. Rev. Française Automat. Informat. Recherche Opérationnelle Sér. Rouge **8**(R-1), 55–61 (1974)

23. Schützenberger, M.P.: Sur certaines opérations de fermeture dans les langages rationnels, pp. 245–253 (1975)
24. Steinberg, B.: On aperiodic relational morphisms. Semigroup Forum **70**(1), 1–43 (2005)
25. Straubing, H.: Aperiodic homomorphisms and the concatenation product of recognizable sets. J. Pure Appl. Algebra **15**(3), 319–327 (1979)
26. Walter, T.: Local divisors in formal languages. Dissertation, Institut für Formale Methoden der Informatik, Universität Stuttgart (2016)
27. Wilke, T.: Classifying discrete temporal properties. In: Meinel, C., Tison, S. (eds.) STACS 1999. LNCS, vol. 1563, pp. 32–46. Springer, Heidelberg (1999). doi:10.1007/3-540-49116-3_3

Overpals, Underlaps, and Underpals

Aayush Rajasekaran[1], Narad Rampersad[2], and Jeffrey Shallit[1(✉)]

[1] School of Computer Science, University of Waterloo, Waterloo,
ON N2L 3G1, Canada
{arajasekaran,shallit}@uwaterloo.ca
[2] Department of Math/Stats, University of Winnipeg, 515 Portage Ave., Winnipeg,
MB R3B 2E9, Canada
n.rampersad@uwinnipeg.ca

Abstract. An overlap in a word is a factor of the form $axaxa$, where x is a (possibly empty) word and a is a single letter; these have been well-studied since Thue's landmark paper of 1906. In this note we consider three new variations on this well-known definition and some consequences.

Keywords: Overlap · Automata · Avoidability in words

1 Introduction

An *overlap* is a word of the form $axaxa$, where x is a (possibly empty) word and a is a single letter. Examples include `alfalfa` in English and `entente` in French. Since Thue's work [2,15,16] in the early 20th century, overlaps and their avoidance have been well-studied in the literature (see, e.g., [12]).

Let μ be the morphism defined by $\mu(0) = 01$ and $\mu(1) = 10$. The Thue-Morse word $\mathbf{t}$ is defined to be the infinite fixed point of μ starting with 0. We have $\mathbf{t} = 0110100110010110\cdots$. We recall two famous results about binary overlaps:

Theorem 1.

(a) The Thue-Morse word $\mathbf{t}$ is overlap-free [2, 16].
(b) The number of binary overlap-free words of length n is $\Omega(n^\alpha)$ and $O(n^\beta)$ for real numbers $1 < \alpha < \beta$ [3, 4, 7].

In this paper we consider three variants of overlaps and study their properties.

2 Definitions and Notation

Throughout, we use the variables a, b, c to denote single letters, and the variables u, v, w, x, y, z to denote words. By $|x|$ we mean the length of a word x, and by x^R we mean its reversal. The empty word is written ε.

If a word w can be written in the form $w = xyz$ for (possibly empty) words x, y, z, then we say that y is a *factor* of w. We say that a word $x = x[1..n]$ has

© Springer International Publishing AG 2017
S. Brlek et al. (Eds.): WORDS 2017, LNCS 10432, pp. 17–29, 2017.
DOI: 10.1007/978-3-319-66396-8_3

period p if $x[i] = x[i + p]$ for $1 \leq i \leq n - p$. We say that a word x is a (p/q)-power, for integers $p > q \geq 1$, if x has period q and length p. For example, the word ionization is a $(10/7)$-power. A 2-power is called a *square*. Finally, we say that a word z *contains* an α-power if z contains a factor x that is a (p/q)-power for some $p/q \geq \alpha$. Otherwise we say that z *avoids* α-powers or is α-power free. We say that a word z avoids $(\alpha + \epsilon)$-powers or is $(\alpha + \epsilon)$-power-free if, for all $p/q > \alpha$, z contains no factor that is a (p/q)-power. By x^ω we mean the infinite word $xxx\cdots$.

We recall the definition of three famous sequences. The *Rudin-Shapiro sequence* $\mathbf{r} = (r_n)_{n\geq 0} = 000100100001110100010\cdots$ is defined by the relations $r_0 = 0$, and $r_{2n} = r_n$, $r_{4n+1} = r_n$, $r_{8n+3} = 1 - r_n$, and $r_{8n+7} = r_{2n+1}$ for $n \geq 0$. The *Fibonacci sequence* $\mathbf{f} = (f_n)_{n\geq 0} = 010010100100101001010\cdots$ is the fixed point of the morphism $\varphi(0) = 01$, $\varphi(1) = 0$. The *Tribonacci sequence* $\mathbf{T} = (T_n)_{n\geq 0} = 010201001020101020\cdots$ is the fixed point of the morphism $\theta(0) = 01$, $\theta(1) = 02$, $\theta(2) = 0$.

3   Overpals

In our first variation, we replace the second occurrence of axa in an overlap with its reversal. Thus, an *overpal* is a word of the form $axax^Ra$, where x^R is the reverse of the (possibly empty) word x and a is a single letter. The English word tartrate contains an occurrence of an overpal corresponding to $a = $ t and $x = $ ar. The *order* of an overpal $axax^Ra$ is defined to be $|ax|$.

We start with some results about binary words.

3.1   The Binary Case

Lemma 2. *Every binary palindrome x of odd length $\ell \geq 7$ contains an occurrence of either aaa, $ababa$, or $abbabba$, for some distinct letters a, b.*

Proof. Let w be an odd-length palindrome of length ≥ 7. Then we can write w in the form $xabcdcbax^R$ for some letters a, b, c, d and x possibly empty. Then a check of all 16 possibilities for a, b, c, d gives the result.

Theorem 3. *A binary word contains an overpal if and only if it contains aaa, $ababa$, or $abbabba$ for letters $a \neq b$.*

Proof. Suppose w contains an overpal $t = axax^Ra$. If $|x| = 0$, then $t = aaa$. If $|x| = 1$, then t is either $aaaaa$ or $ababa$. Otherwise $|x| \geq 2$, so $|t| \geq 7$, and the result follows by Lemma 2. On the other hand, if w contains any of aaa, $ababa$, or $abbabba$, then w contains an overpal.

Theorem 3 allows us to compute the generating function for the number of binary words avoiding overpals.

Corollary 4. *The generating function for the number of binary words avoiding overpals is*

$$\frac{2x^9 + 6x^8 + 8x^7 + 6x^6 + 6x^5 + 5x^4 + 4x^3 + 3x^2 + 2x + 1}{1 - x^2 - x^4}$$

Corollary 4 was apparently first noticed by Colin Barker, in a remark posted at the *On-Line Encyclopedia of Integer Sequences* about sequence A277277.

Proof. We use the DAVID_IAN Maple package [10, 11], implementing the Goulden-Jackson cluster method [6], with the command

```
GJs(0,1,[0,0,0],[1,1,1],[0,1,0,1,0],[1,0,1,0,1],
[0,1,1,0,1,1,0],[1,0,0,1,0,0,1],x);
```

This gives us the above generating function counting the binary words avoiding the patterns *aaa*, *ababa*, and *abbabba*.

Corollary 5. *The number* $\mathrm{ovp}_2(n)$ *of binary words of length n avoiding overpals is, for* $n \geq 6$, *equal to* $a\alpha^n + b\beta^n + c\gamma^n + d\delta^n$ *where*

$$a \doteq 5.096825703528179989223010 \qquad b \doteq 0.008747105471904132213320$$

are the real zeroes of the polynomial $25Z^4 - 300Z^3 + 1240Z^2 - 1840Z + 16$, *and*

$$c = 3 + \frac{\sqrt{5}}{5} + (2\sqrt{5} - 2)^{1/2}i \qquad d = 3 + \frac{\sqrt{5}}{5} - (2\sqrt{5} - 2)^{1/2}i,$$

and $\alpha = ((1 + \sqrt{5})/2)^{1/2} \doteq 1.2720196495140689642524246$, $\beta = -\alpha$, $\gamma = i\alpha^{-1}$, $\delta = -i\alpha^{-1}$.

Proof. From the generating function, we know that $\mathrm{ovp}_2(n)$ satisfies the linear recurrence $\mathrm{ovp}_2(n) = \mathrm{ovp}_2(n-2) + \mathrm{ovp}_2(n-4)$ for $n \geq 10$. Now we use standard techniques to solve this linear recurrence.

Corollary 6. *There are* $\Theta(\alpha^n)$ *binary words of length n containing no overpals.*

We now turn to infinite words avoiding overpals. It is easy to construct a *periodic* binary word avoiding overpals: namely, $(0011)^\omega = 001100110011\cdots$. (To verify this, it suffices to enumerate its subwords of odd length ≤ 7 and check that none of them are of the form *aaa*, *ababa* or *abbabba*.)

Theorem 7. *The lexicographically least infinite binary word that avoids overpals is* $\mathbf{x} := 001(001011)^\omega$.

Proof. To verify that $\mathbf{x}$ avoids overpals, it suffices to enumerate its subwords of odd length ≤ 7. None are of the form *aaa*, *ababa* or *abbabba*.

Suppose there is an infinite binary word **w** that is lexicographically less than **x**, but contains no overpals. Let v be the shortest prefix of **w** such that v is not a prefix of **x**. Suppose $|v| = n$. At position n there must be a 0 in v and **w** and a 1 in **x**. This means there are four possibilities: (i) $v = 000$; (ii) $v = 001(001011)^i 000$ for some $i \geq 0$; (iii) $v = 001(001011)^i 00100$ for some $i \geq 0$; (iv) $v = 001(001011)^i 001010$ for some $i \geq 0$.

In cases (i) and (ii), v ends with the overpal 000, a contradiction. In case (iii), consider the letter at position $n + 1$ of **w**. If it is 0, then $v0$ is a prefix of **w** and ends with 000. If it is 1, then $v1$ is a prefix of **w** and ends with 1001001. Both cases give a contradiction. In case (iv), v ends with the overpal 01010, a contradiction.

We have shown there are ultimately periodic binary words avoiding overpals. We now turn to aperiodic binary words.

Theorem 8. *No (7/3)-power-free binary word contains an overpal.*

Proof. Suppose it did. From Lemma 2 any odd-length palindrome in such a word is of length 1, 3, or 5. A palindrome of length 1 cannot be an overpal. The only overpals of length 3 are 000 and 111, each of which is a cube. Finally, the only overpals of length 5 are 00000 and 01010 and their complements, each of which contains a (7/3)-power.

Corollary 9. *The Thue-Morse word* **t** *contains no overpals.*

Theorem 10. *If $\mu(x)$ contains an overpal, then so does x.*

Proof. Suppose $\mu(x)$ contains an overpal. Then it contains an occurrence of aaa, $ababa$, or $abbabba$. However, it is easy to verify that neither aaa nor $abbabba$ can be the factor of a word that is an image under μ. For $ababa$ to be the factor of $\mu(x)$, it must be that x has the factor aaa, and hence an overpal.

Theorem 11. *The orders of overpals occurring in the Fibonacci word* **f** *are given, for $n \geq 1$, by the n whose Fibonacci representation is accepted by the following automaton.*

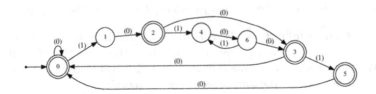

Fig. 1. Automaton accepting orders of overpals in the Fibonacci word

There are infinitely many orders for which there is no overpal factor of **f** *and infinitely many for which there are.*

Proof. We use the automatic theorem-proving software Walnut [9] with the predicate

```
def fiboverpal "?msd_fib (n=0) | Ei (n>=1) & (At (t<=2*n) =>
F[i+t] = F[i+2*n-t]) & F[i]=F[i+n]":
```

Corollary 12. *An overpal of order n exists in the Fibonacci word, for $n \geq 1$, if and only if there exists an integer m such that $n = \lfloor m\alpha + \frac{1}{2} \rfloor$, where $\alpha = (1 + \sqrt{5})/2$.*

Proof. The proof is in six steps.

Step 1: Define the infinite binary word $\mathbf{p} = (p_i)_{i \geq 0}$, where $p_i = 1$ if the Fibonacci representation of i is accepted by the automaton in Fig. 1, and $p_i = 0$ otherwise. Using the usual extension of Cobham's theorem to Fibonacci numeration systems, p is given by the image under the coding τ of the fixed point $f^\omega(0)$, where

$$f(0) = 01 \qquad f(1) = 2 \qquad f(2) = 34$$
$$f(3) = 05 \qquad f(4) = 6 \qquad f(5) = 0 \qquad f(6) = 34$$

and $\tau(0123456) = 1011010$. This is obtained just by reading off the transitions of the automaton, where the image of a state is 1 if the state is accepting, and 0 otherwise.

Step 2: Let $h : \{0,1\}^* \to \{0,1\}^*$ be the morphism sending $1 \to 10110$, $0 \to 110$. A routine induction on n, which we omit, proves that

$$\tau(f^{3n}(0)) = \tau(f^{3n}(2)) = \tau(f^{3n}(3)) = \tau(f^{3n}(3)) = \tau(f^{3n}(4)) = \tau(f^{3n}(6)) = h^n(1)$$
$$\tau(f^{3n}(1)) = h^n(0)$$
$$\tau(f^{3n+3}(5)) = h^n(101)$$

for $n \geq 0$.

Step 3: We now use a result in a paper of Tan and Wen [14]. Define $\pi : \{0,1\}^* \to \{0,1\}^*$ to be the morphism sending $0 \to 1$, $1 \to 0$. Define $\lambda : \{0,1\}^* \to \{0,1\}^*$ to be the morphism corresponding to $\pi \circ h^2 \circ \pi$.

A cutting sequence $K_{q,r}$ is defined as the infinite binary sequence generated by the straight line $y = qx + r$ as it cuts a square grid. See [14] for more on cutting sequences. Let the fixed point of λ be generated by the cutting sequence $K_{\gamma,\beta}$. Tan and Wen give us the slope γ, and the intercept β of this line. We define the additional morphisms $\sigma, \rho : \{0,1\}^* \to \{0,1\}^*$, where σ sends $0 \to 01$, $1 \to 0$, and ρ sends $0 \to 10$, $1 \to 0$.

To get γ, we need to express λ as a composition of $\sigma \circ \pi$, $\rho \circ \pi$ and π. We hence write $\lambda = ((\sigma \circ \pi) \circ \pi \circ (\rho \circ \pi) \circ \pi \circ (\rho \circ \pi) \circ \pi)^2$. Tan and Wen gives the continued fraction expansion of the slope as $\gamma = [0; 1, \overline{1,1,1,1,1,1}] = 1/\alpha$, where $\alpha = \frac{1+\sqrt{5}}{2}$ is the golden ratio.

To get β, we follow Tan and Wen to get the word $u = 010010$ that satisfies $\lambda(01) = u01v$, $\lambda(10) = u10v$, vu is a palindrome, for some word v. We also define $u_n = \lambda^{n-1}(u)\lambda^{n-2}(u) \cdots \lambda(u)u$. Let $|u_n|_0$ denote the number of zeroes in u_n.

The value of β is given by the unique number $x \in [-\gamma, 1 + \gamma)$ that satisfies $e^{2\pi i x} = \lim_{n\to\infty} e^{-2\pi i (|u_n|_0 + 1)\gamma}$. We calculate this value as $\beta = 1 - \sqrt{5}/2$.

Finally, Tan and Wen assert that if the fixed point of λ is given by $K_{\gamma,\beta}$, then the fixed point of h is given by the cutting sequence $K_{1/\gamma, -\beta/\gamma}$. Thus, $h^\omega(1)$ is given by $K_{\alpha, 1-\alpha/2}$.

Step 4: The Sturmian word $s_{e,f} = (s_i)$ is the infinite binary word defined by $s_i = \lfloor e(i + 1) + f \rfloor - \lfloor ei + f \rfloor - \lfloor e \rfloor$. We now use a classical result relating cutting sequences to Sturmian words (e.g., p. 56 of [8]) to conclude that $K_{\alpha, 1-\alpha/2} = s_{1/\alpha, (5-3\alpha)/2}$.

Step 5: We shift this Sturmian word right by 1 position, getting the equality $s_{1/\alpha, (5-3\alpha)/2} = 1 \cdot s_{1/\alpha, -1/(2\alpha)}$.

Step 6: Finally, we use the usual connection between Sturmian words and Beatty sequences (e.g., Lemma 9.1.3 of [1], generalized from characteristic words to the more general setting of Sturmian words) to conclude that $s_{1/\alpha, -1/(2\alpha)} = b_1 b_2 \cdots$, where $b_n = 1$ if and only if there exists an integer $m \geq 1$ such that $n = \lfloor m\alpha + \frac{1}{2} \rfloor$.

Theorem 13. *There are exactly four overpals in the Rudin-Shapiro sequence, and they are given by* 000, 111, 0100010, 1011101.

Proof. We use `Walnut` [9] to find the orders of overpals in the Rudin-Shapiro sequence

```
eval RSOverpal "Ei (n>=1) & (At (t<=2*n) =>
(RS[i+t] = RS[i+2*n-t])) & (RS[i]=RS[i+n])":
```

The only accepted orders are 1 and 2. An exhaustive search yields the result.

3.2   Larger Alphabets

Understanding the words that avoid overpals over large alphabets is more challenging than the binary case. For one thing, there is no analogue of Lemma 2, as the following result shows:

Theorem 14. *Over a ternary alphabet, there are arbitrarily long odd-length palindromes containing no overpals.*

Proof. We know that $\mu^{2n}(0)$ is a palindrome for all $n \geq 0$, and furthermore, since it is a prefix of t, it contains no overpals. Therefore, for all $n \geq 0$, the word $\mu^{2n}(0)2\mu^{2n}(0)$ is a palindrome containing no overpals, and it is of length $2^{2n+1} + 1$.

Theorem 15.

(i) *Every odd-length ternary palindrome of length ≥ 17 contains a $\frac{7}{4}$ power.*

(ii) *There are arbitrarily large odd-length ternary palindromes avoiding $(\frac{7}{4} + \epsilon)$-powers.*

Proof.

(i) It suffices to examine all ternary palindromes of length 17.
(ii) Dejean's word [5] avoids $(\frac{7}{4} + \epsilon)$-powers and contains ternary palindromes of all odd lengths.

Theorem 16. *No infinite ternary word can avoid overpals and $\frac{41}{22}$-powers.*

Proof. We use the usual tree-traversal technique. The tree has 120844 internal nodes, and 241689 leaves. The longest such string is of length 228.

Conjecture 17. There is an infinite ternary word that avoids overpals and $(\frac{41}{22} + \epsilon)$-powers.

4   Underpals

A word is said to be an *underpal* if it is of the form $axbx^Ra$ where x is a (possibly empty) word and a, b are letters with $a \neq b$. An example in English is the word racecar, with $a = $ r, $x = $ ac, and $b = $ e.

Theorem 18. *A word contains an underpal if and only if it contains some word of the form ab^ia with $a \neq b$ and i odd.*

Proof. Let a word contain an underpal $z = axbx^Ra$. Either x ends in b, or it does not end in b.

Case 1: x ends in b. Then either $x = b^l$ for some $l \geq 1$, or $x = ycb^l$ for some word y and letter $c \neq b$.

Case 1a: $x = b^l$. Then $z = ab^ia$ for odd $i = 2l + 1$.

Case 1b: $x = ycb^l$. Then $z = aycb^lbb^lcy^Ra$, which contains cb^ic with odd $i = 2l + 1$.

Case 2: x does not end in b. If $x = \varepsilon$, then $z = ab^ia$, with odd $i = 1$. Otherwise $x = yc$, which gives $z = aycbcy^Ra$, which contains cb^ic with odd $i = 1$.

 Thus, a word that contains an underpal must contain an ab^ia, with odd i, and so a word that avoids such factors must avoid underpals.

 For the converse, suppose w contains $z = ab^ia$ with $a \neq b$ and i odd. Then $z = ab^lbb^la$ for some non-negative integer l. Since z is an underpal with $x = b^l$, the word w contains an underpal.

Theorem 19. *The number of length-n words avoiding underpals over a k-letter alphabet satisfies the recurrence $f_k(0) = 1, f_k(1) = k, f_k(n) = (k - 2)f_k(n - 1) + kf_k(n - 2) + k$.*

Proof. Let Σ be an alphabet with $|\Sigma| = k$. For all p, define $L_{k,p} \subseteq \Sigma^*$ to be the language of words of length p that avoid underpals.

We now define two languages A_1 and A_2 as follows:

$$A_1 = \{ca^{n-1} : a \neq c \in \Sigma\}$$
$$A_2 = \{cx : x = a^l bz \in L_{k,n-1}, a \neq b, a \neq c, b \neq c, a, b, c \in \Sigma, l > 0, z \in \Sigma^*\}$$

Note that $|A_1| = k(k-1)$. Since we exclude unary words of the form a^{n-1}, the number of words of the form $x = a^l bz$ is $|L_{k,n-1}| - k = f_k(n-1) - k$. We thus get that $|A_2| = (k-2)(f_k(n-1) - k)$.

Define $A = A_1 \cup A_2$. All words in A are n-length words avoiding underpals since they must avoid $ab^i a$ with i odd, and so $A \subseteq L_{k,n}$.

Define $D \subseteq L_{k,n-2}$ as follows:

$$D = \{a^l bz \in L_{k,n-2}, a \neq b \in \Sigma, l > 0, l \text{ odd}, z \in \Sigma^*\}.$$

Next, we define the languages B_1, B_2 and B_3 as follows:

$$B_1 = \{ccx : x = a^l bz \in D, c \in \Sigma, c \neq b\}$$
$$B_2 = \{bax : x = a^l bz \in D\}$$
$$B_3 = \{ccx : x \in L_{k,n-2}, x \notin D, c \in \Sigma\}$$

Clearly $|B_1| = (k-1)|D|$, and $|B_2| = |D|$, and $|B_3| = k(|L_{k,n-2}| - |D|)$.

Define $B = B_1 \cup B_2 \cup B_3$. All words in B are n-length words avoiding underpals since they must avoid $ab^i a$ with i odd, and so $B \subseteq L_{k,n}$.

Thus, we get

$$A \cup B \subseteq L_{k,n}. \tag{1}$$

Consider any word $z = d_1 d_2 \cdots d_n \in L_{k,n}$. Note that $d_2 d_3 \cdots d_n \in L_{k,n-1}$ and $d_3 d_4 \cdots d_n \in L_{k,n-2}$. We divide these words z into two cases:

Case 1. $d_1 = d_2$. If $z = d_1 d_1 a^l bx$, for some $a \neq b \in \Sigma$ and even $l \geq 0$, then $z \in B_3$. If $z = d_1 d_1 a^l bx$, for some $a \neq b \in \Sigma$ and odd l, then we consider d_1. If $d_1 \neq b$, then $z \in B_1$. If $d_1 = b$, then z contains $ba^l b$, with odd l, and thus $z \notin L_{k,n}$. If $z = d_1^n$, then $z \in B_3$.

Case 2: $d_1 \neq d_2$. If $z = d_1 d_2^{n-1}$, then $z \in A_1$. If $z = d_1 d_2^l bx$, for some $b \neq d_2 \in \Sigma$ and even $l > 0$, then we consider d_1. If $d_1 \neq b$, then $z \in A_2$. If $d_1 = b$, then $z = d_1 d_2 d_2^{l-1} d_1 x$, where $l-1$ is odd. In this case, $z \in B_2$. If $z = d_1 d_2^l bx$, for some $b \neq d_2 \in \Sigma$ and odd l, then we consider the value of d_1. If $d_1 \neq b$, then $z \in A_2$. If $d_1 = b$, then z contains $bd_2^l b$, with odd l, and thus $z \notin L_{k,n}$. We thus see that for all $z \in L_{k,n}, z \in A \cup B$, and hence

$$L_{k,n} \subseteq A \cup B. \tag{2}$$

Combining Eqs. (1) and (2) gives us $L_{k,n} = A \cup B$, which gives

$$f_k(n) = |L_{k,n}| = |A \cup B|. \tag{3}$$

Since the words in A_1 have exactly two different letters, while those in A_2 have at least 3 different letters, the sets A_1 and A_2 are disjoint.

The sets B_1 and B_2 are disjoint since they disagree on the first two letters. The sets B_1 and B_3 are disjoint since they disagree on the last $n-2$ letters. The sets B_2 and B_3 are disjoint since they disagree on the last $n-2$ letters.

Note that for all words $x = a_1 a_2 \cdots a_n \in A$ we have $a_1 \neq a_2$. The only words $y = b_1 b_2 \cdots b_n \in B$ for which $b_1 \neq b_2$ are in B_2, and are thus of the form $y = baa^l bz$, for some $a \neq b \in \Sigma$ and $z \in \Sigma^*$. Such words y cannot be in A, because A excludes all words with prefix $ba^p b$ for all $a \neq b \in \Sigma$, $p > 0$. This shows that the sets A and B are disjoint.

We have

$$|A| = |A_1| + |A_2| = k(k-1) + (k-2)(f_k(n-1) - k) = (k-2)f_k(n-1) + k. \quad (4)$$

We also have

$$|B| = |B_1| + |B_2| + |B_3| = (k-1)|D| + |D| + k(|L_{k,\,n-2}| - |D|) = kf_k(n-2). \quad (5)$$

Since A and B are disjoint, $|A \cup B| = |A| + |B| = (k-2)f_k(n-1) + kf_k(n-2) + k$. Combining this with Eq. (3) gives us $f_k(n) = (k-2)f_k(n-1) + kf_k(n-2) + k$. Finally, $f_k(0) = 1$, since the empty string avoids underpals, and $f_k(1) = k$, since all strings of length 1 avoid underpals.

Corollary 20. *The number $f_k(n)$ of length-n words avoiding underpals over a k-letter alphabet, for $k \geq 2$ and $n \geq 0$, is given by $f_k(n) = a\alpha^n + b\beta^n + c$, where*

$$\alpha = \frac{k - 2 + \sqrt{k^2 + 4}}{2} \qquad a = \frac{(k-1)(3(k^2+4) + (k+6)\sqrt{k^2+4})}{2(2k-3)(k^2+4)}$$

$$\beta = \frac{k - 2 - \sqrt{k^2 + 4}}{2} \qquad b = \frac{(k-1)(3(k^2+4) - (k+6)\sqrt{k^2+4})}{2(2k-3)(k^2+4)}$$

$$c = \frac{k}{3 - 2k}.$$

Proof. By the usual techniques for handling linear recurrences, we know that $f_k(n) = (k-1)f_k(n-1) + 2f_k(n-2) - kf_k(n-3)$. This means that $f_k(n)$ is expressible as a linear combination of the n'th powers of the zeroes of the polynomial $X^3 + (1-k)X^2 - 2X + k$. Solving the resulting linear system, using Maple as an assistant, gives the result.

Remark 21. For $k = 2$ this simplifies to $f_2(n) = 2^{(n+3)/2} - 2$ for n odd and $f_2(n) = 3 \cdot 2^{n/2} - 2$ for n even.

The *run length encoding* of a binary word is the integer sequence giving the lengths of maximal blocks of 0s and 1s. For example, the run length encoding of 0011101011 is 2, 3, 1, 1, 1, 2.

Theorem 22. *A finite binary word avoids underpals if and only if its run length encoding is of the form $i_1, i_2, \ldots, i_t$ where $i_2, i_3, \ldots, i_{t-1}$ are all even. An infinite binary word that does not end in a^ω for $a \in \{0, 1\}$ avoids underpals if and only if its run length encoding is of the form $i_1, i_2, i_3, \ldots$ where $i_2, i_3, \ldots$ are all even.*

Theorem 23. *The Fibonacci word has underpals of order n for exactly those n accepted by the automaton below (Fig. 2).*

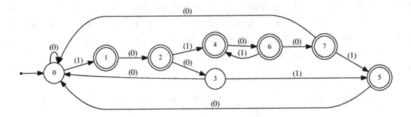

Fig. 2. Automaton accepting orders of underpals in the Fibonacci word

Theorem 24. *The Fibonacci word has both underpals and overpals of order n for exactly those n accepted by the automaton below (Fig. 3).*

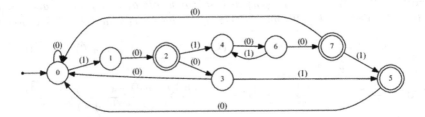

Fig. 3. Automaton accepting orders for which there are both overpals and underpals in the Fibonacci word

Theorem 25. *Every binary word of length ≥ 17 avoiding underpals contains a 4th power.*

Proof. By explicit enumeration of the 1022 binary words of length 17 avoiding underpals.

Theorem 26. *There is an infinite binary word avoiding underpals and avoiding $(4 + \epsilon)$-powers.*

Proof. Let h be the doubling morphism $0 \to 00$ and $1 \to 11$. Applying h to the Thue-Morse word **t** gives a binary word $h(\mathbf{t})$ that contains no underpals and avoids $(4 + \epsilon)$-powers.

Theorem 27. *Every ternary word of length ≥ 6 avoiding underpals has a square.*

Proof. By enumerating all ternary words of length 6 avoiding underpals.

Theorem 28. *There is an infinite ternary word avoiding underpals and $(2+\epsilon)$-powers.*

Proof. Take any infinite squarefree word w over a ternary alphabet $\{0,1,2\}$ and apply the morphism $h : 0 \rightarrow 01, 1 \rightarrow 10, 2 \rightarrow 22$. Then $h(w)$ has no overlaps, overpals, or underpals.

5   Underlaps

In analogy with overlaps, we can define *underlaps*. An *underlap* is a word of the form $axbxa$ with x a (possibly empty) word, and a, b letters with $a \neq b$. Note that x is a bispecial factor of the underlap. An example in English is the word ginning, with $a = g$, $x = in$, and $b = n$.

Theorem 29.

(a) *The only underlaps in the Thue-Morse sequence* t *are*
   $\{010, 101, 0011010, 0101100, 1010011, 1100101\}$.
(b) *The only underlaps in the Fibonacci sequence are* $\{010, 101, 00100\}$.
(c) *The only underlaps in the Rudin-Shapiro sequence are*
   $\{010, 101, 00100, 01110, 10001, 11011, 0001000, 1110111\}$.
(d) *The only underlaps in the Tribonacci sequence are*
   $\{010, 020, 101, 10201, 20102, 001020100\}$.

We now prove a theorem giving the relationship between underlaps and underpals. These concepts actually coincide for binary words.

Theorem 30.

(a) *If z contains an underpal, then it contains an underlap.*
(b) *If z is over a binary alphabet and contains an underlap, then it contains an underpal.*

Proof. (a) Suppose z contains an underpal. Then it can be written in the form $z = uaxbx^Rav$ where $a \neq b$.

Case 1: If x contains some letter $c \neq b$, write $x = ycb^i$ for some $i \geq 0$. Then z contains the word $xbx^R = ycb^i bb^i cy^R$, which contains the word $cb^i bb^i c$, which is an underlap.

Case 2: Otherwise $x = b^i$ for some $i \geq 0$. Then z contains the word $axbx^Ra = ab^i bb^i a$, which is an underlap.

(b) Now suppose z contains an underlap and is over the alphabet $\{0, 1\}$. Then it can be written in the form $z = uaxbxav$ where $a \neq b$.

Case 1: x has no a's. Since x is over a binary alphabet, it must be the case that $x = b^i$ for some $i \geq 0$. Then $axbxa = ab^i bb^i a$, which is an underpal.

Case 2: x has one a. Write $x = b^i a b^j$ for some $i, j \geq 0$. Then $axbxa = ab^i ab^j bb^i ab^j a = ab^i ab^{i+j+1} ab^j a$. If either i (resp., j) is odd, this contains $ab^i a$ (resp., $ab^j a$), which is an underpal. Otherwise i and j are both even, so $i + j + 1$ is odd and $ab^{i+j+1}a$ is an underpal.

Case 3: x has two or more a's. By identifying the first and last occurrences of a, write $x = b^i ayab^j$. Then $axbxa = ab^i ayab^j bb^i ayab^j a = ab^i ayab^{i+j+1} ayab^j a$. If i (resp., j) is odd, this contains $ab^i a$ (resp., $ab^j a$), which is an underpal. Otherwise i and j are both even, so $i + j + 1$ is odd and $ab^{i+j+1}a$ is an underpal.

As an example of a word over the ternary alphabet that contains an underlap but no underpal, consider 001120110.

Theorem 31. *Every binary word of length ≥ 9 has either an overlap or an underlap.*

Proof. It suffices to examine all 512 binary words of length 9.

Theorem 32. *There are exponentially many ternary words avoiding overlaps, underlaps, overpals, and underpals.*

Proof. Take any squarefree word w over a ternary alphabet $\{0, 1, 2\}$ and apply the morphism $h : 0 \to 01, 1 \to 10, 2 \to 22$. Then $h(w)$ has no overlaps, underlaps, overpals, or underpals. Since there are exponentially many squarefree ternary words (the best lower bound known is $\Omega(952^{n/53})$ [13]), the result follows.

References

1. Allouche, J.P., Shallit, J.: Automatic Sequences: Theory, Applications, Generalizations. Cambridge University Press, Cambridge (2003)
2. Berstel, J.: Axel Thue's papers on repetitions in words: a translation, No. 20. In: Publications du Laboratoire de Combinatoire et d'Informatique Mathématique, Université du Québec à Montréal, February 1995
3. Blondel, V.D., Cassaigne, J., Jungers, R.M.: On the number of α-power-free binary words for $2 < \alpha \leq 7/3$. Theoret. Comput. Sci. **410**, 2823–2833 (2009)
4. Cassaigne, J.: Counting overlap-free binary words. In: Enjalbert, P., Finkel, A., Wagner, K.W. (eds.) STACS 1993. LNCS, vol. 665, pp. 216–225. Springer, Heidelberg (1993). doi:10.1007/3-540-56503-5_24
5. Dejean, F.: Sur un théorème de Thue. J. Combin. Theor. Ser. A **13**, 90–99 (1972)
6. Goulden, I., Jackson, D.: An inversion theorem for cluster decompositions of sequences with distinguished subsequences. J. London Math. Soc. **20**, 567–576 (1979)
7. Jungers, R.M., Protasov, V.Y., Blondel, V.D.: Overlap-free words and spectra of matrices. Theor. Comput. Sci. **410**, 3670–3684 (2009)
8. Lothaire, M.: Algebraic Combinatorics on Words, Encyclopedia of Mathematics and its Applications. Cambridge University Press, Cambridge (2011)
9. Mousavi, H.: Automatic theorem proving in Walnut (2016). https://arxiv.org/abs/1603.06017
10. Noonan, J., Zeilberger, D.: DAVID_IAN Maple package (1999). http://www.math.rutgers.edu/~zeilberg/gj.html

11. Noonan, J., Zeilberger, D.: The Goulden-Jackson cluster method: extensions, applications, and implementations. J. Differ. Equ. Appl. **5**, 355–377 (1999)
12. Rampersad, N.: Overlap-free words and generalizations. Ph.D. thesis, University of Waterloo (2008)
13. Sollami, M., Douglas, C.C., Liebmann, M.: An improved lower bound on the number of ternary squarefree words. J. Integer Sequences **19**, 3 (2016). Article 16.6.7
14. Tan, B., Wen, Z.Y.: Invertible substitutions and Sturmian sequences. Eur. J. Comb. **24**, 983–1002 (2003)
15. Thue, A.: Über unendliche Zeichenreihen. Norske vid. Selsk. Skr. Mat. Nat. Kl. **7**, 1–22 (1906). Reprinted in Selected Mathematical Papers of Axel Thue, T. Nagell, editor, Universitetsforlaget, Oslo, 1977, pp. 139–158
16. Thue, A.: Über die gegenseitige Lage gleicher Teile gewisser Zeichenreihen. Norske vid. Selsk. Skr. Mat. Nat. Kl. **1**, 1–67 (1912). Reprinted in Selected Mathematical Papers of Axel Thue, T. Nagell, editor, Universitetsforlaget, Oslo, 1977, pp. 413–478

On Some Interesting Ternary Formulas

Pascal Ochem[1(✉)] and Matthieu Rosenfeld[2]

[1] LIRMM, CNRS, Université de Montpellier, Montpellier, France
ochem@lirmm.fr
[2] LIP, ENS de Lyon, CNRS, UCBL, Université de Lyon, Lyon, France
matthieu.rosenfeld@ens-lyon.fr

Abstract. We show that, up to renaming of the letters, the only infinite ternary words avoiding the formula $ABCAB.ABCBA.ACB.BAC$ (resp. $ABCA.BCAB.BCB.CBA$) have the same set of recurrent factors as the fixed point of $0 \mapsto 012$, $1 \mapsto 02$, $2 \mapsto 1$.

Also, we show that the formula $ABAC.BACA.ABCA$ is 2-avoidable. Finally, we show that the pattern $ABACADABCA$ is unavoidable for the class of C_4-minor-free graphs with maximum degree 3. This disproves a conjecture of Grytczuk.

Keywords: Combinatorics on words · Pattern avoidance

1 Introduction

A *pattern* p is a non-empty finite word over an alphabet $\Delta = \{A, B, C, \ldots\}$ of capital letters called *variables*. An *occurrence* of p in a word w is a non-erasing morphism $h : \Delta^* \to \Sigma^*$ such that $h(p)$ is a factor of w. The *avoidability index* $\lambda(p)$ of a pattern p is the size of the smallest alphabet Σ such that there exists an infinite word over Σ containing no occurrence of p.

A variable that appears only once in a pattern is said to be *isolated*. Following Cassaigne [2], we associate a pattern p with the *formula* f obtained by replacing every isolated variable in p by a dot. The factors between the dots are called *fragments*.

An *occurrence* of a formula f in a word w is a non-erasing morphism $h : \Delta^* \to \Sigma^*$ such that the h-image of every fragment of f is a factor of w. As for patterns, the avoidability index $\lambda(f)$ of a formula f is the size of the smallest alphabet allowing the existence of an infinite word containing no occurrence of f. Clearly, if a formula f is associated with a pattern p, every word avoiding f also avoids p, so $\lambda(p) \leq \lambda(f)$. Recall that an infinite word is *recurrent* if every finite factor appears infinitely many times. If there exists an infinite word over Σ avoiding p, then there exists an infinite recurrent word over Σ avoiding p. This recurrent word also avoids f, so that $\lambda(p) = \lambda(f)$. Without loss of generality, a formula is such that no variable is isolated and no fragment is a factor of

This work was partially supported by the ANR project CoCoGro (ANR-16-CE40-0005).

S. Brlek et al. (Eds.): WORDS 2017, LNCS 10432, pp. 30–35, 2017.
DOI: 10.1007/978-3-319-66396-8_4

another fragment. We say that a formula f is *divisible* by a formula f' if f does not avoid f', that is, there is a non-erasing morphism h such that the image of any fragment of f' by h is a factor of a fragment of f. If f is divisible by f', then every word avoiding f' also avoids f. Let $\Sigma_k = \{0, 1, \ldots, k-1\}$ denote the k-letter alphabet. We denote by Σ_k^n the k^n words of length n over Σ_k.

We say that two infinite words are equivalent if they have the same set of factors. Let b_3 be the fixed point of $0 \mapsto 012$, $1 \mapsto 02$, $2 \mapsto 1$. A famous result of Thue [1,4,5] can be stated as follows:

Theorem 1 [1,4,5]. *Every bi-infinite ternary word avoiding AA, 010, and 212 is equivalent to b_3.*

In Sect. 2, we obtain a similar result for b_3 by forbidding one ternary formula but without forbidding explicit factors in Σ_3^*.

In the remainder of the paper, we discuss a counterexample to a conjecture of Grytczuk stating that every avoidable pattern can be avoided on graphs with an alphabet of size that depends only on the maximum degree of the graph.

2   Formulas Closely Related to b_3

For every letter $c \in \Sigma_3$, $\sigma_c : \Sigma_3^* \mapsto \Sigma_3^*$ is the morphism such that $\sigma_c(a) = b$, $\sigma_c(b) = a$, and $\sigma_c(c) = c$ with $\{a, b, c\} = \Sigma_3$. So σ_c is the morphism that fixes c and exchanges the two other letters.

We consider the following formulas.

- $f_b = ABCAB.ABCBA.ACB.BAC$
- $f_1 = ABCA.BCAB.BCB.CBA$
- $f_2 = ABCAB.BCB.AC$
- $f_3 = ABCA.BCAB.ACB.BCB$
- $f_4 = ABCA.BCAB.BCB.AC.BA$

Theorem 2. *Let $f \in \{f_b, f_1, f_2, f_3, f_4\}$. Every ternary recurrent word avoiding f is equivalent to b_3, $\sigma_0(b_3)$, or $\sigma_2(b_3)$.*

By considering divisibility, we can deduce that Theorem 2 holds for 72 ternary formulas. Since b_3, $\sigma_0(b_3)$, and $\sigma_2(b_3)$ are equivalent to their reverse, Theorem 2 also holds for the 72 reverse ternary formulas.

Proof. For $1 \le i \le 4$, f_b contains an occurrence of f_i. Thus, every word avoiding f_i also avoids f_b. Using Cassaigne's algorithm, we have checked that b_3 avoids f_i. By symmetry, $\sigma_0(b_3)$ and $\sigma_2(b_3)$ also avoid f_i.

Let w be a ternary recurrent word w avoiding f_b. Suppose for contradiction that w contains a square uu. Then there exists a non-empty word v such that $uuvuu$ is a factor of w. Thus, w contains an occurrence of f_b given by the morphism $A \mapsto u, B \mapsto u, C \mapsto v$. This contradiction shows that w is square-free.

An occurrence h of a ternary formula over Σ_3 is said to be *basic* if $\{h(A), h(B), h(C)\} = \Sigma_3$. As it is well-known, no infinite ternary word avoids squares and 012. So, every infinite ternary square-free word contains the 6 factors obtained by letter permutation of 012. Thus, an infinite ternary square-free word contains a basic occurrence of f_b if and only if it contains the same basic occurrence of $ABCAB.ABCBA$. Therefore, w contains no basic occurrence of $ABCAB.ABCBA$.

A computer check shows that the longest ternary words avoiding f_b, squares, 021020120, 102101201, and 210212012 have length 159. So we assume without loss of generality that w contains 021020120.

Suppose for contradiction that w contains 010. Since w is square-free, w contains 20102. Moreover, w contains the factor of 20120 of 021020120. So w contains the basic occurrence $A \mapsto 2$, $B \mapsto 0$, $C \mapsto 1$ of $ABCAB.ABCBA$. This contradiction shows that w avoids 010.

Suppose for contradiction that w contains 212. Since w is square-free, w contains 02120. Moreover, w contains the factor of 021020 of 021020120. So w contains the basic occurrence $A \mapsto 0$, $B \mapsto 2$, $C \mapsto 1$ of $ABCAB.ABCBA$. This contradiction shows that w avoids 212.

Since w avoids squares, 010, and 212, Theorem 1 implies that w is equivalent to b_3. By symmetry, every ternary recurrent word avoiding f_b is equivalent to b_3, $\sigma_0(b_3)$, or $\sigma_2(b_3)$.

3   Avoidability of $ABACA.ABCA$ and $ABAC.BACA.ABCA$

We consider the morphisms $m_a : 0 \mapsto 001$, $1 \mapsto 101$ and $m_b : 0 \mapsto 010$, $1 \mapsto 110$. That is, $m_a(x) = x01$ and $m_b(x) = x10$ for every $x \in \Sigma_2$.

We construct the set S of binary words as follows:

- $0 \in S$.
- If $v \in S$, then $m_a(v) \in S$ and $m_b(v) \in S$.
- If $v \in S$ and v' is a factor of v, then $v' \in S$.

Let $c(n) = |S \cup \Sigma_2^n|$ denote the factor complexity of S. By construction of S,

- $c(3n) = 6c(n)$ for $n \geq 3$,
- $c(3n + 1) = 4c(n) + 2c(n + 1)$ for $n \geq 3$,
- $c(3n + 2) = 2c(n) + 4c(n + 1)$ for $n \geq 2$.

Thus $c(n) = \Theta\left(n^{\ln 6/\ln 3}\right) = \Theta\left(n^{1 + \ln 2/\ln 3}\right)$.

Theorem 3. *Let $f \in \{ABACA.ABCA, ABAC.BACA.ABCA\}$. The set of words u such that u is recurrent in an infinite binary word avoiding f is S.*

Proof. Let R be the set of words u such that u is recurrent in an infinite binary word avoiding $ABACA.ABCA$. Let R' be the set of words u such that u is recurrent in an infinite binary word avoiding $ABAC.BACA.ABCA$. An occurrence of $ABACA.ABCA$ is also an occurrence of $ABAC.BACA.ABCA$, so that $R' \subseteq R$.

Let us show that $R \subseteq S$. We study the small factors of a recurrent binary word w avoiding $ABACA.ABCA$. Notice that w avoid the pattern $ABAAA$ since it contains the occurrence $A \mapsto A$, $B \mapsto B$, $C \mapsto A$ of $ABACA.ABCA$. Since w contains recurrent factors only, w also avoids AAA.

A computer check shows that the longest binary words avoiding $ABACA.ABCA$, AAA, 1001101001, and 0110010110 have length 53. So we assume without loss of generality that w contains 1001101001.

Suppose for contradiction that w contains 1100. Since w avoids AAA, w contains 011001. Then w contains the occurrence $A \mapsto 01, B \mapsto 1, C \mapsto 0$ of $ABACA.ABCA$. This contradiction shows that w avoids 1100.

Since w contains 0110, the occurrence $A \mapsto 0, B \mapsto 1, C \mapsto 1$ of $ABACA.ABCA$ shows that w avoids 01010. Similarly, w contains 1001 and avoids 10101.

Suppose for contradiction that w contains 0101. Since w avoids 01010 and 10101, w contains 001011. Moreover, w avoids AAA, so w contains 10010110. Then w contains the occurrence $A \mapsto 10, B \mapsto 0, C \mapsto 1$ of $ABACA.ABCA$. This contradiction shows that w avoids 0101.

A binary word is a factor of the m_a-image of some binary word if and only if it avoids $\{000, 111, 0101, 1100\}$. Indeed, both kinds of binary words are characterized by the same Rauzy graph with vertex set $\Sigma_2^3 \backslash \{000, 111\}$. So w is the m_a-image of some binary word.

Obviously, the image by a non-erasing morphism of a word containing a formula also contains the formula. Thus, the pre-image of w by m_a also avoids $ABACA.ABCA$. This shows that $R \subseteq S$.

Let us show that $S \subseteq R'$, that is, every word in S avoids $ABAC.BACA.ABCA$. We suppose for contradiction that a finite word $w \in S$ avoids $ABAC.BACA.ABCA$ and that $m_a(w)$ contains an occurrence h of $ABAC.BACA.ABCA$.

The word $m_a(w)$ is of the form $\diamond 01 \diamond 01 \diamond 01 \diamond 01 \ldots$ Thus, in $m_a(w)$:

- Every factor 00 is in position 0 (mod 3).
- Every factor 01 is in position 1 (mod 3).
- Every factor 11 is in position 2 (mod 3).
- Every factor 10 is in position 0 or 2 (mod 3), depending on whether a factor $1 \diamond 0$ is 100 or 110.

We say that a factor s is *gentle* if either $|s| \geq 3$ or $s \in \{00, 01, 11\}$. By previous remarks, all the occurrences of the same gentle factor have the same position modulo 3.

First, we consider the case such that $h(A)$ is gentle. This implies that the distance between two occurrences of $h(A)$ is 0 (mod 3). Because the repetitions $h(ABA)$, $h(ACA)$, and $h(ABCA)$ are contained in the formula, we deduce that

- $|h(AB)| = |h(A)| + |h(B)| \equiv 0 \pmod 3$.
- $|h(AC)| = |h(A)| + |h(C)| \equiv 0 \pmod 3$.
- $|h(ABC)| = |h(A)| + |h(B)| + |h(C)| \equiv 0 \pmod 3$.

This gives $|h(A)| \equiv |h(B)| \equiv |h(C)| \equiv 0 \pmod 3$. Clearly, such an occurrence of the formula in $m_a(w)$ implies an occurrence of the formula in w, which is a contradiction.

Now we consider the case such that $h(B)$ is gentle. If $h(CA)$ is also gentle, then the factors $h(BACA)$ and $h(BCA)$ imply that $|h(A)| \equiv 0 \pmod 3$. Thus, $h(A)$ is gentle and the first case applies. If $h(CA)$ is not gentle, then $h(CA) = 10$, that is, $h(C) = 1$ and $h(A) = 0$. Thus, $m_a(w)$ contains both $h(BAC) = h(B)01$ and $h(BCA) = h(B)10$. Since $h(B)$ is gentle, this implies that 01 and 10 have the same position modulo 3, which is impossible.

The case such that $h(C)$ is gentle is symmetrical. If $h(AB)$ is gentle, then $h(ABAC)$ and $h(ABC)$ imply that $|h(A)| \equiv 0 \pmod 3$. If $h(AB)$ is not gentle, then $h(A) = 1$ and $h(B) = 0$. Thus, $m_a(w)$ contains both $h(ABC) = 01h(C)$ and $h(BAC) = 10h(C)$. Since $h(C)$ is gentle, this implies that 01 and 01 have the same position modulo 3, which is impossible.

Finally, if $h(A)$, $h(B)$, and $h(C)$ are not gentle, then the length of the three fragments of the formula is $2|h(A)|+|h(B)|+|h(C)| \leq 8$. So it suffices to consider the factors of length at most 8 in S to check that no such occurrence exists.

This shows that $S \subseteq R'$. Since $R' \subseteq R \subseteq S \subseteq R'$, we obtain $R' = R = S$, which proves Theorem 3.

4   A Counter-Example to a Conjecture of Grytczuk

Grytczuk [3] has considered the notion of pattern avoidance on graphs. This generalizes the definition of nonrepetitive coloring, which corresponds to the pattern AA. Given a pattern p and a graph G, $\lambda(p, G)$ is the smallest number of colors needed to color the vertices of G such that every non-intersecting path in G induces a word avoiding p.

We think that the natural framework is that of directed graphs, and we consider only non-intersecting paths that are oriented from a starting vertex to an ending vertex. This way, $\lambda(p) = \lambda\left(p, \overrightarrow{P}\right)$ where $\overrightarrow{P}$ is the infinite oriented path with vertices v_i and arcs $\overrightarrow{v_i v_{i+1}}$, for every $i \geq 0$. The directed graphs that we consider have no loops and no multiple arcs, since they do not modify the set of non-intersecting oriented paths. However, opposite arcs (i.e., digons) are allowed. Thus, an undirected graph is viewed as a symmetric directed graph: for every pair of distinct vertices u and v, either there exists no arc between u and v, or there exist both the arcs $\overrightarrow{uv}$ and $\overrightarrow{vu}$. Let P denote the infinite undirected path. We are nitpicking about directed graphs because, even though $\lambda\left(AA, \overrightarrow{P}\right) = \lambda(AA, P) = 3$, there exist patterns such that $\lambda\left(p, \overrightarrow{P}\right) < \lambda(p, P)$. For example, $\lambda(ABCACB) = \lambda\left(ABCACB, \overrightarrow{P}\right) = 2$ and $\lambda(ABCACB, P) = 3$.

We do not attempt the hazardous task of defining a notion of avoidance for formulas on graphs.

A conjecture of Grytczuk [3] says that for every avoidable pattern p, there exists a function g such that $\lambda(p, G) \leq g(\Delta(G))$, where G is an undirected graph and $\Delta(G)$ denotes its maximum degree. Grytczuk [3] obtained that his conjecture holds for doubled patterns.

As a counterexample, we consider the pattern $ABACADABCA$ which is 2-avoidable by the result in the previous section. Of course, $ABACADABCA$ is not doubled because of the variable D. Let us show that $ABACADABCA$ is unavoidable on the infinite oriented graph $\overrightarrow{G}$ with vertices v_i and arcs $\overrightarrow{v_i v_{i+1}}$ and $\overrightarrow{v_{100i} v_{100i+2}}$, for every $i \geq 0$. Notice that $\overrightarrow{G}$ is obtained from $\overrightarrow{P}$ by adding the arcs $\overrightarrow{v_{100i} v_{100i+2}}$. Suppose that $\overrightarrow{G}$ is colored with k colors. Consider the factors in the subgraph $\overrightarrow{P}$ induced by the paths from $v_{300ik+1}$ to $v_{300ik+200k+1}$, for every $i \geq 0$. Since these factors have bounded length, the same factor appears on two disjoint such paths p_l and p_r (such that p_l is on the left of p_r). Notice that p_l contains $2k + 1$ vertices with index $\equiv 1 \pmod{100}$. By the pigeon-hole principle, p_l contains three such vertices with the same color a. Thus, p_l contains an occurrence of $ABACA$ such that $A \mapsto a$ on vertices with index $\equiv 1 \pmod{100}$. The same is true for p_r. In $\overrightarrow{G}$, the occurrences of $ABACA$ in p_l and p_r imply an occurrence of $ABACADABCA$ since we can skip an occurrence of the variable A in p_l thanks to some arc of the form $\overrightarrow{v_{100j} v_{100j+2}}$.

This shows that $ABACADABCA$ is unavoidable on $\overrightarrow{G}$, which has maximum degree 3.

References

1. Berstel, J.: Axel Thue's papers on repetitions in words: a translation, vol. 20. Publications du LACIM. Université du Québec à Montréal (1994)
2. Cassaigne, J.: Motifs évitables et régularité dans les mots. Ph.D. thesis, Université Paris VI (1994)
3. Grytczuk, J.: Pattern avoidance on graphs. Discrete Math. **307**(11–12), 1341–1346 (2007)
4. Thue, A.: Über unendliche Zeichenreihen. Norske Vid. Selsk. Skr. I. Mat. Nat. Kl. Christiania **7**, 1–22 (1906)
5. Thue, A.: Über die gegenseitige Lage gleicher Teile gewisser Zeichenreihen. Norske Vid. Selsk. Skr. I. Mat. Nat. Kl. Christiania **10**, 1–67 (1912)

Minimal Forbidden Factors of Circular Words

Gabriele Fici[✉], Antonio Restivo, and Laura Rizzo

Dipartimento di Matematica e Informatica, Università di Palermo,
Via Archirafi 34, Palermo, Italy
{gabriele.fici,antonio.restivo}@unipa.it, rizzolaura88@gmail.com

Abstract. Minimal forbidden factors are a useful tool for investigating properties of words and languages. Two factorial languages are distinct if and only if they have different (antifactorial) sets of minimal forbidden factors. There exist algorithms for computing the minimal forbidden factors of a word, as well as of a regular factorial language. Conversely, Crochemore et al. [IPL, 1998] gave an algorithm that, given the trie recognizing a finite antifactorial language M, computes a DFA of the language having M as set of minimal forbidden factors. In the same paper, they showed that the obtained DFA is minimal if the input trie recognizes the minimal forbidden factors of a single word. We generalize this result to the case of a circular word. We also discuss combinatorial properties of the minimal forbidden factors of a circular word. Finally, we characterize the minimal forbidden factors of the circular Fibonacci words.

Keywords: Minimal forbidden factor · Circular word · L-AUTOMATON

1 Introduction

Minimal forbidden factors are a useful combinatorial tool in several areas, ranging from symbolic dynamics to string processing. They have many applications, e.g. in text compression (where they are also known as *antidictionaries*) [1], in bioinformatics (where they are also known under the name *minimal absent words*) [2,3], etc. The theory of minimal forbidden factors is well developed, both from the combinatorial and the algorithmic point of view (see, for instance, [1,4–8]). In particular, there exist algorithms for computing the minimal forbidden factors of a single word [3,9–11], as well as of a regular factorial language [5]. Conversely, Crochemore et al. [6], gave an algorithm, called L-AUTOMATON that, given a trie representing a finite antifactorial set M, builds a deterministic automaton recognizing the language L whose set of minimal forbidden factors is M. The automaton built by the algorithm is not, in general, minimal. However, if M is the set of minimal forbidden factors of a single word w, then the algorithm builds the factor automaton of w, i.e., the minimal deterministic automaton recognizing the language of factors of w (see [6]).

The notion of a minimal forbidden factor has been recently extended to the case of circular words [12–14]. A circular word can be seen as a sequence of

© Springer International Publishing AG 2017
S. Brlek et al. (Eds.): WORDS 2017, LNCS 10432, pp. 36–48, 2017.
DOI: 10.1007/978-3-319-66396-8_5

symbols drawn on a circle, where there is no beginning nor end. Although a circular word can be formally defined as an equivalence class of the free monoid under the relation of conjugacy, the fact that in a circular word there is no beginning nor end leads to a less clear definition of notions as prefixes, suffixes, factors. In this paper, we consider the set of factors of a circular word w as the (infinite) set of words that appear as a factor in some power of w. Although this set is infinite, we show that its set of minimal forbidden factors is always finite.

As a main result, we prove that if M is the set of minimal forbidden factors of a circular word, then algorithm L-AUTOMATON with input a trie recognizing M builds the minimal automaton representing the set of factors of the circular word. To this end, we use combinatorial properties of the minimal forbidden factors of a circular word.

Finally, we explore the case of circular Fibonacci words, and give a combinatorial characterization of their minimal forbidden factors.

2   Preliminaires

Let A be a finite alphabet, and let A^* be the free monoid generated by A under the operation of concatenation. The elements of A^* are called *words* over A. The *length* of a word w is denoted by $|w|$. The *empty word*, denoted by ε, is the unique word of length zero and is the neutral element of A^*. If $x \in A$ and $w \in A^*$, we let $|w|_x$ denote the number of occurrences of x in w.

A *prefix* (resp. a *suffix*) of a word w is any word u such that $w = uz$ (resp. $w = zu$) for some word z. A *factor* of w is a prefix of a suffix (or, equivalently, a suffix of a prefix) of w. A prefix/suffix/factor of a word is *proper* if it is nonempty and does not coincide with the word itself. From the definitions, we have that ε is a prefix, a suffix and a factor of any word. An *occurrence* of a factor u in w is a factorization $w = vuz$. An occurrence of u is *internal* if both v and z are nonempty. The set of factors of a word w is denoted by $\mathcal{F}_w$.

The word $\widetilde{w}$ obtained by reading w from right to left is called the *reversal* (or *mirror image*) of w. A *palindrome* is a word w such that $\widetilde{w} = w$. In particular, the empty word is a palindrome.

The *conjugacy* is the equivalence relation over A^* defined by

$$w \sim w' \text{ iff } \exists\, u, v | w = uv, w' = vu.$$

When the word w is conjugate to the word w', we say that w is a *rotation* of w'. An equivalence class $[w]$ of the conjugacy relation is called a *circular word*. A representative of a conjugacy class $[w]$ is called a *linearization* of the circular word $[w]$. Therefore, a circular word $[w]$ can be viewed as the set consisting of all the rotations of a word w.

A word w is *a power* of a word v if there exists a positive integer $k > 1$ such that $w = v^k$. Conversely, w is *primitive* if $w = v^k$ implies $k = 1$. Notice that a word is primitive if and only if any of its rotations is. We can therefore extend the definition of primitivity to circular words straightforwardly. Notice that a word w (resp. a circular word $[w]$) is primitive if and only if $|[w]| = |w|$.

Remark 1. A circular word can be seen as a word drawn on a circle, where there is no beginning and no end. Therefore, the definitions of prefix/suffix/factor lose their meaning for a circular word. In the literature, a factor of a circular word $[w]$ is often defined as a factor of any linearization w of $[w]$. Nevertheless, since there is no beginning nor end, one can define a factor of w as a word that appears as a factor of w^k for some k. We will adopt this point of view in this paper.

2.1 Minimal Forbidden Factors

We now recall some basic facts about minimal forbidden factors. For further details and references, the reader may see [7,12].

A *language* over the alphabet A is a set of finite words over A, that is, a subset of A^*. A language is *factorial* if it contains all the factors of its words. The *factorial closure* of a language L is the language consisting of all factors of the words in L, that is, the language $\mathcal{F}_L = \cup_{w \in L} \mathcal{F}_w$.

The counterparts of factorial languages are antifactorial languages. A language is called *antifactorial* if no word in the language is a proper factor of another word in the language. Dual to the notion of factorial closure, there also exists the notion of *antifactorial part* of a language, obtained by removing the words that are factors of another word in the language.

Definition 2. Given a factorial language L, the (antifactorial) language of *minimal forbidden factors* of L is defined as

$$\mathcal{M}_L = \{aub \mid a, b \in A, aub \notin L, au, ub \in L\}.$$

Every factorial language L is uniquely determined by its (antifactorial) language of minimal forbidden factors $\mathcal{M}_L$, through the equation

$$L = A^* \backslash A^* \mathcal{M}_L A^*. \tag{1}$$

The converse is also true, since by the definition of a minimal forbidden factor we have

$$\mathcal{M}_L = AL \cap LA \cap (A^* \backslash L). \tag{2}$$

The previous equations define a bijection between factorial and antifactorial languages.

In the case of a single word w, the set of minimal forbidden factors of w, that we denote by $\mathcal{M}_w$, is defined as the antifactorial language $\mathcal{M}_{\mathcal{F}_w}$. Indeed, a word aub, with $a, b \in A$ and $u \in A^*$, is a minimal forbidden factor of a word w if $aub \notin \mathcal{F}_w$ and $au, ub \in \mathcal{F}_w$.

For example, consider the word $w = aabbabb$ over the alphabet $A = \{a, b\}$. The set of minimal forbidden factors of w is $\mathcal{M}_w = \{aaa, aba, bbb, baa, babba\}$.

Applying (1) and (2) to the language of factors of a single word, we have that, given two words u and v, $u = v$ if and only if $\mathcal{M}_u = \mathcal{M}_v$, that is, every word is uniquely represented by its set of minimal forbidden factors.

An important property of the minimal forbidden factors of a word w, which plays a crucial role in algorithmic applications, is that their number is linear in the size of w. Let w be a word of length n over an alphabet A of cardinality σ. In [7] it is shown that the total number of minimal forbidden factors of w is smaller than or equal to σn. Actually, $\mathcal{O}(\sigma n)$ is a tight asymptotic bound for the number of minimal forbidden factors of w whenever $2 \leq \sigma \leq n$ [12]. They can therefore be stored on a trie[1], whose number of nodes is linear in the size of the word.

2.2 Automata for Minimal Forbidden Factors

Recall that a *deterministic finite state automaton* (DFA) is a 5-tuple $\mathcal{A} = (Q, A, i, T, \delta)$, where Q is the finite set of states, A is the current alphabet, i is the initial state, T the set of terminal (or final) states, and $\delta : (Q \times A) \mapsto Q$ is the transition function. A word is *recognized* (or *accepted*) by $\mathcal{A}$ if reading w from the initial state one ends in a final state. The language recognized (or accepted) by $\mathcal{A}$ is the set of all words recognized by $\mathcal{A}$. A language is *regular* if it is recognized by some DFA. A DFA $\mathcal{A}$ is *minimal* if it has the least number of states among all the DFA's recognizing the same language as $\mathcal{A}$. The minimal DFA is unique.

It follows from basic closure properties of regular languages that the bijection between factorial and antifactorial languages expressed by (1) and (2) preserves regularity, that is, a factorial language is regular if and only if its language of minimal forbidden factors is.

The *factor automaton* of a word w is the minimal DFA recognizing the (finite) language $\mathcal{F}_w$. The factor automaton of a word of length n has less than $2n$ states, and can be built in $\mathcal{O}(n)$ time and space by an algorithm that also constructs the *failure function* of the automaton [15]. The failure function of a state p (different from the initial state) is a link to another state q defined as follows. Let u be a nonempty word and $p = \delta(i, u)$. Then $q = \delta(i, u')$, where u' is the longest suffix of u for which $\delta(i, u) \neq \delta(i, u')$. It can be shown that this definition does not depend on the particular choice of u [6]. An example of a factor automaton is displayed in Fig. 1.

In [5], the authors gave a quadratic-time algorithm to compute the set of minimal forbidden factors of a regular factorial language L. However, computing the minimal forbidden factors of a single word can be done in linear time in the length of the word. Algorithm MF-TRIE, described in [6] and presented in Fig. 2, builds the trie of the set $\mathcal{M}_w$ having as input the factor automaton of w, together with its failure function. Moreover, the states of the output trie recognizing the set $\mathcal{M}_w$ are the same as those of the factor automaton of w, plus some sink states, which are the terminal states with no outgoing edges, corresponding to the minimal forbidden factors. An example is given in Fig. 3.

More recently, other algorithms have been introduced to compute the minimal forbidden factors of a word. The computation of minimal forbidden factors

---

[1] A *trie* representing a finite language L is a tree-like deterministic automaton recognizing L, where the set of states is the set of prefixes of words in L, the initial state is the empty word ε, the set of final states is a set of *sink* states, and the set of transitions is $\{(u, a, ua) | a \in A\}$.

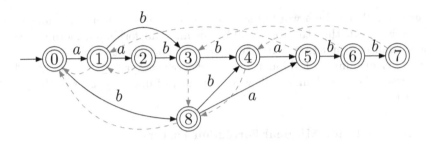

Fig. 1. The factor automaton of the word $w = aabbabb$. It is the minimal DFA recognizing $\mathcal{F}_w$. Dashed edges correspond to the failure function links.

MF-TRIE (factor automaton $\mathcal{A} = (Q, A, i, T, \delta)$ and its failure function f)
1. **for** each state $p \in Q$ in width-first search from i **and** each $a \in A$
2.   **if** $\delta(p, a)$ undefined **and** $(p = i$ **or** $\delta(f(p), a)$ defined)
3. $\delta'(p, a) \leftarrow$ new sink;
4.   **else**
5.   **if** $\delta(p, a) = q$ **and** q not already reached
6. $\delta'(p, a) \leftarrow q$;
7. **return** $(Q, A, i, \{sinks\}, \delta')$;

Fig. 2. Algorithm MF-TRIE. It takes as input the factor automaton of a word w and builds the trie of the set $\mathcal{M}_w$.

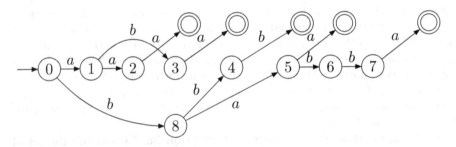

Fig. 3. The trie of the set $\mathcal{M}_w = \{aaa, aba, bbb, baa, babba\}$ of minimal forbidden factors of the word $w = aabbabb$ output by algorithm MF-TRIE when the input is the factor automaton of w. The edges are labeled by single letters for convenience.

based on the construction of suffix arrays was considered in [9]; although this algorithm has a linear-time performance in practice, the worst-case time complexity is $\mathcal{O}(n^2)$. New $\mathcal{O}(n)$-time and $\mathcal{O}(n)$-space suffix-array-based algorithms were presented in [3,10,11]. A more space-efficient solution to compute all minimal forbidden factors in time $\mathcal{O}(n)$ was also presented in [16].

We have described algorithms for computing the set of minimal forbidden factors of a given factorial language. We are now describing an algorithm performing the reverse operation. Let M be an antifactorial language. We let $L(M)$

L-AUTOMATON (trie $\mathcal{T} = (Q, A, i, T, \delta')$)
1. **for** each $a \in A$
2.   **if** $\delta'(i, a)$ defined
3.   set $\delta(i, a) = \delta'(i, a)$;
4.   set $f(\delta(i, a)) = i$;
5.   **else**
6.   set $\delta(i, a) = i$;
7. **for** each state $p \in Q \setminus \{i\}$ in width-first search **and** each $a \in A$
8.   **if** $\delta'(p, a)$ is defined
9.   set $\delta(p, a) = \delta'(p, a)$;
10.   set $f(\delta(p, a)) = \delta(f(p), a)$;
11.   **else if** $p \notin T$
12.   set $\delta(p, a) = \delta(f(p), a)$;
13.   **else**
14.   set $\delta(p, a) = p$;
15. **return** $(Q, A, i, Q \setminus T, \delta)$;

Fig. 4. Algorithm L-AUTOMATON. It builds an automaton recognizing the language $L(M)$ of words avoiding an antifactorial language M on the input trie $\mathcal{T}$ accepting M.

denote the (factorial) language avoiding M, that is, the language of all the words that do not contain any word of M as a factor. Clearly, from Eqs. (1) and (2), we have that $L(M)$ is the unique language whose set of minimal forbidden factors is M, i.e., the unique language L such that $\mathcal{M}_L = M$.

For a finite antifactorial language M, algorithm L-AUTOMATON [6] builds a DFA recognizing $L(M)$. It is presented in Fig. 4. The algorithm runs in linear time in the size of the trie storing the words of M. It uses a failure function f defined in a way analogous to the one used for building the factor automaton.

The algorithm can be applied for retrieving a word from its set of minimal forbidden factors, and this can be done in linear time in the length of the word, since the size of the trie of minimal forbidden factors of a word is linear in the length of the word. Notice that even if M is finite, the language $L(M)$ can be finite or infinite. Moreover, if $L(M)$ is finite, it can be the language of factors of a single word or of a set of words.

Algorithm L-AUTOMATON builds an automaton recognizing the language $L(M)$ of words avoiding a given antifactorial language M, but this automaton is not, in general, minimal. However, the following result holds [6]:

Theorem 3. *If M is the set of the minimal forbidden factors of a finite word w, then the automaton output from algorithm L-AUTOMATON on the input trie recognizing M, after removing sink states, is the factor automaton of w, i.e., it is minimal.*

To see that the minimality described in the previous theorem does not hold in general, consider for instance the antifactorial language $M = \{aa, ba\}$. It can be easily checked that the automaton output from algorithm L-AUTOMATON, after removing sink states, has three states, while the minimal automaton of the language $L(M) = \{b^n | n \geq 0\} \cup \{ab^n | n \geq 0\}$ has only two states.

We will prove in the next section that this minimality property still holds true in the case of minimal forbidden factors of a circular word.

3   Minimal Forbidden Factors of a Circular Word

Given a word w, the language *generated* by w is the language $w^* = \{w^k | k \geq 0\} = \{\varepsilon, w, ww, www, \ldots\}$. Analogously, the language L^* generated by $L \subset A^*$ is the set of all possible concatenations of words in L, i.e., $L^* = \{\varepsilon\} \cup \{w_1 w_2 \cdots w_n | w_i \in L \text{ for } i = 1, 2, \ldots, n\}$.

Let w be a word of length at least 2. The language w^* generated by w is not a factorial language, nor is the language generated by all the rotations of w. Nevertheless, if we take the factorial closure of the language generated by w, then of course we get a factorial language $\mathcal{F}_{w^*}$. Now, if z is conjugate to w, then although w and z generate different languages, the factorial closures of the languages they generate coincide, i.e., $\mathcal{F}_{w^*} = \mathcal{F}_{z^*}$. Moreover, for any power w^k of w, $k > 0$, one clearly has $\mathcal{F}_{w^*} = \mathcal{F}_{(w^k)^*}$.

Based on the previous discussion, and on Remark 1, we give the following definition: We let the set of *factors of a circular word* $[w]$ be the (factorial) language $\mathcal{F}_{w^*}$, where w is any linearization of $[w]$. By the previous remark, this definition is independent of the particular choice of the linearization. Moreover, we can suppose that $[w]$ is a primitive circular word.

The set of minimal forbidden factors of the circular word $[w]$ is defined as the set $\mathcal{M}_{\mathcal{F}_{w^*}}$ of minimal forbidden factors of the language $\mathcal{F}_{w^*}$, where w is any linearization of $[w]$. We already showed that this is independent from the particular choice of the linearization. To simplify the notation, in the remainder of this paper we will let $\mathcal{M}_{[w]}$ denote the set of minimal forbidden factors of the circular word $[w]$.

For instance, if $[w] = [aabbabb]$, then we have

$$\mathcal{M}_{[w]} = \{aaa, aba, bbb, aabbaa, babbab\}.$$

Notice that $\mathcal{M}_{[w]}$ does not coincide with the set of minimal forbidden factors of the factorial closure of the language of all the rotations of w (see [12] for a comparison between the two definitions).

Although $\mathcal{F}_{w^*}$ is an infinite language, the set $\mathcal{M}_{[w]} = \mathcal{M}_{\mathcal{F}_{w^*}}$ of minimal forbidden factors of $[w]$ is always finite. More precisely, we have the following structural lemma.

Lemma 4. *Let* $[w]$ *be a circular word and* w *any linearization of* $[w]$*. Then*

$$\mathcal{M}_{[w]} = \mathcal{M}_{ww} \cap A^{\leq |w|}. \tag{3}$$

Proof. If aub, with $a, b \in A$ and $u \in A^*$, is an element in $\mathcal{M}_{ww} \cap A^{\leq |w|}$, then clearly $aub \in \mathcal{M}_{\mathcal{F}_{w^*}} = \mathcal{M}_{[w]}$.

Conversely, let aub, with $a, b \in A$ and $u \in A^*$, be an element in $\mathcal{M}_{[w]} = \mathcal{M}_{\mathcal{F}_{w^*}}$. Then $aub \notin \mathcal{F}_{w^*}$, while $au, ub \in \mathcal{F}_{w^*}$. So, there exists some letter $\bar{b}$ different from b such that $au\bar{b} \in \mathcal{F}_{w^*}$ and a letter $\bar{a}$ different from a such that $\bar{a}ub \in \mathcal{F}_{w^*}$. Therefore, $au, \bar{a}u, ub, u\bar{b} \in \mathcal{F}_{w^*}$. It is readily verified that any word of length at least $|w| - 1$ cannot be extended to the right nor to the left by different letters in $\mathcal{M}_{\mathcal{F}_{w^*}}$. Hence $|aub| \leq |w|$. Since au and ub are factors of some rotation of w, we have $au, ub \in \mathcal{F}_{ww}$, whence $aub \in \mathcal{M}_{ww}$. □

The equality (3) was first introduced as the definition for the set of minimal forbidden factors of a circular word in [14].

About the number of minimal forbidden factors of a circular words we have the following bounds.

Lemma 5. *Let* $[w]$ *be a circular word over the alphabet* A *and let* $A(w)$ *be the set of letters of* A *that occur in* w. *Then*

$$|A| \leq |\mathcal{M}_{[w]}| \leq |A| + (n-1)|A(w)| - n. \tag{4}$$

Proof. The inequality $|A| \leq |\mathcal{M}_{[w]}|$ follows from the fact that for each letter $a \in A$ there exists an integer $n_a > 0$ such that $a^{n_a} \in \mathcal{M}_{[w]}$. For the upper bound, we first observe that the minimal forbidden factors of length 1 of $[w]$ are precisely the elements of $A \backslash A(w)$. We now count the minimal forbidden factors of length greater than one. Recall by Lemma 4 that $\mathcal{M}_{[w]} = \mathcal{M}_{ww} \cap A^{\leq |w|}$. Let $ww = w_1 w_2 \cdots w_{2n}$. Consider a position i in ww such that $n \leq i < 2n$. We claim that there are at most $|A|$ distinct elements of $\mathcal{M}_{[w]}$ of length greater than one whose longest proper prefixes have an occurrence ending in position i. Indeed, by contradiction, let $b \in A$ such that there exist $ub, vb \in \mathcal{M}_{[w]}$ and both u and v occur in ww ending in position i. This implies that ub and vb are one suffix of another, against the minimality of the minimal forbidden factors. Since the letter b must be different from the letter of ww occurring in position $i + 1$, we therefore have that the number of minimal forbidden factors obtained for i ranging from n to $2n - 1$ is at most $n(|A(w)| - 1)$. For i such that $1 \leq i < n$ (resp. $i = 2n$), if an element $ub \in \mathcal{M}_{[w]}$, $b \in A$, is such that u has an occurrence in ww ending in position i, then u has also an occurrence ending in position $i + n$ (resp. n), so it has already been counted. Hence,

$$|\mathcal{M}_{[w]}| \leq |A| - |A(w)| + n(|A(w)| - 1) = |A| + (n-1)|A(w)| - n. \quad □$$

We now give a result analogous to Theorem 3 in the case of circular words.

Theorem 6. *If* M *is the set of the minimal forbidden factors of a primitive circular word* $[w]$, *then the automaton output from algorithm* L-AUTOMATON *on the input trie* $\mathcal{T}$ *recognizing* M, *after removing sink states, is the minimal automaton recognizing the language* $\mathcal{F}_{w^*}$ *of factors of* $[w]$.

Proof. Let $\mathcal{A} = (Q, A, i, Q \backslash T, \delta)$ be the automaton output by algorithm L-AUTOMATON with input the trie $\mathcal{T}$ recognizing the set of the minimal forbidden factors of a circular word $[w]$. Let $w = w_1 w_2 \cdots w_n$ be a linearization of $[w]$. The automaton $\mathcal{A}$ recognizes the language $\mathcal{F}_{w^*}$ since its input recognizes the language $\mathcal{M}_{[w]} = \mathcal{M}_{\mathcal{F}_{w^*}}$. To prove that $\mathcal{A}$ is minimal, we have to prove that any two states are distinguishable. Suppose by contradiction that there are two nondistinguishable states $p, q \in Q$. By construction, p and q are respectively associated with two proper prefixes, v_p and v_q, of words in $\mathcal{M}_{\mathcal{F}_{w^*}}$, which, by Lemma 4, is equal to $\mathcal{M}_{ww} \cap A^{\leq |w|}$. Therefore, v_p and v_q are factors of w^* of length $\leq |w|$. Hence, they are both factors of w^2. Let us then write $w^2 = xv_p y = x'v_q y'$, with x and x' of minimal length.

Suppose first that there exists i such that xv_p and $x'v_q$ both end in $w_1 w_2 \cdots w_i$. Then they are one suffix of another. Since p and q are nondistinguishable, there exists a word z such that $xv_p z$ and $x'v_q z$ end in a sink state, that is, are elements of $\mathcal{M}_{[w]}$. This is a contradiction since $\mathcal{M}_{[w]}$ is an antifactorial set and $xv_p z$ and $x'v_q z$ are one suffix of another.

Suppose now that xv_p ends in $w_1 w_2 \cdots w_i$ and $x'v_q$ ends in $w_1 w_2 \cdots w_j$ for $i \neq j$. Since p and q are nondistinguishable, for any word u one has that that $v_p u \in \mathcal{F}_{w^*}$ if and only if $v_q u \in \mathcal{F}_{w^*}$. Since $\mathcal{F}_{w^*}$ is a factorial language, we therefore have that there exists a word z of length $|w|$ such that $v_p z$ and $v_q z$ are both in $\mathcal{F}_{w^*}$. But this implies that $z = w_{i+1} w_{i+2} \cdots w_i = w_{j+1} w_{j+2} \cdots w_j$, and this leads to a contradiction since w is primitive and therefore all its rotations are distinct.   □

4   Circular Fibonacci Words and Minimal Forbidden Factors

In this section, we illustrate the combinatorial results discussed in the previous section in the special case of the circular Fibonacci words.

The sequence $(f_n)_{n \geq 1}$ of Fibonacci words is defined recursively by $f_1 = b$, $f_2 = a$ and $f_n = f_{n-1} f_{n-2}$ for $n > 2$. The length of the word f_n is the Fibonacci number F_n.

Let us recall some well-known properties of the Fibonacci words. For every $n \geq 3$, one can write $f_n = u_n ab$ if n is odd or $f_n = u_n ba$ if n is even, where u_n is a palindrome. Moreover, since $f_n = f_{n-1} f_{n-2}$ and the words u_n are palindromes, one has that for every $n \geq 5$

$$f_n = u_n xy = u_{n-1} yx u_{n-2} xy = u_{n-2} xy u_{n-1} xy \qquad (5)$$

for letters x, y such that $\{x, y\} = \{a, b\}$. The first few Fibonacci words f_n and the first few words u_n are shown in Table 1.

Recall that a *bispecial factor* of a word w over the alphabet $A = \{a, b\}$ is a word v such that av, bv, va, vb are all factors of w. From basic properties of Fibonacci words, it can be proved that for every $n \geq 4$ the set of bispecial factors of the word f_n is $\{u_3, u_4, \ldots, u_{n-1}\}$, while the set of bispecial factors of the word $f_n f_n$ is $\{u_3, u_4, \ldots, u_n\}$.

The words f_n (as well as the words $f_n f_n$) are *balanced*, that is, for every pair of factors u and v of the same length, one has $||u|_a - |v|_a| \leq 1$ (and therefore also $||u|_b - |v|_b| \leq 1$).

Table 1. The first few Fibonacci words f_n and the first few words u_n.

$$f_1 = b$$
$$f_2 = a$$
$$f_3 = ab \qquad\qquad\qquad u_3 = \varepsilon$$
$$f_4 = aba \qquad\qquad\qquad u_4 = a$$
$$f_5 = abaab \qquad\qquad\qquad u_5 = aba$$
$$f_6 = abaababa \qquad\qquad u_6 = abaaba$$
$$f_7 = abaababaabaab \qquad u_7 = abaababaaba$$

Table 2. The first few elements of the sequences $\hat{f}_n$ and $\hat{g}_n$.

$$\hat{f}_3 = aa \qquad\qquad\qquad \hat{g}_3 = bb$$
$$\hat{f}_4 = bab \qquad\qquad\qquad \hat{g}_4 = aaa$$
$$\hat{f}_5 = aabaa \qquad\qquad\qquad \hat{g}_5 = babab$$
$$\hat{f}_6 = babaabab \qquad\qquad \hat{g}_6 = aabaabaa$$
$$\hat{f}_7 = aabaababaabaa \qquad \hat{g}_7 = babaababaabab$$

Let us now define the sequence of words $(\hat{f}_n)_{n\geq 3}$ by $\hat{f}_n = au_na$ if n is odd, $\hat{f}_n = bu_nb$ if n is even. These words are known as *singular words*. Analogously, we can define the sequence of words $(\hat{g}_n)_{\geq 3}$ by $\hat{g}_n = bu_nb$ if n is odd, $\hat{g}_n = au_na$ if n is even. For every n, the word $\hat{g}_n$ is obtained from the word $\hat{f}_n$ by changing the first and the last letter. The first few values of the sequences $\hat{f}_n$ and $\hat{g}_n$ are shown in Table 2.

We will now describe the structure of the sets of minimal forbidden factors of circular Fibonacci words in terms of the words $\hat{f}_n$ and $\hat{g}_n$.

The first few sets $\mathcal{M}_{[f_n]}$ are displayed in Table 3. We have $\mathcal{M}_{[f_1]} = \mathcal{M}_{[b]} = \{a\}$, $\mathcal{M}_{[f_2]} = \mathcal{M}_{[a]} = \{b\}$ and $\mathcal{M}_{[f_3]} = \mathcal{M}_{[ab]} = \{aa, bb\}$. The following theorem gives a characterization of the sets $\mathcal{M}_{[f_n]}$ for $n \geq 4$.

Table 3. The first few sets of minimal forbidden factors of the circular Fibonacci words.

$$\mathcal{M}_{[f_1]} = \{a\}$$
$$\mathcal{M}_{[f_2]} = \{b\}$$
$$\mathcal{M}_{[f_3]} = \{aa, bb\}$$
$$\mathcal{M}_{[f_4]} = \{bb, aaa, bab\}$$
$$\mathcal{M}_{[f_5]} = \{bb, aaa, aabaa, babab\}$$
$$\mathcal{M}_{[f_6]} = \{bb, aaa, babab, aabaabaa, babaabab\}$$
$$\mathcal{M}_{[f_7]} = \{bb, aaa, babab, aabaabaa, aabaababaabaa, babaababaabab\}$$

Theorem 7. *For every $n \geq 4$, $\mathcal{M}_{[f_n]} = \{\hat{g}_3, \hat{g}_4, \ldots, \hat{g}_n, \hat{f}_n\}$.*

Proof. By Lemma 4, $\mathcal{M}_{[f_n]} = \mathcal{M}_{f_n f_n} \cap A^{\leq |f_n|}$. Let xuy, $u \in A^*$, $x, y \in A$, be in $\mathcal{M}_{f_n f_n} \cap A^{\leq |f_n|}$. Then xu has an occurrence in $f_n f_n$ followed by letter $\bar{y}$, the complement of y, and uy has an occurrence in $f_n f_n$ preceded by letter $\bar{x}$, the complement of x. Therefore, u is a bispecial factor of the word $f_n f_n$, hence $u \in \{u_3, u_4, \ldots, u_n\}$. Thus, an element in $\mathcal{M}_{[f_n]}$ is of the form $\alpha u_i \beta$ for some $3 \leq i \leq n$ and $\alpha, \beta \in A$.

Claim. The singular word $\hat{f}_n$ is a minimal forbidden factor of the word $f_n f_n$.

Proof: Let $\hat{f}_n = x u_n x$, $x \in A$. The word $u_n x$ appears in $f_n f_n$ only as a prefix of one of the two occurrences of f_n, so it appears in $f_n f_n$ only preceded by the letter $\bar{x}$ different from x, hence $x u_n x$ cannot be a factor of $f_n f_n$. Finally, the word $x u_n$ appears as a factor in $f_n f_n$ since from (5) one can write

$$f_n f_n = u_n x y u_n x y = u_{n-1} y \ x u_{n-2} x y u_{n-1} y \ u_{n-1} y x u_{n-2} x y \tag{6}$$
$$= u_{n-1} y \ x u_n y \ u_{n-1} y x u_{n-2} x y.$$

Claim. The singular word $\hat{f}_n$ is a factor of the word $f_{n+1} f_{n+1}$.

Proof: The first letter of $\hat{f}_n$ is equal to the last letter of f_{n+1} and, by removing the first letter from $\hat{f}_n$, one obtains a prefix of f_{n+1}. Hence, $\hat{f}_n$ is a factor of the word $f_{n+1} f_{n+1}$.

Claim. For every $3 \leq i \leq n$, the word $\hat{g}_i$ is a minimal forbidden factor of the word $f_n f_n$.

Proof: From the previous claim, it follows that for every $3 \leq i \leq n$, the word $\hat{f}_i$ is factor of the word $f_n f_n$. Therefore $\hat{g}_i$ cannot be a factor of $f_n f_n$ otherwise the word $f_n f_n$ would not be balanced. Since removing the first or the last letter from the word $\hat{g}_i$ one obtains a factor of the word $\hat{f}_i$, the claim is proved.

Finally, from (5) and (6), for every $3 \leq i \leq n$, the words $x u_i y$ and $y u_i x$ are factors of $f_n f_n$. This completes the proof.   □

Notice that, by Lemma 4, for any circular word $[w]$, one has that $|w|$ is an upper bound on the length of the minimal forbidden factors of $[w]$. The previous theorem shows that this bound is indeed tight. However, the maximum length of a minimal forbidden factor of a circular word $[w]$ is not always equal to $|w|$. For example, for $w = aabbab$ one has $\mathcal{M}_{[w]} = \{aaa, bbb, aaba, abab, babb, bbaa\}$.

Corollary 8. *For every $n \geq 2$, the cardinality of $\mathcal{M}_{[f_n]}$ is $n - 1$.*

By Theorem 6, if $\mathcal{T}$ is the trie recognizing the set $\{\hat{g}_3, \hat{g}_4, \ldots, \hat{g}_n, \hat{f}_n\}$, then algorithm L-AUTOMATON on the input trie $\mathcal{T}$ builds the minimal deterministic automaton recognizing $\mathcal{F}_{f_n^*}$. Since the automaton output by algorithm L-AUTOMATON has the same set of states of the input trie $\mathcal{T}$ after removing sink states, and since removing the last letter from each word $\hat{g}_i$ results in a prefix of $\hat{f}_i$, we have that the factor automaton of the circular Fibonacci word $[f_n]$, that is, the minimal automaton recognizing $\mathcal{F}_{f_n^*}$, has exactly $2F_n - 1$ states (see Fig. 5 for an example).

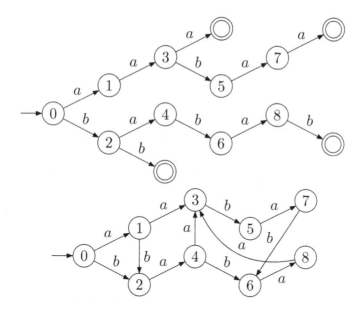

Fig. 5. The trie $\mathcal{T}$ recognizing the set $\mathcal{M}_{[f_5]}$ (top), and the automaton output by algorithm L-AUTOMATON on the input trie $\mathcal{T}$ after removing sink states (bottom), which is the minimal automaton recognizing $\mathcal{F}_{f_5^*}$. It has $9 = 2F_5 - 1$ states.

5 Conclusions and Open Problems

We proved that the automaton built by algorithm L-AUTOMATON on the input trie recognizing the set of minimal forbidden factors of a circular word is minimal. More generally, it would be interesting to characterize those antifactorial languages for which algorithm L-AUTOMATON builds a minimal automaton.

References

1. Crochemore, M., Mignosi, F., Restivo, A., Salemi, S.: Text compression using antidictionaries. In: Wiedermann, J., Boas, P.E., Nielsen, M. (eds.) ICALP 1999. LNCS, vol. 1644, pp. 261–270. Springer, Heidelberg (1999). doi:10.1007/3-540-48523-6_23
2. Chairungsee, S., Crochemore, M.: Using minimal absent words to build phylogeny. Theor. Comput. Sci. **450**, 109–116 (2012)
3. Barton, C., Héliou, A., Mouchard, L., Pissis, S.P.: Linear-time computation of minimal absent words using suffix array. BMC Bioinform. **15**, 388 (2014)
4. Béal, M., Mignosi, F., Restivo, A., Sciortino, M.: Forbidden words in symbolic dynamics. Adv. Appl. Math. **25**(2), 163–193 (2000)
5. Béal, M., Crochemore, M., Mignosi, F., Restivo, A., Sciortino, M.: Computing forbidden words of regular languages. Fundam. Inform. **56**(1–2), 121–135 (2003)
6. Crochemore, M., Mignosi, F., Restivo, A.: Automata and forbidden words. Inf. Process. Lett. **67**, 111–117 (1998)

7. Mignosi, F., Restivo, A., Sciortino, M.: Words and forbidden factors. Theor. Comput. Sci. **273**(1–2), 99–117 (2002)
8. Fici, G., Mignosi, F., Restivo, A., Sciortino, M.: Word assembly through minimal forbidden words. Theor. Comput. Sci. **359**(1), 214–230 (2006)
9. Pinho, A.J., Ferreira, P., Garcia, S.P.: On finding minimal absent words. BMC Bioinform. **10**(1), 137 (2009)
10. Fukae, H., Ota, T., Morita, H.: On fast and memory-efficient construction of an antidictionary array. In: ISIT, pp. 1092–1096. IEEE (2012)
11. Barton, C., Heliou, A., Mouchard, L., Pissis, S.P.: Parallelising the computation of minimal absent words. In: Wyrzykowski, R., Deelman, E., Dongarra, J., Karczewski, K., Kitowski, J., Wiatr, K. (eds.) PPAM 2015. LNCS, vol. 9574, pp. 243–253. Springer, Cham (2016). doi:10.1007/978-3-319-32152-3_23
12. Crochemore, M., Fici, G., Mercas, R., Pissis, S.P.: Linear-time sequence comparison using minimal absent words and applications. In: Kranakis, E., Navarro, G., Chávez, E. (eds.) LATIN 2016. LNCS, vol. 9644. Springer, Heidelberg (2016). doi:10.1007/978-3-662-49529-2_25
13. Ota, T., Morita, H.: On antidictionary coding based on compacted substring automaton. In: ISIT, pp. 1754–1758. IEEE (2013)
14. Ota, T., Morita, H.: On a universal antidictionary coding for stationary ergodic sources with finite alphabet. In: ISITA, pp. 294–298. IEEE (2014)
15. Crochemore, M., Hancart, C.: Automata for matching patterns. In: Handbook of Formal Languages. Springer 399–462(1997)
16. Belazzougui, D., Cunial, F., Kärkkäinen, J., Mäkinen, V.: Versatile succinct representations of the bidirectional Burrows-Wheeler transform. In: Bodlaender, H.L., Italiano, G.F. (eds.) ESA 2013. LNCS, vol. 8125, pp. 133–144. Springer, Heidelberg (2013). doi:10.1007/978-3-642-40450-4_12

A de Bruijn Sequence Construction by Concatenating Cycles of the Complemented Cycling Register

Daniel Gabric and Joe Sawada[(✉)]

University of Guelph, Guelph, Canada
{dgabric,jsawada}@uoguelph.ca

Abstract. We present a new de Bruijn sequence construction based on co-necklaces and the complemented cycling register (CCR). A co-necklace is the lexicographically smallest string in an equivalence class of strings induced by the CCR. We prove that a concatenation of the cycles of the CCR forms a de Bruijn sequence when the cycles are ordered in colexicographic order with respect to their co-necklace representatives. We also give an algorithm that produces the de Bruijn sequence in $O(1)$-time per bit. Finally, we prove that our construction has a discrepancy bounded above by $2n$.

1 Introduction

Let $\mathbf{B}(n)$ be the set of binary strings of length n. It is well known that the pure cycling register, which takes a binary string and outputs its first bit, partitions $\mathbf{B}(n)$ into equivalence classes under rotation. The lexicographically smallest representative of each equivalence class is called a necklace. For $n = 5$, the eight necklace equivalence classes are listed in columns as follows:

00000	00001	00011	00101	00111	01011	01111	11111.
	00010	00110	01010	01110	10110	11110	
	00100	01100	10100	11100	01101	11101	
	01000	11000	01001	11001	11010	11011	
	10000	10001	10010	10011	10101	10111	

The first string in each equivalence class is its necklace representative and the necklaces are listed from left to right in lexicographic order. For each equivalence class, observe that the string obtained by concatenating the first bit from each string yields the longest aperiodic prefix of the necklace representative. Now consider the string of length $2^5 = 32$ obtained by concatenating these highlighted bits (top down, then left to right):

$$0\ 00001\ 00011\ 00101\ 00111\ 01011\ 011111.$$

Amazingly, when considered cyclicly, this constructed string contains every string in $\mathbf{B}(5)$ as a substring exactly once. Strings with this property for a given n are called de Bruijn sequences.

© Springer International Publishing AG 2017
S. Brlek et al. (Eds.): WORDS 2017, LNCS 10432, pp. 49–58, 2017.
DOI: 10.1007/978-3-319-66396-8_6

In this paper, we show a similar property with respect to the complemented cycling register (CCR), which takes a binary string and outputs the complement of the first bit. The CCR partitions $\mathbf{B}(n)$ into equivalence classes of size up to $2n$. We call the lexicographically smallest string in each such equivalence class a co-necklace. For $n = 5$, the four co-necklace equivalence classes are listed in columns as follows:

00000	00010	00100	01010
00001	00101	01001	10101.
00011	01011	10011	
00111	10111	00110	
01111	01110	01101	
11111	11101	11011	
11110	11010	10110	
11100	10100	01100	
11000	01000	11001	
10000	10001	10010	

Observe that the co-necklace representative is positioned at the top of each class, and the classes are ordered in lexicographic order with respect to the co-necklaces. Within each equivalence class, each successive string is a left rotation of the string above it after complementing the final bit. For each equivalence class, let α denote the co-necklace and observe that the string obtained by concatenating together the first bit from each string yields the longest aperiodic prefix of $\alpha\bar{\alpha}$, where $\bar{\alpha}$ denotes the complement of α. We call such a string a *cycle of the CCR*. Consider the string obtained by concatenating these cycles of the CCR (top down, then left to right):

$$0000011111 \ 0001011101 \ 0010011011 \ 01.$$

This string is not a de Bruijn sequence since it contains the substring 11010 twice. It also contains the substring 10100 twice and is missing 01010 and 10101. However, observe what happens if we list the equivalence classes in colexicographic order with respect to the co-necklace representatives. This listing is obtained by swapping the second and third classes from the previous lexicographic order example. Using this colexicographic order and concatenating the CCR cycles together yields the following de Bruijn sequence

$$0000011111 \ 0010011011 \ 0001011101 \ 01.$$

Main result: The main result of this paper is to provide a simple proof that this construction using co-necklaces and CCR cycles, for arbitrary n, yields a de Bruijn sequence. We provide an algorithm to generate the co-necklaces of length n in colexicographic order that runs in $O(n)$-amortized time per string. This allows us to generate the de Bruijn sequence in $O(1)$-time per it. Additionally, we demonstrate that the constructed de Bruijn sequences have a very nice property: they have discrepancy bounded above by $2n$.

In the next subsection, we provide some history of de Bruijn sequence constructions and the complemented cycling register. Then in Sect. 2, we present formal definitions of our key objects and notation. In Sect. 3, we prove our main result which includes implementation details and an analysis. Finally in Sect. 4 we discuss an interesting property of our constructed de Bruijn sequences.

1.1 History

The lexicographic necklace concatenation approach was first presented by Fredricksen and Maiorana [11]. An algorithm to generate necklaces in lexicographic order [13] was shown to run in $O(1)$-amortized time per necklace [19] which means the corresponding de Bruijn sequence can be generated in $O(1)$-amortized time per bit. Rather surprisingly, it was only recently discovered that a similar algorithm works for necklaces in colexicographic order [6]. Since necklaces can also be generated in colexicographic order in $O(1)$-time [21], the corresponding de Bruijn sequence can also be generated in $O(1)$-amortized time per bit.

Other methods for generating de Bruijn sequences of order n include greedy approaches and successor-based approaches. A major drawback of the greedy approaches [2,7,10,17] is they all require an exponential amount of space. The first successor-rule based approach was by Fredricksen [12] for the lexicographically smallest de Bruijn sequence, which also happens to be equivalent to the lexicographic necklace concatenation approach. Additional successor based approaches [8,9,16] generate de Bruijn sequences requiring $O(n^2)$-time per bit. In particular, the constructions by Etzion [9] and Huang [16] are also based on the CCR. A more recent construction [20] based on the PCR requires $O(n)$-time to compute each successive bit, and an optimization allows the entire sequence to be generated in $O(1)$-amortized time per bit.

There is a well known correspondence between co-necklaces of order n and necklaces of order n that contain an odd number of 1s. A discussion from [4] describes how representatives from these equivalence classes can be generated in $O(1)$-amortize time per string. However, there is no efficient algorithm known to generate the lexicographically smallest representatives, the co-necklaes, in either lexicographic or colexicographic order. The enumeration sequence for co-necklaces A000016 was one of the first listed in the Online Encyclopedia of Integer Sequences [1].

2 Background and Definitions

A *necklace* is the lexicographically smallest string in an equivlance class of strings under rotation. Let $\bar{\alpha}$ be the complement of the string α. We say that α is a *co-necklace* if $\alpha\bar{\alpha}$ is a necklace. Let **coneck**(α) denote the set containing all length n substrings of the circular string $\alpha\bar{\alpha}$. For example,

coneck$(00000) = \{00000, 00001, 00011, 00111, 01111, 11111, 11110, 11100, 11000, 10000\}$.

Let $\mathbf{coN}(n)$ be the set of all co-necklaces of length n, so $\mathbf{coN}(5) = \{00000, 00010, 00100, 01010\}$. Clearly every distinct **coneck** set contains a co-necklace, which is it's lexicographically least element. It is also well known that $\{\mathbf{coneck}(\alpha) \mid \alpha \in \mathbf{coN}(n)\}$ is a partition of $\mathbf{B}(n)$ [4,9,15,18]. The number of co-necklaces of length n is the same as the number of cycles of the CCR (with respect to n) as well as the number of necklaces of length n with an odd number of 1s and is given by the formula [14]

$$|\mathbf{coN}(n)| = \frac{1}{2n} \sum_{\substack{\text{odd } d|n}} \phi(d) 2^{n/d}, \tag{1}$$

where ϕ is Euler's totient function.

Given a non-empty subset $\mathbf{S}$ of $\mathbf{B}(n)$, a *universal cycle* for $\mathbf{S}$ is a sequence of length $|\mathbf{S}|$ that contains every string in $\mathbf{S}$ as a substring exactly once when the string is viewed circularly. A universal cycle is a *de Bruijn* sequence in the case that $\mathbf{S} = \mathbf{B}(n)$.

The *aperiodic prefix* of α denoted by $ap(\alpha)$ is the shortest prefix $a_1 a_2 \cdots a_i, i \in \{1, 2, \ldots, n\}$ such that $\alpha = (a_1 a_2 \cdots a_i)^{\frac{n}{i}}$. We say α is *periodic* if $|ap(\alpha)| < |\alpha|$ and is *aperiodic* if $|ap(\alpha)| = |\alpha|$.

Lemma 1. *If $\alpha = a_1 a_2 \cdots a_n$ is a binary string and $\alpha \overline{\alpha}$ is periodic, then $\frac{|\alpha \overline{\alpha}|}{|ap(\alpha \overline{\alpha})|}$ is odd.*

Proof. The proof is by contradiction. Assume $\alpha \overline{\alpha}$ is periodic and $\frac{|\alpha \overline{\alpha}|}{|ap(\alpha \overline{\alpha})|} = 2k$ is even. Then $\alpha \overline{\alpha} = (ap(\alpha \overline{\alpha}))^{2k} = (ap(\alpha \overline{\alpha}))^k (ap(\alpha \overline{\alpha}))^k = \alpha \alpha$, a contradiction. $\square$

Lemma 2. *If $\alpha = a_1 a_2 \cdots a_n$ is a binary string and $\beta = ap(\alpha \overline{\alpha})$, then $\beta = a_1 a_2 \cdots a_i \overline{a_1 a_2 \cdots a_i}$ for some $1 \le i \le n$.*

Proof. If $\alpha \overline{\alpha}$ is aperiodic, then $\beta = \alpha \overline{\alpha} = a_1 a_2 \cdots a_n \overline{a_1 a_2 \cdots a_n}$. If $\alpha \overline{\alpha}$ is periodic, then $ap(\alpha \overline{\alpha}) = a_1 a_2 \cdots a_j$ for some $j \le n$ and by Lemma 1 the value j must be even and $\alpha \overline{\alpha} = (a_1 a_2 \cdots a_j)^{2k+1} = (a_1 a_2 \cdots a_j)^k (a_1 a_2 \cdots a_j)(a_1 a_2 \cdots a_j)^k$. Thus, it follows that $a_1 a_2 \cdots a_{j/2} = \overline{a_{j/2+1} \cdots a_{j-1} a_j}$. $\square$

Let $\alpha = a_1 a_2 \cdots a_n$ and $\beta = b_1 b_2 \cdots b_n$ be two distinct binary strings of equal length. Then α comes before β in *colexicographic* (colex) order if $a_i < b_i$ for the largest i where $a_i \ne b_i$.

Lemma 3. *If $\alpha = a_1 a_2 \cdots a_n$ and $\beta = b_1 b_2 \cdots b_n$ are consecutive co-necklaces in colex order, where α comes before β and j is the smallest index where $b_j = 1$, then $a_{j+1} a_{j+2} \cdots a_n = b_{j+1} b_{j+2} \cdots b_n$.*

Proof. The proof is by contradiction. Suppose $a_{j+1} a_{j+2} \cdots a_n \ne b_{j+1} b_{j+2} \cdots b_n$. Then there exists some largest $i > j$ such that $a_i \ne b_i$. Since α comes before β in colex order, $a_i = 0$ and $b_i = 1$. However, since β is a co-necklace, then $\gamma = 0^{i-1} 1 b_{i+1} b_{i+2} \cdots b_n$ will also be a co-necklace, since 0^i is the largest run of 0's in γ. But this means γ comes between α and β in colex order, which is a contradiction. Thus $a_{j+1} a_{j+2} \cdots a_n = b_{j+1} b_{j+2} \cdots b_n$. $\square$

3   de Bruijn Sequence Construction

In this section we present a de Bruijn sequence construction obtained by concatenating the cycles of the CCR. The construction generalizes to produce universal cycles for certain subsets of $\mathbf{B}(n)$.

Let $\alpha_1, \alpha_2, \ldots, \alpha_m$ be the first m co-necklaces of length n in colex order. Let

$$\mathcal{U}_{m,n} = ap(\alpha_1\overline{\alpha_1})ap(\alpha_2\overline{\alpha_2}) \cdots ap(\alpha_m\overline{\alpha_m}).$$

When $m = |\mathbf{coN}(n)|$, let $\mathcal{DB}_n = \mathcal{U}_{m,n}$.

Theorem 4. Let $\alpha_1, \alpha_2, \ldots, \alpha_m$ be the first m co-necklaces of order n in colex order, where $m \geq 1$. Then $\mathcal{U}_{m,n}$ is a universal cycle for $\bigcup_{k=1}^{m} \mathbf{coneck}(\alpha_k)$ and $\overline{\alpha_m}$ is a suffix of $\mathcal{U}_{m,n}$.

Proof. The proof is by induction. In the base case when $m = 1$, $\alpha_1 = 0^n$. Clearly $ap(\alpha_1\overline{\alpha_1}) = 0^n 1^n$ is a universal cycle that contains all the strings in $\mathbf{coneck}(0^n)$, and $\overline{\alpha_1} = 1^n$ is its suffix. For $m \geq 1$, assume $\mathcal{U}_{m,n}$ is a universal cycle for $\bigcup_{k=1}^{m} \mathbf{coneck}(\alpha_k)$ with suffix $\overline{\alpha_m}$. Consider $\mathcal{U}_{m+1,n} = \mathcal{U}_{m,n}ap(\alpha_{m+1}\overline{\alpha_{m+1}})$, where $\alpha_m = a_1 a_2 \cdots a_n$ and $\alpha_{m+1} = 0^j 1 b_{j+2}b_{j+3} \cdots b_n$ where $j+1$ is the smallest index where $b_{j+1} = 1$. Let $\beta = ap(\alpha_{m+1}\overline{\alpha_{m+1}}) = 0^j 1 b_{j+2}b_{j+3} \cdots b_{|\beta|}$. First we show that $\overline{\alpha_{m+1}}$ is a suffix of $\mathcal{U}_{m+1,n}$. If $\alpha_{m+1}\overline{\alpha_{m+1}}$ is aperiodic, then by definition $\overline{\alpha_{m+1}}$ is a suffix of $\mathcal{U}_{m+1,n}$. If $\alpha_{m+1}\overline{\alpha_{m+1}}$ is periodic, we know that $\overline{\alpha_m}$ appears as a suffix of $\mathcal{U}_{m,n}$ by the inductive hypothesis. Also by Lemma 3 we see that $\overline{a_{j+2}a_{j+3}\cdots a_n} = \overline{b_{j+2}b_{j+3}\cdots b_n}$, and this implies that a suffix of $\overline{\alpha_m}$ is $(b_1 b_2 \cdots b_{|\beta|})^k$ where $\alpha_{m+1}\overline{\alpha_{m+1}} = (b_1 b_2 \cdots b_{|\beta|})^{2k+1}$ (a result of Lemma 1). Lemma 2 tells us that $\overline{\alpha_{m+1}}$ is a suffix of $\overline{\alpha_m}\beta = (b_1 b_2 \cdots b_{|\beta|})^{k+1}$, so $\overline{\alpha_{m+1}}$ is a suffix of $\mathcal{U}_{m+1,n}$. Now we prove that $\mathcal{U}_{m+1,n}$ is a universal cycle for $\bigcup_{k=1}^{m+1} \mathbf{coneck}(\alpha_k)$. By the inductive hypothesis, $\mathcal{U}_{m+1,n}$ will contain all the strings in $\bigcup_{k=1}^{m} \mathbf{coneck}(\alpha_k)$ except for possibly the strings $\{\overline{a_2 a_3 \cdots a_n}0, \overline{a_3 a_4 \cdots a_n}00, \ldots, \overline{a_n}0^{n-1}\}$ which were involved in the wraparound. First, we show they still exist as substring in the cyclic $\mathcal{U}_{m+1,n}$. By Lemma 3, $\overline{a_{j+2}a_{j+3}\cdots a_n} = \overline{b_{j+2}b_{j+3}\cdots b_n}$. Because we already showed that $\overline{\alpha_{m+1}}$ is a suffix of $\mathcal{U}_{m+1,n}$, this implies that each string in $\{\overline{a_{j+2}a_{j+3}\cdots a_n}0^{j+1}, \overline{a_{j+3}a_{j+4}\cdots a_n}0^{j+2}, \ldots, \overline{a_n}0^{n-1}\}$ occurs as a substring in the wrap-around of the cyclic $\mathcal{U}_{m+1,n}$. Furthermore, the strings $\{\overline{a_2 a_3 \cdots a_n}0, \overline{a_3 a_4 \cdots a_n}00, \ldots, \overline{a_{j+1}\cdots a_n}0^j\}$ exist within $\mathcal{U}_{m+1,n}$ because β has prefix 0^j. Finally, we show that all strings in $\mathbf{coneck}(\alpha_{m+1})$ occur as a substring in $\mathcal{U}_{m+1,n}$. Those that are not trivially substrings of β occur either in the wrap-around or have their prefix as a suffix in $\mathcal{U}_{m,n}$ and suffix in a prefix of β. The latter case covers each string in $\{\overline{b_{j+2}b_{j+3}\cdots b_n}0^j 1, \overline{b_{j+3}b_{j+4}\cdots b_n}0^j 1 b_{j+2}, \ldots, \overline{b_{x+1}\cdots b_n}0^j 1 b_{j+2}\cdots b_x\}$, where $x = n$ if $|\beta| > n$ and $x = |\beta|$ otherwise. Since the length n suffix of $\mathcal{U}_{m+1,n}$

is $\overline{\alpha_{m+1}} = \overline{b_1 b_2 \cdots b_n}$, and $\alpha_1 = 0^n$, the strings $\{1^{j-1} 0 \overline{b_{j+2} \cdots b_n} 0, 1^{j-2} 0$ $\overline{b_{j+2} \cdots b_n} 00 \ldots, 0 \overline{b_{j+2} \cdots b_n} 0^j\}$ occur in the wraparound of $\mathcal{U}_{m+1,n}$. Thus, every string in $\bigcup_{k=1}^{m+1} \mathbf{coneck}(\alpha_k)$ appears as a substring in the cyclic string $\mathcal{U}_{m+1,n}$. Therefore since the length of $\mathcal{U}_{m+1,n}$ is equal to $|\bigcup_{k=1}^{m+1} \mathbf{coneck}(\alpha_k)|$, $\mathcal{U}_{m+1,n}$ is a universal cycle for $\bigcup_{k=1}^{m+1} \mathbf{coneck}(\alpha_k)$. $\qquad\square$

Corollary 5. *$\mathcal{DB}_n$ is a de Bruijn sequence of order n.*

3.1 Efficient Implementation

In order to construct $\mathcal{DB}_n$, we must first generate co-necklaces in colex order. A naïve algorithm will consider all strings $\alpha \in \mathbf{B}(n)$ in colex order and test if $\alpha \overline{\alpha}$ is a necklace. Such a necklace test can be computed in $O(n)$-time [3]. Since there are $\Theta(2^n/n)$ co-necklaces of length n by Eq. 1 this approach will result in each co-necklace being generated in $O(n^2)$-amortized time. We will present an algorithm that improves this method by a factor of n.

Our strategy is to apply a standard recursive algorithm to generate strings in colex order, building the global string $\alpha = a_1 a_2 \cdots a_n$ from right to left one bit at a time. Such an algorithm is given in Algorithm 1 with the following modifications optimized for co-necklace generation.

- Keep track of the length of the current run of 0s in the parameter $curZero$.
- Keep track of the longest substring of the form 0^* in the parameter $maxZero$.
- Terminate the recursion when the length of the remaining prefix of α to be completed, given by parameter t, is less than or equal to $maxZero$. This is because a longest run of 0s must be at the start of any co-necklace.[1] The algorithm can be further optimized by keeping track of the current run of the form 1^*. At this point, the prefix $a_1 a_2 \cdots a_t$ is set to 0^t and then we test if $\alpha \overline{\alpha}$ is a co-necklace using the boolean function IsNecklace.

The function Print outputs the string passed as input and the function Max returns the larger of its two integer inputs. After initializing $a_n = 0$, since all co-necklaces end with 0, the initial call to generate all co-necklaces of length n is GenerateConeck$(n-1, 1, 1)$.

Algorithm 1. An algorithm to generate all co-necklaces of length n in colex order.

```
1: procedure GenerateConeck(t, curZero, maxZero)
2:     if t ≤ maxZero then
3:         a₁a₂···aₜ ← 0ᵗ
4:         if IsNecklace(αᾱ) then Print(αᾱ)
5:     else
6:         aₜ ← 0
7:         GenerateConeck(t − 1, curZero + 1, Max(curZero + 1, maxZero))
8:         aₜ ← 1
9:         GenerateConeck(t − 1, 0, maxZero)
```

---

[1] Because a longest run of the form 0^* or 1^* must be at the start of a co-necklace, the algorithm can be further optimized by keeping track of the longest current run of the form 1^*. However, it will not affect the asymptotic analysis.

Recall that IsNECKLACE can be implemented in $O(n)$-time and also note that setting the prefix $a_1 a_2 \cdots a_t$ also requires at most $O(n)$-time. Thus, since the recursion always has a branch factor of two, the total work done by the algorithm will be $O(n)$ times the number of strings α generated before the necklace test. An example computation tree for $n = 6$ is shown in Fig. 1. Each such string $\alpha = a_1 a_2 \cdots a_n$ is constructed to have a longest run of 0s at the start of the string and $a_n = 0$. Now observe that $a_n a_1 a_2 \cdots a_{n-1}$ is the prefix of a necklace (called a prenecklace). In particular, $a_n a_1 a_2 \cdots a_{n-1}$ is clearly a necklace since the longest number of 0s occurs uniquely at the start of the string. The number of prenecklaces of length n is known to be $\Theta(2^n/n)$ [4], which in turn is proportional to the number of co-necklaces of length n as mentioned earlier. Thus, the total work done by the algorithm is bounded by $O(n)$ times the number of co-necklaces of length n.

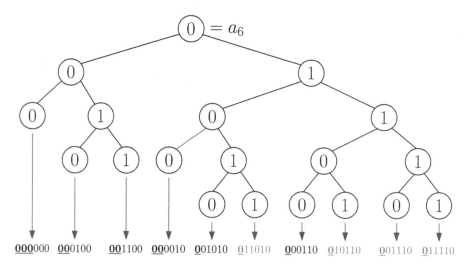

Fig. 1. Computation tree for GENERATECONECK for $n = 6$. Observe that the six co-necklaces in **coN**(6) are listed in colex order: 000000, 000100, 001100, 000010, 001010, 000110.

Theorem 6. *The set of all co-necklaces of length n can be listed in colex order in $O(n)$-amortized time per string.*

To apply the algorithm GENERATECONECK to construct $\mathcal{DB}_n$, we only need to determine $p = |ap(\alpha\bar{\alpha})|$ for each co-necklace α, then pass $a_1 a_2 \cdots a_p$ to the function PRINT instead of $\alpha\bar{\alpha}$. A complete C implementation to generate $\mathcal{DB}_n$ is provided in the appendix. Since the value p can be computed in $O(n)$ time, we obtain the following corollary.

Corollary 7. *The de Bruijn sequence* $\mathcal{DB}_n$ *can be constructed in* $O(1)$-*amortized time per bit.*

It remains an open problem to generate co-necklaces in colex (or lexicographic) order in $O(1)$-amortized time per string.

4   Discrepancy

In this section we focus on a measure studied by Cooper and Heitsch [5] known as the discrepancy of a de Bruijn sequence. In particular, they show that the discrepancy of the lexicographically smallest de Bruijn sequence, which happens to be obtained via the necklace concatenation approach discussed earlier, is $\Theta(\frac{2^n \log n}{n})$.

Let α be a binary string. Let $diff(\alpha)$ denote the absolute difference between the number of 1s and number of 0s in α. Thus $diff(011011) = 2$ because there are four 1s and two 0s, and the absolute difference is 2. The *discrepancy* of α, denoted $D(\alpha)$, is defined to be the maximum value of $diff(\beta)$ over all substrings β of α. For example, the discrepancy of 101001110110 is 4 because $diff(111011) = 4$ and 111011 is a substring that results in the maximal difference.

Theorem 8. *For all* $n \geq 1$, $n \leq D(\mathcal{DB}_n) < 2n$.

Proof. Clearly n is the lower bound for the discrepancy of any de Bruijn sequence, since they all must contain the substring 0^n. All substrings contained within $ap(\alpha\overline{\alpha})$ for a co-necklace α clearly have difference less than n, unless $\alpha = 0^n$ in which case the difference is n. Any other substring will be of the form

$$\sigma \; ap(\alpha_i\overline{\alpha_i}) \; ap(\alpha_{i+1}\overline{\alpha_{i+1}}) \; \cdots \; ap(\alpha_j\overline{\alpha_j})\tau$$

for some $i \leq j$ where σ is a suffix of $ap(\alpha_{i-1}\overline{\alpha_{i-1}})$ and τ is a prefix of $ap(\alpha_{j+1}\overline{\alpha_{j+1}})$. Thus since $diff(ap(\alpha\overline{\alpha})) = 0$, the difference of the substring must be less than $2n$.   □

In fact, $D(\mathcal{DB}_n)$ approaches $2n$ as n gets large. Consider the substring of $\mathcal{DB}_n$ starting from $\overline{\alpha_1} = 1^n$ and ending with $\alpha_j = 0^i(1^{i-1}0)^r$ where $i(r+1) = n$. Note that α_j is a co-necklace and $\alpha_j\overline{\alpha_j}$ is aperiodic, so such a substring exists. The difference of the string between this prefix and suffix is 0, as outlined in the proof of Theorem 8. Thus, the difference of the entire substring is given by $n + diff(\alpha_j)$ since α_j has more 1s than 0s. Since $diff(\alpha_j) = r(i-1) - r - i$, by solving for $r = n/i - 1$ we get $diff(\alpha_j) = \frac{i-2}{i}(n) - 2i + 2$. Thus considering i to be any constant, as n goes to infinity $D(\mathcal{DB}_n)$ approaches $n + \frac{i-2}{i}n$.

Appendix - C Code

```c
#include <stdio.h>
#define MAX(a,b) (a)>(b)?(a):(b)
int N,b[1000];

//--------------------------------------------------------------
// Returns 0 if string is not necklace, and index of
// longest aperiodic prefix if is necklace
//--------------------------------------------------------------
int IsNecklace(int b[], int n) {
    int i, p=1;

    for (i=2; i<=n; i++) {
        if (b[i-p] > b[i]) return 0;
        if (b[i-p] < b[i]) p = i;
    }
    if (n % p != 0) return 0;
    return p;
}

void Gen(int t, int curZero, int maxZero){
    int i,p;
    if(t <= maxZero){
        for (i=N+1;i<=2*N;++i) b[i]=1-b[i-N];
        p = IsNecklace(b,2*N);
        for (i=1; i<=p; i++) printf("\%d", b[i]);
    }
    else {
        b[t]=0;
        Gen(t-1,curZero+1,MAX(curZero+1,maxZero));
        b[t]=1;
        Gen(t-1,0,maxZero);
    }
}

int main(){
    printf("Enter N:");scanf("\%d",&N);
    b[N] = 0;
    Gen(N-1,1,1);
    printf("\n");
    return 0;
}
```

References

1. The On-Line Encyclopedia of Integer Sequences (2010). https://oeis.org, sequence A000016

2. Alhakim, A.: A simple combinatorial algorithm for de Bruijn sequences. Am. Math. Monthly **117**(8), 728–732 (2010). http://www.jstor.org/stable/10.4169/000298910x515794
3. Booth, K.S.: Lexicographically least circular substrings. Inf. Process. Lett. **10**(4/5), 240–242 (1980)
4. Cattell, K., Ruskey, F., Sawada, J., Serra, M., Miers, C.: Fast algorithms to generate necklaces, unlabeled necklaces, and irreducible polynomials over GF(2). J. Algorithms **37**(2), 267–282 (2000)
5. Cooper, J., Heitsch, C.: The discrepancy of the lex-least de Bruijn sequence. Discrete Math. **310**, 1152–1159 (2010)
6. Dragon, P.B., Hernandez, O.I., Williams, A.: The grandmama de Bruijn sequence for binary strings. In: Kranakis, E., Navarro, G., Chávez, E. (eds.) LATIN 2016. LNCS, vol. 9644, pp. 347–361. Springer, Heidelberg (2016). doi:10.1007/978-3-662-49529-2_26
7. Eldert, C., Gray, H., Gurk, H., Rubinoff, M.: Shifting counters. AIEE Trans. **77**, 70–74 (1958)
8. Etzion, T., Lempel, A.: Construction of de Bruijn sequences of minimal complexity. IEEE Trans. Inf. Theory **30**(5), 705–709 (1984)
9. Etzion, T.: Self-dual sequences. J. Comb. Theory Ser. A **44**(2), 288–298 (1987). http://www.sciencedirect.com/science/article/pii/0097316587900355 http://www.sciencedirect.com/science/article/pii/0097316587900355
10. Ford, L.: A cyclic arrangement of M-tuples. Report No. P-1071, Rand Corporation, Santa Monica, 23 April 1957
11. Fredricksen, H., Maiorana, J.: Necklaces of beads in k colors and k-ary de Bruijn sequences. Discrete Math. **23**, 207–210 (1978)
12. Fredricksen, H.: Generation of the Ford sequence of length 2^n, n large. J. Comb. Theory Ser. A **12**(1), 153–154 (1972). http://www.sciencedirect.com/science/article/pii/009731657290091X
13. Fredricksen, H., Kessler, I.: An algorithm for generating necklaces of beads in two colors. Discrete Math. **61**(2), 181–188 (1986). http://www.sciencedirect.com/science/article/pii/0012365X86900890
14. Golomb, S.W.: Shift Register Sequences. Aegean Park Press, Laguna Hills (1981)
15. Hauge, E.R.: On the cycles and adjacencies in the complementary circulating register. Discrete Math. **145**(1), 105–132 (1995). http://www.sciencedirect.com/science/article/pii/0012365X9400057P
16. Huang, Y.: A new algorithm for the generation of binary de Bruijn sequences. J. Algorithms **11**(1), 44–51 (1990). http://www.sciencedirect.com/science/article/pii/019667749090028D
17. Martin, M.H.: A problem in arrangements. Bull. Am. Math. Soc. **40**(12), 859–864 (1934)
18. Mayhew, G.L., Golomb, S.W.: Characterizations of generators for modified de Bruijn sequences. Adv. Appl. Math. **13**(4), 454–461 (1992). http://www.sciencedirect.com/science/article/pii/019688589290021N
19. Ruskey, F., Savage, C., Wang, T.M.Y.: Generating necklaces. J. Algorithms **13**, 414–430 (1992)
20. Sawada, J., Williams, A., Wong, D.: A surprisingly simple de Bruijn sequence construction. Discrete Math. **339**, 127–131 (2016)
21. Sawada, J., Williams, A., Wong, D.: Necklaces and Lyndon words in colexicographic and reflected Gray code order (2017). Submitted manuscript

On Words with the Zero Palindromic Defect

Edita Pelantová[1] and Štěpán Starosta[2(✉)]

[1] Department of Mathematics, Faculty of Nuclear Sciences and Physical Engineering,
Czech Technical University in Prague, Prague, Czech Republic
edita.pelantova@fjfi.cvut.cz
[2] Department of Applied Mathematics, Faculty of Information Technology,
Czech Technical University in Prague, Prague, Czech Republic
stepan.starosta@fit.cvut.cz

Abstract. We study the set of finite words with zero palindromic defect, i.e., words rich in palindromes. This set is factorial, but not recurrent. We focus on description of pairs of rich words which cannot occur simultaneously as factors of a longer rich word.

Keywords: Palindrome · Palindromic defect · Rich words

1 Introduction

In [14], Droubay, Justin and Pirillo observed that the number of distinct palindromes occurring in a finite word w of length n does not exceed $n + 1$. This upper bound motivated Brlek, Hamel, Nivat, and Reutenauer to define in [9] the notion *palindromic defect* $D(w)$ of a finite word w as the difference of the upper bound $n + 1$ and the actual number of palindromic factors occurring in w. One can say that the palindromic defect measures the number of "missing" palindromic factors in the given word. A word with zero palindromic defect is usually shortly called *rich* or *full*.

For an infinite word $\mathbf{u}$ the palindromic defect $D(\mathbf{u})$ is naturally defined as the supremum of the set $\{D(w)\colon w \text{ is a factor of } \mathbf{u}\}$. Many classes of words with the defect zero have been found, for example Sturmian words, words coding symmetrical interval exchange and complementary symmetric Rote words (see [2,7,15]).

Palindromic defect is actively studied in the last decade. During these years many nice properties of words with zero defect have been brought into light. Some of them have been already proved, some of them are formulated as conjectures and are still open. Neither the basic question "What is the number of rich words of a given length?" has been answered. This question is extremely interesting as the set of rich words is a very naturally defined factorial language which has superpolynomial and subexponential growth as was shown in [17] by C. Guo, J. Shallit and A.M. Shur and in [30] by J. Rukavicka, respectively.

This article consists of three parts. In the first part, we present relevant known results. In the last part we give a list of open questions connected to the palindromic defect and we also recall a narrow connection to the well known

© Springer International Publishing AG 2017
S. Brlek et al. (Eds.): WORDS 2017, LNCS 10432, pp. 59–71, 2017.
DOI: 10.1007/978-3-319-66396-8_7

conjecture of Hof, Knill, and Simon. The middle part contains a new result. It is devoted to so-called compatible words, i.e., to the pairs of finite rich words which can occur simultaneously as factors of a longer rich word. We believe that our result may help to characterize words w with the following property: $D(w) = 1$ and $D(u) = 0$ for each proper factor u of w. A characterization of these words seems to be the missing point in answering several open questions.

2   Preliminaries

2.1   Basic Notations and Definitions

Let $\mathcal{A}$ be a finite set, called an *alphabet*. Its elements are called *letters*. A *finite word* w is an element of $\mathcal{A}^n$ for $n \in \mathbb{N}$. The *length* of w is n and is denoted $|w|$. The set of all finite words over $\mathcal{A}$ is denoted $\mathcal{A}^*$. An *infinite word* over $\mathcal{A}$ is an infinite sequence of letters from $\mathcal{A}$.

A finite word w is a *factor* of a finite or infinite word v if there exist words p and s such that v is a concatenation of p, w, and s, denoted $v = pws$. The word p is said to be a *prefix* and s a *suffix* of v. The set of all factors of a word $\mathbf{u}$ is the *language of* $\mathbf{u}$ and is denoted $\mathcal{L}(\mathbf{u})$. All factors of $\mathbf{u}$ of length n are denoted by $\mathcal{L}_n(\mathbf{u})$.

An *occurrence* of $w = w_0 w_1 \cdots w_{n-1} \in \mathcal{A}^n$ in a word $v = v_0 v_1 v_2 \ldots$ is an index i such that $v_i \cdots v_{i+n-1} = w$. A factor w is *unioccurrent* in v if there is exactly one occurrence of w in v. A *complete return word* of a factor w (in v) is a factor f (of v) containing exactly two occurrences of w such that w is its prefix and also its suffix. For instance, the word 010011010 is a complete return word of 010.

The *reversal* or mirror mapping assigns to a word $w \in \mathcal{A}^*$ the word $\widetilde{w}$ with the letters reversed, i.e.,

$$\widetilde{w} = w_{n-1} w_{n-2} \cdots w_1 w_0 \quad \text{where } w = w_0 w_1 \cdots w_{n-1} \in \mathcal{A}^n.$$

A word is *palindrome* if $w = \widetilde{w}$. We say that a language $\mathcal{L} \subset \mathcal{A}^*$ is *closed under reversal* if for all $w \in \mathcal{L}$ we have $\widetilde{w} \in \mathcal{L}$.

Given an infinite word $\mathbf{u}$, its *factor complexity* $\mathcal{C}_\mathbf{u}(n)$ is the count of its factors of length n:

$$\mathcal{C}_\mathbf{u}(n) = \#\mathcal{L}_n(\mathbf{u}) \quad \text{for all } n \in \mathbb{N}.$$

Let $\mathrm{Pal}(\mathbf{u})$ be the set of all palindromic factors of the infinite word $\mathbf{u}$. The *palindromic complexity* $\mathcal{P}_\mathbf{u}(n)$ of $\mathbf{u}$ is given by

$$\mathcal{P}_\mathbf{u}(n) = \#(\mathcal{L}_n(\mathbf{u}) \cap \mathcal{P}(\mathbf{u})) \quad \text{for all } n \in \mathbb{N}.$$

We omit the subscript $\mathbf{u}$ if there is no confusion.

2.2   Fixed Points of Morphisms and Their Properties

A *morphism* φ is a mapping $\mathcal{A}^* \to \mathcal{B}^*$ where $\mathcal{A}$ and $\mathcal{B}$ are alphabets such that for all $v, w \in \mathcal{A}^*$ we have $\varphi(vw) = \varphi(v)\varphi(w)$ (it is a homomorphism of the monoids $\mathcal{A}^*$ and $\mathcal{B}^*$). Its action is extended to $\mathcal{A}^{\mathbb{N}}$: if $\mathbf{u} = u_0 u_1 u_2 \ldots \in \mathcal{A}^{\mathbb{N}}$, then

$$\varphi(\mathbf{u}) = \varphi(u_0)\varphi(u_1)\varphi(u_2) \ldots \in \mathcal{B}^{\mathbb{N}}.$$

If φ is an endomorphism of $\mathcal{A}^*$, we may find its fixed point, i.e., a word $\mathbf{u}$ such that $\varphi(\mathbf{u}) = \mathbf{u}$. We are interested mainly in the case of $\mathbf{u}$ being infinite. A morphism $\varphi : \mathcal{A}^* \to \mathcal{A}^*$ is *primitive* if there exists an integer k such that for every $a, b \in \mathcal{A}$ the letter b occurs in $\varphi^k(a)$.

Two morphisms $\varphi, \psi : \mathcal{A}^* \to \mathcal{B}^*$ are *conjugate* if there exists a word $w \in \mathcal{B}^*$ such that

$$\forall a \in \mathcal{A}, \varphi(a)w = w\psi(a) \quad \text{or} \quad \forall a \in \mathcal{A}, w\varphi(a) = \psi(a)w.$$

If φ is primitive, then the languages of fixed points of φ and ψ are the same.

A morphism $\psi : \mathcal{A}^* \to \mathcal{B}^*$ is of *class P* if $\psi(a) = pp_a$ for all $a \in \mathcal{A}$ where p and p_a are both palindromes (possibly empty). A morphism φ is of class P' if it is conjugate to a morphism of class P.

The following examples illustrate the last few notions.

Example 1. Let $\varphi : \{a, b\}^* \to \{a, b\}^*$ be determined by $\varphi : \begin{matrix} a \mapsto abab, \\ b \mapsto aab. \end{matrix}$ The fixed point of φ is

$$\mathbf{u} = \lim_{k \to +\infty} \varphi^k(a) = \underbrace{abab}_{\varphi(a)} \underbrace{aab}_{\varphi(b)} \underbrace{abab}_{\varphi(a)} \underbrace{aab}_{\varphi(b)} \underbrace{abab}_{\varphi(a)} \ldots$$

The morphism φ is of class P' since it is conjugate to ψ given by $\psi : \begin{matrix} a \mapsto abab, \\ b \mapsto aba. \end{matrix}$ Indeed, we have $ab\varphi(a) = \psi(a)ab$ and $ab\varphi(b) = \psi(b)ab$. To see that ψ is of class P, i.e., it is of the form $a \mapsto pp_a$ and $b \mapsto pp_b$, it suffices to set $p = aba$, $p_a = b$ and $p_b = \varepsilon$. The fixed point of ψ is

$$\mathbf{v} = \lim_{k \to +\infty} \psi^k(a) = \underbrace{abab}_{\psi(a)} \underbrace{aba}_{\psi(b)} \underbrace{abab}_{\psi(a)} \underbrace{aba}_{\psi(b)} \underbrace{abab}_{\psi(a)} \ldots$$

We have $\mathcal{L}(\mathbf{u}) = \mathcal{L}(\mathbf{v})$.

Example 2. The two famous examples of infinite words, the Thue–Morse word $\mathbf{t}$ and the Fibonacci word $\mathbf{f}$, are both fixed points of a morphism.

The word $\mathbf{t}$ is fixed by the morphism φ_{TM} determined by $\varphi_{TM}(0) = 01$ and $\varphi_{TM}(1) = 10$. Note that this morphism in fact has two fixed points, one being the other one after replacing 0 with 1 and 1 with 0. The word $\mathbf{t}$ as given above is the fixed points starting in 0.

The word $\mathbf{f}$ is fixed by the morphism φ_F defined by $\varphi_F(0) = 01$ and $\varphi_F(1) = 0$.

An (infinite) fixed point of a morphism of class P' clearly contains infinitely many palindromes which is one motivation for this notion. Class P is introduced in [19] in the context of discrete Schrödinger operators.

3 The Study of Palindromic Defect

3.1 Characterizations of Words with the Zero Defect

We start by giving some of the known characterizations of infinite rich words.

Theorem 3. *For an infinite word* **u** *with language closed under reversal the following statements are equivalent:*

1. $D(\mathbf{u})$ *is zero [9];*
2. *any prefix of* **u** *has a unioccurrent longest palindromic suffix [14];*
3. *for any palindromic factor w of* **u**, *every complete return word of w is a palindrome [16];*
4. *for any factor w of* **u**, *every factor of* **u** *that contains w only as its prefix and $\widetilde{w}$ only as its suffix is a palindrome [16];*
5. *for each $n \in N$ we have $\mathcal{C}(n+1) - \mathcal{C}(n) + 2 = \mathcal{P}(n) + \mathcal{P}(n+1)$ [11].*

We generalized the previous theorem to infinite words with finite palindromic defect, see [3,26]. In particular, we showed that an infinite word has a finite palindromic defect $D(\mathbf{u})$ if and only if the equality $\mathcal{C}(n+1) - \mathcal{C}(n) + 2 = \mathcal{P}(n) + \mathcal{P}(n+1)$ is valid for all $n \in N$ up to finitely many exceptions. A surprising observation that these exceptional indices allow to determine the value of the palindromic defect was made by Brlek and Reutenauer. In [8] they proved for infinite periodic words and conjectured for general words the following equality

$$2D(\mathbf{u}) = \sum_{n=0}^{+\infty} \Big(\mathcal{C}_{\mathbf{u}}(n+1) - \mathcal{C}_{\mathbf{u}}(n) + 2 - \mathcal{P}_{\mathbf{u}}(n+1) - \mathcal{P}_{\mathbf{u}}(n) \Big). \tag{1}$$

The conjecture was confirmed in [4] where we showed the following theorem.

Theorem 4. *Equation* (1) *is true for any infinite word* **u** *whose language is closed under reversal.*

Besides these general properties, many examples of words with zero or finite palindromic defect were found:

- In [12,27], another characterizations of rich words are given.
- In [13], the relation of rich words to so-called periodic-like words is exhibited.
- Links to another class of words, trapezoidal words, are shown in [24].
- Words coding symmetric interval exchange transformations are rich by [2].
- In [7], the authors show that words coding rotation on the unit circle with respect to partition consisting of two intervals are rich.
- In [29], the authors show a connection of rich words with the Burrows–Wheeler transform.
- In [32], we show that morphic images of episturmian words, a known class of rich words, produces a word with finite palindromic defect.
- The articles [20,28,31] exhibit more examples of words with finite palindromic defect (along with some examples of words with finite generalized palindromic defect).

3.2   Palindromic Defect of Fixed Points of Morphisms

We now focus on words that are fixed by a morphism with the assumption that their language is closed under reversal. The main motivation to study their palindromic defect is the following conjecture.

Conjecture 5 (Zero defect conjecture [6]). Let **u** be an aperiodic fixed point of a primitive morphism having its language closed under reversal. We have $D(\mathbf{u}) = 0$ or $D(\mathbf{u}) = +\infty$.

The Thue–Morse word **t** and the Fibonacci word **f** are examples of aperiodic fixed points of a primitive morphism (see Example 2) having their language closed under reversal. We have $D(\mathbf{f}) = 0$ and $D(\mathbf{t}) = +\infty$.

Counterexamples to the conjecture were given in [1,10]. Thus, the current statement of the conjecture is not true. There still might some refinement of the current statement that is valid as there are many witnesses and the found counterexamples seem to have some specific properties. Indeed, in [22] we prove that the conjecture is true for a special class of morphisms. A morphism φ is *marked* if there exists two morphisms φ_1 and φ_2, both being conjugate to φ, such that

$$\{\text{last letter of } \varphi_1(a) \colon a \in \mathcal{A}\} = \{\text{first letter of } \varphi_2(a) \colon a \in \mathcal{A}\} = \mathcal{A}.$$

In other words, the set of the last letters of the images of letters by φ_1 is the whole alphabet $\mathcal{A}$ and the set of the first letters of the images of letters by φ_2 is also the whole alphabet $\mathcal{A}$.

For instance, $\varphi = \varphi_{TM} \colon 0 \mapsto 01, 1 \mapsto 10$ is marked (here $\varphi = \varphi_1 = \varphi_2$). For $\varphi = \varphi_F \colon 0 \mapsto 01, 1 \mapsto 0$ we have $\varphi = \varphi_1$ and $\varphi_2 \colon 0 \mapsto 10, 1 \mapsto 0$. Thus, φ_F is also marked.

If a morphism φ is conjugate to no other morphism except for φ itself, then we say that φ is *stationary*. In other words, a morphism φ is stationary if the longest common prefix and the longest common suffix of φ-images of all letters are both empty words.

In [22] we show the following theorems:

Theorem 6. *Let φ be a primitive marked morphism and let **u** be its fixed point with finite palindromic defect. If all complete return words of all letters in **u** are palindromes or φ is not stationary, then $D(\mathbf{u}) = 0$.*

Moreover, the binary alphabet allows for all of the assumptions to be dropped:

Theorem 7. *If $\mathbf{u} \in \mathcal{A}^{\mathbb{N}}$ is a fixed point of a primitive morphism over binary alphabet and $D(\mathbf{u}) < +\infty$, then $D(\mathbf{u}) = 0$ or **u** is periodic.*

We thus confirm that for a large class of fixed points of morphisms, their palindromic defect is either zero or infinite.

3.3   Enumeration of Rich Words

Let $R_d(n)$ denote the number of rich words of length n over an alphabet with d elements. As we have already mentioned, there is no closed-form formula for $R_d(n)$.

In [34], Vesti gives a recursive lower bound on $R_d(n)$ and an upper bound on $R_2(n)$. Both these estimates seem to be very rough.

In [17], Guo, Shallit and Shur constructed for each n a large set of binary rich words of length n. They show that for any two sequences of integers $0 \le n_1 \le n_2 \le \cdots \le n_k$ and $0 \le m_1 \le m_2 \le \cdots \le m_k$ satisfying $n = \sum_{i=1}^{k} n_k + \sum_{i=1}^{k} m_k$, the word $a^{n_1} b^{m_1} a^{n_1} b^{m_1} \cdots a^{n_k} b^{m_k}$ of length n is rich. This construction gives, currently, the best lower bound on the number of binary rich words, namely $R_2(n) \ge \frac{C^{\sqrt{n}}}{p(n)}$ where $p(n)$ is a polynomial and the constant $C \sim 37$. They also conjectured that $R_2(n) = \Theta\left(\frac{n}{g(n)}\right)^{\sqrt{n}}$ for some infinitely growing function $g(n)$.

The best upper bound is provided by Rukavicka in [30]. He shows that $R_d(n)$ has a subexponential growth on any alphabet. More precisely, for any cardinality d of the alphabet $\lim_{n \to \infty} \sqrt[n]{R_d(n)} = 1$. The result uses a specific factorization of a rich word into distinct rich palindromes, called UPS-factorization (Unioccurrent Palindromic Suffix factorization).

4   Compatible Pairs

The set of rich words is a factorial language but it is not recurrent. Let us recall that a language $\mathcal{L} \subset \mathcal{A}^*$ is *recurrent* if for any two words $u, v \in \mathcal{L}$ there exists $w \in \mathcal{L}$ such that u is a prefix of w and v is a suffix of w. Using results of Glen et al. [16], Vesti in [34] formulated a sufficient condition which prevents two rich words u, v to be simultaneously factors of another rich word. His proposition uses the notion of *longest palindromic suffix* of a factor u, denoted lps(u) and *longest palindromic prefix* of a factor u, denoted lpp(u). We say that two finite words are *compatible* if there exists a rich words having these two words as factors.

Proposition 8. *Let u and v be two words such that*

$$u \ne v, \quad u, v \ \text{rich}, \quad \mathrm{lpp}(u) = \mathrm{lpp}(v) \quad \text{and} \quad \mathrm{lps}(u) = \mathrm{lps}(v). \qquad (2)$$

If a word w contains factors u and v, then w is not rich, i.e., u and v are not compatible.

We give an example which demonstrates that a word w can be non-rich without containing factors u and v satisfying (2).

Example 9. Consider the word $w = 11010011$, which is not rich. In fact, it is a factor of the Thue–Morse word. As pointed out in [5], the length 8 is the shortest length of a non-rich binary word.

Table 1 depicts all non-empty rich factors u of w together with the pairs (lpp(u), lps(u)). The map $u \mapsto (\mathrm{lpp}(u), \mathrm{lps}(u))$ is injective. In other words, no pair of factors u, v of the non-rich word $w = 11010011$ satisfies (2).

Table 1. All non-empty rich factors u of w from Example 9 together with the pairs $(\mathrm{lpp}(u), \mathrm{lps}(u))$.

u	$(\mathrm{lpp}(u), \mathrm{lps}(u))$		u	$(\mathrm{lpp}(u), \mathrm{lps}(u))$
1	$(1,1)$		0	$(0,0)$
11	$(11,11)$		01	$(0,1)$
110	$(11,0)$		010	$(010,010)$
1101	$(11,101)$		0100	$(010,00)$
11010	$(11,010)$		01001	$(010,1001)$
110100	$(11,00)$		010011	$(010,11)$
1101001	$(11,1001)$		100	$(1,00)$
10	$(1,0)$		1001	$(1001,1001)$
101	$(101,101)$		10011	$(1001,11)$
1010	$(101,010)$		00	$(00,00)$
1010	$(101,010)$		001	$(00,1)$
10100	$(101,00)$		0011	$(00,11)$
101001	$(101,1001)$		011	$(0,11)$
1010011	$(101,11)$			

Let us formulate another sufficient condition for non-richness of a word w.

Proposition 10. *Let u and v be two words satisfying*

$$u \neq \tilde{v}, \quad u, v \;\; rich, \quad \mathrm{lps}(u) = \mathrm{lpp}(v) \quad and \quad \mathrm{lps}(v) = \mathrm{lpp}(u). \tag{3}$$

If a word w contains factors u and v, then w is not rich.

Proof. First we show (by contradiction) that the assumption (3) gives

$$u, \tilde{u} \notin \mathcal{L}(v) \cup \mathcal{L}(\tilde{v}) \quad and \quad v, \tilde{v} \notin \mathcal{L}(u) \cup \mathcal{L}(\tilde{u}). \tag{4}$$

As the roles of v and u are symmetric, we have to discuss the following two cases:

(1) $u \in \mathcal{L}(v)$:
As v is rich, $\mathrm{lps}(v)$ is unioccurrent in v. Since $\mathrm{lps}(v) = \mathrm{lpp}(u)$, we have that $\mathrm{lpp}(u)$ occurs only as a suffix of v. Since $u \in \mathcal{L}(v)$, necessarily $u = \mathrm{lpp}(u)$ and thus u is a palindrome. It follows that $u = \mathrm{lps}(u) = \mathrm{lpp}(v) = \mathrm{lps}(v)$. Richness of v implies that $\mathrm{lpp}(v)$ and $\mathrm{lps}(v)$ are unioccurrent in v and consequently v is a palindrome satisfying $v = \mathrm{lpp}(v) = u = \tilde{u}$, which is a contradiction.
(2) $\tilde{u} \in \mathcal{L}(v)$:
Since $\mathrm{lps}(v) = \mathrm{lps}(\tilde{u})$ is unioccurrent in v, we have that $\tilde{u}$ occurs only as a suffix of v. Similarly, as $\mathrm{lpp}(v) = \mathrm{lpp}(\tilde{u})$ is unioccurrent in v, we get that $\tilde{u}$ occurs only as a prefix of v. It implies $v = \tilde{u}$, which is again a contradiction.

Obviously, the assumption (3) implies that u and v are not palindromes.

To prove the proposition itself (again by contradiction), we assume that w is rich and let f denote the shortest factor of w such that f contains as its factor u or $\tilde{u}$ and f contains as its factor v or $\tilde{v}$. Without loss of generality and due to (4), we have to discuss the following two cases:

(1) u is a proper prefix and v is a proper suffix of f:
The word $\mathrm{lps}(f)$ is not longer than v; otherwise, we obtain a contradiction with the choice of f as the shortest factor with the given property. Thus $\mathrm{lps}(f) = \mathrm{lps}(v)$. Similarly, $\mathrm{lpp}(f) = \mathrm{lpp}(u)$. It means that $\mathrm{lps}(f)$ is not unioccurrent in f—a contradiction.

(2) u is a proper prefix and $\widetilde{v}$ is a proper suffix of f:
By the same argument as before, $\mathrm{lps}(f) = \mathrm{lps}(\widetilde{v}) = \mathrm{lpp}(v)$. It means that $\mathrm{lpp}(v) = \mathrm{lps}(u)$ occurs as a suffix of f and also as a suffix of u. Since u is a proper prefix of f, the factor $\mathrm{lpp}(v) = \mathrm{lps}(f)$ occurs in f twice—a contradiction with the richness of f.

Example 11. We consider again the non-rich word $w = 11010011$. It contains the factors $u = 11010$, $v = 010011$ such that $\mathrm{lpp}(u) = 11 = \mathrm{lps}(v)$ and $\mathrm{lps}(u) = 010 = \mathrm{lpp}(v)$. Also the pairs $u' = 1101001$, $v' = 10011$ and $u'' = 110100$, $v'' = 0011$ satisfy (3).

We show that a pair of factors with the property (3) occurs in each non-rich word.

Proposition 12. *If w be is a non-rich word, then w has two factors u and v such that*

$$u \neq \widetilde{v}, \quad u, v \ \text{rich}, \quad \mathrm{lps}(u) = \mathrm{lpp}(v) \quad \text{and} \quad \mathrm{lps}(v) = \mathrm{lpp}(u).$$

Proof. As w is not rich, it contains a complete return word r to a palindrome p such that r is not a palindrome. Let r be the shortest non-palindromic return word in w to a palindrome. Denote by t the first letter of r and find the longest q such that tq is a prefix of r and $\widetilde{q}t$ is a suffix of v. Clearly, p is a prefix of tq and p is a suffix of $\widetilde{q}t$. Let us denote x and y the letters such that tqx is a prefix of r and $y\widetilde{q}t$. Obviously, $x \neq y$.

– If q is empty, then r is a non-palindromic complete return word to the letter t, i.e., the letter t does not occur in the factor f given by $r = tft$, i.e., $f = t^{-1}rt^{-1}$. Choose $z \in \{x, y\}$ such that $z \neq t$ and put
$u :=$ the shortest prefix of r which ends with the letter z and
$v :=$ the shortest suffix of r which starts with the letter z.
In particular, both letters z and t are unioccurrent in u and also in v. It means that $\mathrm{lpp}(u) = t = \mathrm{lps}(v)$ and $\mathrm{lps}(u) = z = \mathrm{lpp}(v)$. One of the words u and v has length 2 and the second one is longer than 2. It implies that $u \neq \widetilde{v}$.

– Let us assume that $q \neq \varepsilon$. The word $f = t^{-1}rt^{-1}$ has a prefix qx and a suffix $y\widetilde{q}$. First we show

Claim: Occurrences of q and $\widetilde{q}$ in f alternate and moreover each factor of f starting with q and ending with $\widetilde{q}$ without other occurrences of q and $\widetilde{q}$ is a palindrome.

Proof of the claim: Let w' be arbitrary suffix of f such that $|w'| > |\widetilde{q}|$ and w' has a prefix q. Clearly, f has a suffix $\widetilde{q}$ and thus $\widetilde{q}$ is a suffix of w' as well. Let us denote $p' = \mathrm{lpp}(q)$. Since q is rich, p' is unioccurrent in q. But p' occurs in w' at least twice, as $\widetilde{q}$ is a suffix of w'. Let us denote r' a complete return word to p' in w'. From minimality of r, the complete return word r' to p' is a palindrome. Therefore, w' has prefixes p', q and r', their lengths satisfy $|p'| \leq |q| < |r'|$. It implies that $\widetilde{q}$ is a suffix of the palindrome r' and thus the first occurrence of q in w' is followed by the occurrence of $\widetilde{q}$.

Since f is not a palindrome, the previous claim implies that q and $\widetilde{q}$ occur also as inner factors of f. It means that there exists a palindromic factor, say w'', of the word f such that $\widetilde{q}$ is a prefix and q is a suffix of w'' and $|w''| > |q|$. Let z denote the letter satisfying that $\widetilde{q}z$ is a prefix of w''. Obviously, zq is a suffix of w''. Let us stress that $z \neq t$, otherwise r would not be a complete return word to the palindrome p. The letter z enables us to identify the factors v and u announced in the proposition. Put

$u :=$ the shortest prefix of $r = tft$ which ends with $\widetilde{q}z$
$v :=$ the shortest suffix of $r = tft$ which starts with zq.

To prove $\mathrm{lpp}(u) = \mathrm{lps}(v)$, we apply the simple observation: If a word s' is a prefix of a word s and $\mathrm{lpp}(s)$ is a prefix of s', then $\mathrm{lpp}(s) = \mathrm{lpp}(s')$.
   In our situation: $p = \mathrm{lpp}(r) = \mathrm{lpp}(u)$. Analogously, $p = \mathrm{lps}(r) = \mathrm{lps}(v)$.
   To show $\mathrm{lps}(u) = \mathrm{lpp}(v)$, we use a simple consequence of the claim: Any occurrence of ℓq in r, where ℓ is a letter with $\ell \neq t$, is preceded with an occurrence of $\widetilde{q}\ell$. Therefore, our definition of u guarantees that $\mathrm{lps}(u)$ is not longer than $\widetilde{q}z$, i.e., $\mathrm{lps}(u) = \mathrm{lps}(\widetilde{q}z)$. By the same reason, $\mathrm{lpp}(v) = \mathrm{lpp}(zq)$. As $\mathrm{lpp}(zq) = \mathrm{lps}(\widetilde{q}z)$, the equality $\mathrm{lps}(u) = \mathrm{lpp}(v)$ is proven.
   Obviously, $u \neq \widetilde{v}$. Otherwise, we have a contradiction with the assumption that tq is the longest prefix of r such that $\widetilde{q}t$ is a suffix of r.
   The last proof has an interesting direct consequence on a binary alphabet. It is based on the fact that the case $q = \varepsilon$ is not possible on a binary alphabet and the second case implies that q is not a palindrome. We state this consequence of the construction in the second case as the following corollary.

Corollary 13. *Let $w \in \{0,1\}^*$ be a binary word. The word w is not rich if and only if there exists a non-palindromic word q such that*

$$0q0, 1q1, 0\widetilde{q}1, 1\widetilde{q}0 \in \mathcal{L}(w).$$

5   Open Questions and Related Problems

We finish this article with a list of open questions that we deem important in further understanding of the structure of rich words (and more generally, words with finite palindromic defect).

– The subexponential upper bound on the number of rich words $R_d(n)$ of length n over d letters is based on the statement that any rich word of length n can be factorized into at most $c\frac{n}{\ln n}$ distinct palindromes. In fact, the number of palindromes is exaggerated, as the factorization does not take into consideration that each of the palindromes is rich as well. Any asymptotic improvement of the bound $c\frac{n}{\ln n}$ would improve the upper bound on $R_d(n)$.

- To our knowledge, there are no result on morphisms preserving the set of rich words. Such a class of morphisms preserving richness would allow to construct a set of class other than the set constructed in [17] to obtain a lower bound on $R_2(n)$. In particular, any fixed point of a primitive morphism which preserves the set of rich words must be rich as well. In this point of view the following question is also important.
- Theorem 6 confirms the validity of the zero defect conjecture only for marked morphisms φ satisfying the following assumption: all complete return words of all letters in **u** are palindromes or φ is not stationary. We have no example that this peculiar assumption is really needed.
- We do not know how to decide whether two rich words u and v are factors of a common rich word w. The related task is to identify a minimal non-rich word, i.e., to look for a word which is not rich but any its proper factor is rich.

Primitive morphisms that preserve the set of rich words are included in a larger set of morphisms having infinitely many palindromic factors in their fixed points. An infinite word having infinitely many palindromic factors is usually called *palindromic*. A very useful property of morphisms in this larger set is given by the following conjecture.

Conjecture 14 (Class P conjecture [19]). Let **u** be a palindromic fixed point of a primitive morphism φ. There exists a morphism of class P' such that its fixed point has the same language as **u**.

The original statement of the conjecture in [19] is ambiguous and allows for more interpretations, see also [21] or [18]. The above given statement of Conjecture 14 follows from two results. First, for binary alphabet the question is solved by B. Tan in [33]: if a fixed point of a primitive morphism φ over a binary alphabet contains infinitely many palindromes, then φ or φ^2 is of class P'. Second, in [23], S. Labbé shows that the analogy of the previous result cannot be generalized for multiliteral alphabet: there exists a word **w** over ternary alphabet which is a palindromic fixed point of a primitive morphism and not being fixed by any morphism of class P'. However, the authors of [18] note that the language of the word **w** may indeed be generated by a morphism of class P.

At this moment only partial answers to Conjecture 14 are known: as already mentioned, the binary case is solved ([33]); for larger alphabets an affirmative answer is provided only for some special classes of morphisms.

In [25], we confirm the conjecture for morphisms fixing a codings a non-degenerate exchange of 3 intervals. In [21], the authors prove the validity of the conjecture for marked morphisms. Moreover, they show that a power of the marked morphism itself is in class P'. The technique and results used in the proofs of the latter fact is crucial in showing the defect conjecture for marked morphisms in [22].

Palindromicity of a fixed point $\mathbf{u}$ is linked to the symmetry of the language $\mathcal{L}(\mathbf{u})$, namely the closedness under reversal. One direction of this connection is trivial: If a fixed point of a primitive morphism contains infinitely many palindromes, then its language is closed under reversal. The non-trivial converse is shown in [21] for marked morphisms. The mentioned results and computer experiments lead to the formulation of the following conjecture.

Conjecture 15. Let $\varphi : \mathcal{A}^* \to \mathcal{A}^*$ be a primitive morphism having a fixed point $\mathbf{u}$. Its language $\mathcal{L}(\mathbf{u})$ is closed under reversal if and only if $\mathbf{u}$ is palindromic.

A proof in full generality of this conjecture has applications in algorithmic analysis of the language of a given morphism. Specifically, it allows for an efficient test whether the language of a fixed point is closed under reversal. For marked primitive morphisms, such an algorithm may be devised based on the following results of [21]:

1. Every marked morphism has a so-called well-marked power (see [21] for a definition). If the fixed point of the morphism is palindromic, then this power is of class P'.
2. Conjecture 15 is true for marked morphisms.

Overall, closedness under reversal of the language generated by a marked primitive morphism is equivalent to palindromicity of the language which is equivalent to the well-marked power being in class P'. Therefore, given a marked primitive morphism, the test whether the language it generates is closed under reversal consists of find the well-marked power and checking if this power is in class P'. Since both these tasks can be performed efficiently in a straightforward manner, the whole test can be easily executed.

In the view of this special case, Conjecture 15 may be seen as a first step to provide an efficient test of closedness under reversal for the language generated by any primitive morphism for which the class P conjecture holds.

Acknowledgements. The authors acknowledge financial support by the Czech Science Foundation grant GAČR 13-03538S.

References

1. Bašić, B.: On highly potential words. Eur. J. Combin. **34**(6), 1028–1039 (2013)
2. Baláži, P., Masáková, Z., Pelantová, E.: Factor versus palindromic complexity of uniformly recurrent infinite words. Theoret. Comput. Sci. **380**(3), 266–275 (2007)
3. Balková, L., Pelantová, E., Starosta, Š.: Infinite words with finite defect. Adv. Appl. Math. **47**(3), 562–574 (2011)
4. Balková, L., Pelantová, E., Starosta, Š.: Proof of the Brlek-Reutenauer conjecture. Theoret. Comput. Sci. **475**, 120–125 (2013)

5. Massé, A.B., Brlek, S., Frosini, A., Labbé, S., Rinaldi, S.: Reconstructing words from a fixed palindromic length sequence. In: Ausiello, G., Karhumäki, J., Mauri, G., Ong, L. (eds.) TCS 2008. IIFIP, vol. 273, pp. 101–114. Springer, Boston, MA (2008). doi:10.1007/978-0-387-09680-3_7

6. Blondin Massé, A., Brlek, S., Garon, A., Labbé, S.: Combinatorial properties of ƒ-palindromes in the Thue-Morse sequence. Pure Math. Appl. 19(2–3), 39–52 (2008)

7. Blondin Massé, A., Brlek, S., Labbé, S., Vuillon, L.: Palindromic complexity of codings of rotations. Theoret. Comput. Sci. 412(46), 6455–6463 (2011)

8. Brlek, S., Reutenauer, C.: Complexity and palindromic defect of infinite words. Theoret. Comput. Sci. 412(4–5), 493–497 (2011)

9. Brlek, S., Hamel, S., Nivat, M., Reutenauer, C.: On the palindromic complexity of infinite words. Int. J. Found. Comput. Sci. 15(2), 293–306 (2004)

10. Bucci, M., Vaslet, E.: Palindromic defect of pure morphic aperiodic words. In: Proceedings of the 14th Mons Days of Theoretical Computer Science (2012)

11. Bucci, M., De Luca, A., Glen, A., Zamboni, L.Q.: A connection between palindromic and factor complexity using return words. Adv. Appl. Math. 42(1), 60–74 (2009)

12. Bucci, M., De Luca, A., Glen, A., Zamboni, L.Q.: A new characteristic property of rich words. Theoret. Comput. Sci. 410(30–32), 2860–2863 (2009)

13. Bucci, M., Luca, A., Luca, A.: Rich and periodic-like words. In: Diekert, V., Nowotka, D. (eds.) DLT 2009. LNCS, vol. 5583, pp. 145–155. Springer, Heidelberg (2009). doi:10.1007/978-3-642-02737-6_11

14. Droubay, X., Justin, J., Pirillo, G.: Episturmian words and some constructions of de Luca and Rauzy. Theoret. Comput. Sci. 255(1–2), 539–553 (2001)

15. Droubay, X., Pirillo, G.: Palindromes and Sturmian words. Theoret. Comput. Sci. 223(1–2), 73–85 (1999)

16. Glen, A., Justin, J., Widmer, S., Zamboni, L.Q.: Palindromic richness. Eur. J. Combin. 30(2), 510–531 (2009)

17. Guo, C., Shallit, J., Shur, A.M.: Palindromic rich words and run-length encodings. Inf. Process. Lett. 116(12), 735–738 (2016)

18. Harju, T., Vesti, J., Zamboni, L.Q.: On a question of Hof, Knill and Simon on palindromic substitutive systems. Monatsh. Math. 179(3), 379–388 (2016)

19. Hof, A., Knill, O., Simon, B.: Singular continuous spectrum for palindromic Schrödinger operators. Commun. Math. Phys. 174, 149–159 (1995)

20. Jajcayová, T., Pelantová, E., Starosta, Š.: Palindromic closures using multiple antimorphisms. Theoret. Comput. Sci. 533, 37–45 (2014)

21. Labbé, S., Pelantová, E.: Palindromic sequences generated from marked morphisms. Eur. J. Combin. 51, 200–214 (2016)

22. Labbé, S., Pelantová, E., Starosta, Š.: On the zero defect conjecture. Eur. J. Comb. 62, 132–146 (2017)

23. Labbé, S.: A counterexample to a question of Hof, Knill and Simon. Electron. J. Combin. 21 (2014). Paper #P3.11

24. de Luca, A., Glen, A., Zamboni, L.Q.: Rich, sturmian, and trapezoidal words. Theoret. Comput. Sci. 407(1), 569–573 (2008)

25. Masáková, Z., Pelantová, E., Starosta, Š.: Exchange of three intervals: substitutions and palindromicity. Eur. J. Combin. 62, 217–231 (2017)

26. Pelantová, E., Starosta, Š.: Languages invariant under more symmetries: overlapping factors versus palindromic richness. Discret. Math. 313, 2432–2445 (2013)

27. Pelantová, E., Starosta, Š.: Palindromic richness for languages invariant under more symmetries. Theor. Comput. Sci. 518, 42–63 (2014)

28. Pelantová, E., Starosta, Š.: Constructions of words rich in palindromes and pseudopalindromes. Discret. Math. Theoret. Comput. Sci. **18**(3), 1–26 (2016)
29. Restivo, A., Rosone, G.: Balancing and clustering of words in the Burrows-Wheeler transform. Theoret. Comput. Sci. **412**(27), 3019–3032 (2011)
30. Rukavicka, J.: On number of rich words (2017). Preprint available at arXiv:1701.07778
31. Starosta, Š.: Generalized Thue-Morse words and palindromic richness. Kybernetika **48**(3), 361–370 (2012)
32. Starosta, Š.: Morphic images of episturmian words having finite palindromic defect. Eur. J. Combin. **51**, 359–371 (2016)
33. Tan, B.: Mirror substitutions and palindromic sequences. Theoret. Comput. Sci. **389**(1–2), 118–124 (2007)
34. Vesti, J.: Extensions of rich words. Theoret. Comput. Sci. **548**, 14–24 (2014)

Equations Enforcing Repetitions
Under Permutations

Joel D. Day, Pamela Fleischmann$^{(\boxtimes)}$, Florin Manea, and Dirk Nowotka

Institut für Informatik, Christian-Albrechts-Universität zu Kiel, Kiel, Germany
{jda,fpa,flm,dn}@informatik.uni-kiel.de

Abstract. The notion of repetition of factors in words is central to combinatorics on words. A recent generalisation of this concept considers repetitions under permutations: give an alphabet Σ and a morphism or antimorphism f on Σ^*, whose restriction to Σ is a permutation, w is an $[f]$-repetition if there exists $\gamma \in \Sigma^*$ such that $w = f^{i_1}(\gamma)f^{i_2}(\gamma) \cdots f^{i_k}(\gamma)$, for some $k \geq 2$. In this paper, we extend a series of classical repetition enforcing word equations to this general setting to obtain a series of word equations whose solutions are $[f]$-repetitions.

1 Introduction

The study of repetitive sequences in words is one of the central topics of combinatorics on words, with applications in e.g., pattern matching and stringology in general, data compression, bioinformatics (see [10,13]). Part of the investigations on this topic deal with repetition enforcing relations or equations. Basically, a repetition enforcing relation for words is a relation, or a statement, that holds only for words that can be expressed as repetitions (i.e., repeated concatenation) of some (other) word. For instance, it is well known (see, e.g., [12]) that a word w is a factor, other than prefix of suffix, of the word ww if and only if $w \in \{t\}^+$ for some shorter word t, i.e., w is a repetition. Another prominent example of a repetition enforcing statement is the Theorem of Fine and Wilf [6] (FWT): if $\alpha = u^\ell$ and $\beta = v^k$ and α and β share a common prefix of length at least $|u| + |v| - \gcd(|u|, |v|)$, then both u and v are repetitions of some word t, i.e., $u, v \in \{t\}^+$. The equation of Lyndon and Schützenberger [14] (LSE) is an example of a repetition enforcing equation: if $u^\ell = v^m w^n$ holds, for some words u, v, w and $\ell, m, n \geq 2$, then there exists a word t such that $u, v, w \in \{t\}^+$, so u, v, w are repetitions of the same root.

Pseudo-repetitions were introduced [4,5], as a generalisation of classical repetitions, inspired by molecular biology. A word w is a *pseudo-repetition* (more precisely, f-repetition) if it equals a repeated concatenation of one of its prefixes t and its image $f(t)$ under some morphism or antimorphism (for short "anti-/morphism") f, thus $w \in t\{t, f(t)\}^+$. To fit the biological motivation, in [5] f was defined as an antimorphic involution (i.e., $f^2(w) = w$ for all words w). More

---

D. Nowotka—Research supported by DFG grant NO 872/3-2 (jda, dn), DFG grant MA 5725/1-2 (flm), BMBF HPSV grant 01IH15006A (fpa).

S. Brlek et al. (Eds.): WORDS 2017, LNCS 10432, pp. 72–84, 2017.
DOI: 10.1007/978-3-319-66396-8_8

interesting to us, if f is not restricted, pseudo-repetitions generalise not only repetitions (when f is the identity morphism), but also palindromes (when f is the mirror image); both these concepts are central in combinatorics on words, so their generalisations are of intrinsic theoretical interest. Initial results (see [3,5]) concerned generalisations of the FWT, of the LSE, and of other repetition enforcing results to the setting of f-repetitions for antimorphic involutions f. For instance, Czeizler et al. [3] introduced a different generalisation of LSE. They considered equations of the form $u_1 u_2 \cdots u_\ell = v_1 v_2 \cdots v_m w_1 w_2 \cdots w_n$, where $u_i \in \{u, \theta(u)\}$ for all $1 \leq i \leq \ell$, $v_j \in \{v, \theta(v)\}$ for all $1 \leq j \leq m$, and $w_k \in \{w, \theta(w)\}$ for all $1 \leq k \leq n$, and studied under which conditions $u, v, w \in \{t, \theta(t)\}^+$ yield for some word t. That is, they studied the case when u, v, w are generalised repetitions (more precisely, θ-repetitions). A complete characterisation of the conditions under which the aforementioned equation has only θ-repetitive solutions was obtained in [17].

Going a step further, the case of f-repetitions (over an alphabet Σ) for an anti-/morphism f that acts as a permutation on Σ (anti-/morphic permutation) was considered in [15]; where a series of results in the style of the FWT were given. Introduced in [15], but only briefly studied in that paper, was also a more general notion of repetition, that we will call here an $[f]$-repetition. If f is an anti-/morphic permutation and $w = f^{i_1}(\gamma) f^{i_2}(\gamma) \cdots f^{i_k}(\gamma)$, for some $k \geq 2$, then w is called $[f]$-repetition of root γ. A variant of the FWT was shown for $[f]$-repetitions in the case when f is a morphism. This notion also appears in a series of papers regarding avoidability of patterns under anti-/morphic permutations: in [16] the avoidability of patterns of the form $\pi^i(x) \pi^j(x) \pi^k(x)$, i.e., $[\pi]$-cubes, for π a variable that can be replaced by anti-/morphic permutations, was studied, while in [2] the avoidability of general $[\pi]$-repetitions was considered. Finally, algorithmic problems like deciding whether a word is an $[f]$-repetition [7,8] or whether a word contains $[f]$-repetitions [1,9,18] for different types of functions (including anti-/morphic permutations) were investigated. However, we are not aware of any algorithmic results regarding $[f]$-repetitions.

In this paper we analyse a series of $[f]$-repetition enforcing word equations, for an anti-/morphic permutation f. We first analyse the morphic case, and we show that a series of classical repetition enforcing equations are extendible to this more general setting. For instance we show that both $f^a(x) f^b(y) = f^c(y) f^d(x)$ and $f^a(u) f^b(u) = x f^c(u) y$ with $x, y \neq \varepsilon$ enforce x, y resp. u to be an $[f]$-repetition each with one root.

Our main result is an extension of the LSE: if $f^{i_1}(u) \ldots f^{i_r}(u) f^{j_1}(v) \ldots$ $f^{j_s}(v) = f^{k_1}(w) \ldots f^{k_t}(w)$ for some $r, s, t \geq 2$, then u, v, w are $[f]$-repetitions of the same root t. These results complement the generalised FWT obtained in this setting in [15]. In the case when f is antimorphic, we show that the equation $f^a(u) f^b(u) = x f^c(u) y$ may have solutions which are not $[f]$-repetitions. Thus, following the results of [3,5], we characterise exactly the equations $f^{a_1}(u) f^{a_2}(u) f^{a_3}(u) = x f^{b_1}(u) f^{b_2}(u) y$, with $x, y \neq \varepsilon$, whose solutions are $[f]$-repetitions. We use this characterisation to show a result in the style of the FWT and to define a class of extensions of the LSE that only have solutions which are $[f]$-repetitions.

The paper is organised as follows: we first give the basic definitions and recall some preliminary results, then we present the results for the case when f is a morphic permutation, and finally we present the results for when f is an antimorphic permutation. Due to space restrictions some proofs are omitted.

2   Preliminaries

Let $\mathbb{N}$ be the set of natural numbers, $\mathbb{N}_0 = \mathbb{N} \cup \{0\}$ and $\mathbb{N}_{\geq k} = \{x \in \mathbb{N}_0 \mid x \geq k\}$. For $n \in \mathbb{N}$, $[n]$ denotes $\{1, \ldots, n\}$ and $[n]_0 = [n] \cup \{0\}$. For the set of integers $\mathbb{Z}$, $a \equiv_k b$ holds if and only if $a, b \in \mathbb{Z}$ have the same remainder modulo $k \in \mathbb{N}$ and $\mathbb{Z}_k$ denotes the quotient ring of integers modulo k. For $m, n \in \mathbb{N}$, let $\gcd(m, n)$ denote their greatest common divisor.

Let Σ be a finite alphabet. In this paper Σ^* denotes the set of all words over Σ, ε the *empty word*, $\Sigma^+ := \Sigma^* \backslash \{\varepsilon\}$, and for the word's length $|w|$, $\Sigma^{\leq k} := \{w \in \Sigma^* \mid |w| \leq k\}$. For two words u and v, set $d_{u,v} := \gcd(|u|, |v|)$.

For some words x, y, u is a *factor* of w, if $w = xuy$; u is a *prefix* of w if $x = \varepsilon$ and a *suffix* if $y = \varepsilon$. A word u is said to occur strictly inside another word w if u is a factor of w, other than a prefix or a suffix. Moreover, $w = u^{-1}v$, whenever $v = uw$. The powers of w are defined recursively by $w^0 = \varepsilon$, $w^n = ww^{n-1}$ for $n \geq 1$. If w cannot be expressed as a power of another word, then w is said to be *primitive*.

We say that $f : \Sigma^* \to \Sigma^*$ is a morphism (resp., antimorphism) if $f(xy) = f(x)f(y)$ (resp., $f(xy) = f(y)f(x)$) for any words $x, y \in \Sigma^*$. Note that, to define an anti-/morphism it is enough to define $f(a)$ for all $a \in \Sigma$. If f is a bijective morphism (resp., antimorphism), then we call f a morphic (resp., antimoprhic) permutation. If f is a permutation of Σ then $\mathrm{ord}(f)$ denotes the smallest positive number such that $f^{\mathrm{ord}(f)}(a) = a$ for all $a \in \Sigma$. If f is a morphic permutation then $f^{\mathrm{ord}(f)}(w) = w$ and if f is an antimorphic permutation then $f^{2\mathrm{ord}(f)}(w) = w$, for all $w \in \Sigma^*$. This leads to the fact that the exponents of an anti-/morphic permutation f can be considered to be elements of $\mathbb{Z}_{\mathrm{ord}(f)-1}$ resp. $\mathbb{Z}_{2\mathrm{ord}(f)-1}$, i.e. f^{a-b} is for all $a, b \in \mathbb{Z}$ a well-defined iteration of f.

For an anti-/morphic permutation f, a word $w \in \Sigma^*$ is said to be an $[f]$-*repetition* if there exists $t \in \Sigma^+$, $k \geq 2$, and $i_1, \ldots, i_k \in \mathbb{Z}$ such that $w = f^{i_1}(t)f^{i_2}(t) \cdots f^{i_k}(t)$. In this case, t is called the root of the $[f]$-repetition w. If w is not an $[f]$-repetition, then w is $[f]$-*primitive*. For instance, the word $w = abcaab$ is $[\mathrm{Id}_\Sigma]$-primitive, where Id_Σ is the identical morphism on Σ, and $[f]$-primitive for some morphism or antimorphism f with $f(a) = b$, $f(b) = a$ and $f(c) = c$. However, for the morphism $f(a) = c$, $f(b) = a$ and $f(c) = b$, $w = abf(ab)ab = abcaab$, thus, w is an $[f]$-power in this setting; $abbcab = abf^2(ab)ab$ is also an $[f]$-repetition as well.

In the following, several classical repetition enforcing results are recalled. The first one is folklore (see, e.g., [12]). The next three are classical results of Fine and Wilf and Lyndon and Schützenberger, respectively.

Theorem 1 (1-in-2). *A word $w \in \Sigma^*$ is a repetition iff w occurs strictly inside ww.*

Theorem 2 (Fine and Wilf [6]). *Let $u, v \in \Sigma^*$. If two words $\alpha \in u\{u, v\}^*$ and $\beta \in v\{u, v\}^*$ have a common prefix of length at least $|u| + |v| - d_{u,v}$, then u and v are powers of a common word of length $d_{u,v}$. The bound $|u| + |v| - d_{u,v}$ is optimal.*

Theorem 3 (Lyndon and Schützenberger [14]). *Let $u, v, w \in \Sigma^*$. Then $uv = vw$ if and only if there exist words $p, q \in \Sigma^*$, such that $u = (pq)^i, w = (qp)^i$, and $v = (pq)^j p$ for some $i, j \geq 0$ and pq is primitive.*

Theorem 4 (Lyndon and Schützenberger [14]). *If $u^\ell = v^m w^n$ for some words $u, v, w \in \Sigma^*$ and $\ell, m, n \geq 2$, then $u, v, w \in \{t\}^*$ for some word $t \in \Sigma^*$.*

Theorem 2 was extended in [15] for $[f]$-repetitions.

Theorem 5. *Let $u, v \in \Sigma^*$ and $f : \Sigma^* \to \Sigma^*$ be a morphic permutation with $\mathrm{ord}(f) = k + 1$. Let $S(u, v) = \{u, f(u), \dots, f^k(u), v, f(v), \dots, f^k(v)\}^*$. If two words $\alpha \in uS(u, v)$ and $\beta \in vS(u, v)$ have a common prefix of length at least $|u| + |v| - d_{u,v}$, then there exists a $t \in V^*$, such that $u, v \in t\{t, f(t), \dots, f^k(t)\}^*$.*

Theorem 3 was extended in the setting of anti-/morphic involutions in [5]. Theorem 4 was extended for $[f]$-repetitions where f is an antimorphic involution in a series of papers that culminated in [17], where a full characterisation of the triples (ℓ, m, n) for which $u_1 u_2 \cdots u_\ell = v_1 v_2 \cdots v_m w_1 w_2 \cdots w_n$, where $u_i \in \{u, f(u)\}$ for all $1 \leq i \leq \ell$, $v_j \in \{v, f(v)\}$ for all $1 \leq j \leq m$, and $w_k \in \{w, f(w)\}$ for all $1 \leq k \leq n$, has only solutions which are $[f]$-repetitions was given.

3 The Morphic Case

In this section some well known equations which only have repetitions as solutions are generalised to equations whose solutions are repetitions under morphic permutations. These results are used to ultimately show that a version of Theorem 4 holds for $[f]$-repetitions in the case that f is a morphic permutation.

Some basic lemmas are first established, which provide some fundamental combinatorial tools for proving the later results. They focus on two very well-known equations, namely $xy = xy$ and $xy = yz$ with $x, y, z \in \Sigma^+$, and describe their solutions in this more general setting.

Lemma 6. *Let $x, y \in \Sigma^+$, f a morphic permutation on Σ^*, and $a, b, c, d \in [\mathrm{ord}(f)]_0$ with $f^a(x)f^b(y) = f^c(y)f^d(x)$. Then there exists a $t \in \Sigma^+$ such that x, y are $[f, t]$-repetitions.*

Proof. Theorem 5 can be applied for $\alpha = f^a(x)f^b(y)$, $\beta = f^c(y)f^d(x)$, $u = f^a(x)$, and $v = f^c(y)$. Clearly, α and β have a common prefix of length $|\alpha| = |\beta| = |u| + |v|$, it follows that there exists t such that $f^a(x), f^c(y) \in t\{t, f(t), \dots, f^{\mathrm{ord}(f)-1}(t)\}^*$. Consequently, x, y are $[f, t]$-repetitions. $\square$

While Lemma 6 provides a direct analogy to the standard setting, for which the "repetition-enforcing" nature of the equation is folklore, it is also possible to provide the following more specific insight which is essential to the proofs.

Lemma 7. *Let* $x, y \in \Sigma^+$ *with* $x = x_1 x_2$ *such that* $|x_1| = |x_2|$, *f a morphic permutation on* Σ^*, *and* $a, b, c \in [\mathrm{ord}(f)]_0$ *with*

$$y x_1 x_2 = f^a(x_1) f^b(x_2) f^c(y)$$

Then there exists a $t \in \Sigma^+$ *such that* $x, y, f^a(x_1) f^b(x_2)$ *are* $[f, t]$-*repetitions.*

Lemma 8. *Let* $x, y \in \Sigma^+$, *f a morphic permutation on* Σ^* *and* $a, b, c, d \in [\mathrm{ord}(f)]_0$. *The equation*

$$f^a(x) f^b(y) = f^c(y) f^d(z)$$

holds if and only if there exist $u, v \in \Sigma^*$, $i, s, r, q \in \mathbb{N}_0$ *with*

$$x = uv, \quad z = f^q(v) f^{q+r}(u), \quad \text{and} \quad y = f^{s+r}(uv) \dots f^{s+ir}(uv) f^{s+(i+1)r}(u).$$

If Theorem 3 holds for three words x, y, z (i.e., $xy = yz$) then the words x and z are *conjugate*, $x \sim z$ for short. It is well known that the conjugacy relation is an equivalence relation. When working in the setting of equations under morphic permutations, this relation can be extended to f-*conjugacy*. For a morphic permutation f, the words $x, y \in \Sigma^*$ are said to be f-*conjugate* (written $x \sim_f y$) if there exist $a, b, c, d \in [\mathrm{ord}(f)]_0$ such that $f^a(x) f^b(y) = f^c(y) f^d(z)$ – so if they satisfy the equation addressed in Lemma 8. It can be seen that $x \sim_f y$ follows from $x \sim y$. More interestingly however, while $\sim_f$ is symmetrical and reflexive, it is not transitive (unless f is the identical morphism). Accordingly, $\sim_f$ is an equivalence if and only if f is the identity morphism.

The following lemma extends another fundamental result mentioned in Theorem 1, and may be proved by reducing to the case considered by Lemma 6 (see Fig. 1).

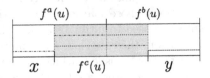

Fig. 1. x, y reoccur each twice within $f^c(y)$, such that - except for permutation application - the pattern $xy = yx$ occurs (shown by the dotted and dashed lines), so Lemma 6 may be applied.

Lemma 9. *Let* $f : \Sigma^* \to \Sigma^*$ *be a morphic permutation and* $a, b, c \in [\mathrm{ord}(f)]_0$. *If* $u \in \Sigma^*$ *is* $[f]$-*primitive, then for* $x, y \in \Sigma^*$ *with*

$$f^a(u) f^b(u) = x f^c(u) y$$

either $x = \varepsilon$ *or* $y = \varepsilon$ *follows.*

Lemma 7 can be extended in a similar fashion.

Lemma 10. *Let* $f : \Sigma^* \to \Sigma^*$ *be a morphic permutation and* $a, b, c, d \in [\text{ord}(f)]_0$. *If* $u \in \Sigma^*$ *is* $[f]$-*primitive and* $u = u_1 u_2$, *with* $|u_1| = |u_2|$, *then for* $x, y \in \Sigma^*$ *with*

$$f^a(u)f^b(u) = xf^c(u_1)f^d(u_2)y$$

either $x = \varepsilon$ *or* $y = \varepsilon$ *follows.*

In the rest of this section it will be shown that Lyndon and Schützenberger's result can be reproven without any additional restrictions in the setting of repetitions under morphic permutations. In this setting, the LSE is defined for words $u, v, w \in \Sigma^+$ by

$$f^{a_1}(u) \dots f^{a_r}(u)f^{c_1}(v) \dots f^{c_s}(v) = f^{b_1}(w) \dots f^{b_t}(w), \tag{1}$$

for $r, s, t \in \mathbb{N}_{\geq 2}$, $a_i, b_k, c_j \in [\text{ord}(f)]_0$, $i \in [r]$, $k \in [s]$, $j \in [t]$, and a morphic permutation f on Σ^*. For simplicity, the following notations are sometimes used: $\alpha_1 = f^{a_1}(u) \dots f^{a_r}(u)$, $\alpha_2 = f^{c_1}(v) \dots f^{c_s}(v)$, and $\beta = f^{b_1}(w) \dots f^{b_t}(w)$. The intention is to show that there exists a word t such that $u, v, w \in \{t, f(t), \dots, f^{\text{ord}(f)-1}(t)\}^*$, and thus that the equation, when augmented by the presence of morphic permutations, remains a repetition enforcing relation.

In order to show that indeed u, v, and w are $[f]$-repetitions with the same root under these conditions, the proof is divided into various cases. To begin with, the cases in which the us and vs "fit" exactly inside the ws and vice-versa are given.

Lemma 11. *If Eq. 1 holds for* $r, s, t \geq 2$, *and* $|u| \mid |w|$ *or* $|v| \mid |w|$ *holds, then* u, v, w *are* $[f]$-*repetitions.*

Lemma 12. *If Eq. 1 holds for* $r, s, t \geq 2$, *and* $|w| \mid |u|$ *or* $|w| \mid |v|$ *holds, then* u, v, w *are* $[f]$-*repetitions.*

The following lemma demonstrates how, in some cases, the extension of the FWT (Theorem 5), may be applied. This is straightforward if the theorem may be applied from both endpoints, in opposite directions, showing first that u and w share an f-root and then that v and w share an f-root. In fact, as the lemma states, it is sufficient to be able to apply the theorem in just one direction – although this requires more effort to prove.

Lemma 13. *In Eq. 1, if* α_1 *and* β *have a common prefix of length at least* $|w| + |u| - d_{u,w}$ *or* α_2 *and* β *have common suffix of length at least* $|w| + |v| - d_{v,w}$, *then* u, v, *and* w *are* $[f]$-*repetitions.*

Following from Lemma 13 (see also Fig. 2), it is now possible to show that the equation is repetition enforcing provided r, s and t are large enough.

Theorem 14. *If Eq. 1 holds with* r, s, t *at least 3 or with* $t \geq 4$ *and* r, s *at least 2, then one of the conditions of Lemma 13 also holds. Hence, there exists a word* t *such that* $u, v, w \in \{t, f(t), \dots, f^{\text{ord}(f)-1}(t)\}^*$.

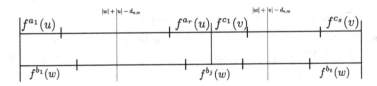

Fig. 2. If the parts of permutations of u and the one permutation of v are each long enough, Theorem 5 can be applied from either side.

Recalling the original LSE (Theorem 4), the claim is *nearly* proven for the permutation setting as well. It remains to prove the cases for r, s, t being 2, respectively. In order to accomplish this, the following auxiliary result is needed.

Lemma 15. *Let w_1, w_2, be in Σ^+, f a permutation, and $a, b, c, d \in \mathbb{N}$. If $w_1 f^a(w_2) = f^b(x) f^c(w_2) f^d(x)$ holds, then there exists a suffix x' of a permutation of x and there exists $n_1, \ldots, n_r \in \mathbb{N}$ for $r \in \mathbb{N}$ with*

$$w_2 = x' f^{n_1}(x) \ldots f^{n_r}(x).$$

Remark 16. For $f^a(w_1) w_2 = f^b(x) f^c(w_1) f^d(x)$ an analogous result can be obtained.

The case that $r = s = t = 2$ is one of the most straightforward remaining cases, and is addressed first.

Lemma 17. *If Eq. 1 holds for $r = s = t = 2$, then u, v, and w are $[f]$-repetitions.*

Proof. Consider w.l.o.g. $2|u| > |w|$. Choose $u_1, u_2 \in \Sigma^+$ with $f^{a_2}(u) = u_1 u_2$ such that $u_1 \in \text{Suff}(f^{b_1}(w))$ and $u_2 \in \text{Pref}(f^{b_2}(w))$. This implies $f^{b_1}(w) = f^{a_1-a_2}(u_1) f^{a_1-a_2}(u_2) u_1$ and

$$f^{b_2-b_1+a_1-a_2}(u_1) f^{b_2-b_1+a_1-a_2}(u_2) f^{b_2-b_1}(u_1) = f^{b_2}(w) = u_2 f^{c_1}(v) f^{c_2}(v).$$

By this $2|u_1| = 2|v|$ follows. Moreover, since $f^{b_2-b_1}(u_1)$ and $f^{c_2}(v)$ are suffixes of $f^{b_2}(w)$, $f^{b_2-b_1}(u_1) = f^{c_2}(v)$ holds. Substituting this result in $f^{b_2}(w)$ leads to

$$f^{b_2-b_1+a_1-a_2}(u_1) f^{b_2-b_1+a_1-a_2}(u_2) = u_2 f^{c_1-c_2+b_2-b_1}(u_1)$$

By Lemma 6 follows the existence of a $\gamma \in \Sigma^*$ such that u_1, u_2 are $[f, \gamma]$-repetitions and consequently u, v, and w are $[f]$-repetitions as well (Fig. 3).

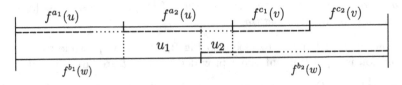

Fig. 3. In the case of $r = s = t = 2$ the pattern $u_1 u_2 = u_2 u_1$ - neglecting the permutations - occurs in the second w.

Lemma 18. *If Eq. 1 holds for $t = 3$ and $r = s = 2$ then u, v, and w are $[f]$-repetitions. Similarly, f Eq. 1 holds for $r = 2$ and $s = t = 3$, then u, v, and w are $[f]$-repetitions.*

Lemma 19. *If Eq. 1 holds for $s = t = 2$ and $r \geq 3$ then u, v, and w are $[f]$-repetitions.*

Lemma 20. *If Eq. 1 holds for $t = 2$ and $r, s \geq 3$ then u, v, and w are $[f]$-repetitions.*

From the preceding lemmas, we can conclude with the following main result.

Theorem 21. *If Eq. 1 holds for $t, r, s \geq 2$ then u, v, and w are $[f]$-repetitions.*

4   The Antimorphic Case

Firstly, note that the results of Lemma 9 do not hold in the case of antimorphic permutations.

Remark 22. Consider the equation $f^a(w)f^b(w) = xf^c(w)y$, where f is an antimorphism. The following counterexamples show that this equation is not repetition enforcing, no matter the values of a, b, c; in all cases, $\Sigma = \{\mathsf{a}, \mathsf{b}\}$ and f is the mirror image on Σ^*. Let $w = \mathsf{aaba}$, which is not a $[f]$-repetition. However, for a even and b, c odd, $f^a(w)f^b(w) = \mathsf{aabaabaa} = \mathsf{aab}f^c(w)\mathsf{a}$ holds. By Lemma 25, it follows immediately that the equation $f^a(w)f^b(w) = xf^c(w)y$ when a is odd and b, c even, a, c odd and b even, and a, c even and b odd, may have solutions which are not $[f]$-repetitions. If a, b are even and c odd, for the same w and f, we have $f^a(w)f^b(w) = \mathsf{aabaaaba} = \mathsf{a}f^c(w)\mathsf{aba}$. Again, it is an immediate consequence that when a, b are odd and c is even then $f^a(w)f^b(w) = xf^c(w)y$ may have solutions which are not $[f]$-repetitions.

Following the ideas of [11], an extension of Lemma 9 to the antimorphic case can be obtained by considering equations of the form

$$f^{a_1}(w)f^{a_2}(w)f^{a_3}(w) = xf^{b_1}(w)f^{b_2}(w)y$$

for f antimorphic permutation on Σ, $a_1, a_2, a_3, b_1, b_2 \in \mathbb{N}_0$, $w, x, y \in \Sigma^*$, $0 < |x|, |y| < |w|$. Our goal is to identify under which restrictions on a_1, a_2, a_3, b_1, b_2 the equation above enforces (for some f that makes the equality between the sides of the equation hold) that x, y, w are $[f]$-repetitions.

The main difference to the morphic case is that when iterating an antimorphic permutation f, $f^i(w)$ preserves the order of letters of w when i is even, and reverses it when i is odd; in the morphic case, the order was preserved for all exponents. Therefore, it seems a good approach to classify the equations considered above by the parity of their exponents. In the following, e (from even) and o (from odd) are used for 0 and 1 resp., for convenience. Moreover for each number a let $\bar{a}$ denote its residue class modulo 2.

Definition 23. *Define the set of all these equations by*

$$\mathcal{E} := \{f^{a_1}(w)f^{a_2}(w)f^{a_3}(w) = xf^{b_1}(w)f^{b_2}(w)y \mid f \text{ antimorphic permutation,}$$
$$w, x, y \in \Sigma^*, 0 < |x|, |y| < |w|, a_1, a_2, a_3, b_1, b_2 \in \mathbb{N}_0\}.$$

The equations

$$E : f^{a_1}(w)f^{a_2}(w)f^{a_3}(w) = xf^{b_1}(w)f^{b_2}(w)y \text{ and}$$
$$E' : f^{a'_1}(w)f^{a'_2}(w)f^{a'_3}(w) = x'f^{b'_1}(w)f^{b'_2}(w)y'$$

are called equivalent *($E \sim E'$) if $(\overline{a_1}, \overline{a_2}, \overline{a_3}, \overline{b_1}, \overline{b_2}) = (\overline{a'_1}, \overline{a'_2}, \overline{a'_3}, \overline{b'_1}, \overline{b'_2})$ A class of equivalent equations will be denoted by the quintuple $(\overline{a_1}, \overline{a_2}, \overline{a_3} \mid \overline{b_1}, \overline{b_2})$. Such a class (resp., quintuple) is called* repetition enforcing *if every equation in the class it defines has only solutions which are $[f]$-repetitions.*

Remark 24. Note that $\sim$ is an equivalence relation. Thus, the quotient set $\mathcal{E}/\sim$ is well defined. Since the elements of $\mathcal{E}$ are defined by five parameters, which are further reduced by the factorization w.r.t. $\sim$ to their canonical representative from $\mathbb{Z}_2$, then $\mathcal{E}$ has only 32 elements. Since all equivalent equations are associated to the same quintuple of elements of Z_2, these quintuples can be used as canonical representatives for the classes of $\mathcal{E}/\sim$.

In order to further group together classes of equation, it is worth noting the following.

Lemma 25. *Let f be an antimorphic permutation on Σ. Consider the equations*

$$f^{i_1}(w) \cdots f^{i_k}(w) = xf^{j_1}(w) \cdots f^{j_{k-1}}(w)y, 0 < |x|, |y| < |w| \tag{2}$$

$$f^{i_1-1}(u) \cdots f^{i_k-1}(u) = xf^{j_1-1}(u) \cdots f^{j_{k-1}-1}(u)y, 0 < |x|, |y| < |u| \tag{3}$$

$$f^{i_k+1}(w) \cdots f^{i_1+1}(w) = f(y)f^{j_{k-1}+1}(w) \cdots f^{j_1+1}(w)f(x), 0 < |x|, |y| < |w| \tag{4}$$

$$f^{i_k}(u) \cdots f^{i_1}(u) = f(y)f^{j_{k-1}}(u) \cdots f^{j_1}(u)f(x), 0 < |x|, |y| < |u| \tag{5}$$

All the solutions of equation (i) are $[f]$-repetitions if and only if all solutions of equation (j) are $[f]$-repetitions, with $1 \le i, j \le 4$.

Following the ideas of Lemma 25, it makes sense to define the following relation.

Definition 26. *Two elements $E = (a_1, a_2, a_3 \mid b_1, b_2)$ and $E' = (a'_1, a'_2, a'_3 \mid b'_1, b'_2)$ of $\mathcal{E}/\sim$ are called* dual *($E_1 \circ\!\!-\!\!\circ E_2$) if either $E = E'$ or one of the following cases holds (all the sums below are done in Z_2):*

1. $(a'_1, a'_2, a'_3 \mid b'_1, b'_2) = (a_3 + 1, a_2 + 1, a_1 + 1 \mid b_2 + 1, b_1 + 1)$ *(equating to the application of f to E)*
2. $(a'_1, a'_2, a'_3 \mid b'_1, b'_2) = (a_1 + 1, a_2 + 1, a_3 + 1 \mid b_1 + 1, b_2 + 1)$ *(equating to $f^z(w) = f^{z-1}(f(w))$ for $z \in Z_2$)*

3. $(a_1', a_2', a_3' \mid b_1', b_2') = (a_3, a_2, a_1 \mid b_2, b_1)$ *(equating to the application of 1. and 2.)*

Remark 27. Since $\multimap$ is also an equivalence relation the above mentioned 32 classes can be reduced to the following 10 classes of $(\mathcal{E}/\!\sim)/\!\multimap$

(1) $[(e,e,e \mid e,e)]$		$[(e,e,e \mid e,o)]$ (2)
(3) $[(e,e,e \mid o,o)]$		$[(e,e,o \mid e,e)]$ (4)
(5) $[(e,e,o \mid e,o)]$		$[(e,e,o \mid o,e)]$ (6)
(7) $[(e,e,o \mid o,o)]$		$[(e,o,e \mid e,e)]$ (8)
(9) $[(e,o,e \mid e,o)]$		$[(o,e,o \mid e,e)]$ (10)

The following lemma is a direct consequence of Lemma 25.

Lemma 28 (Duality Lemma). *Let C be a class of $(\mathcal{E}/\!\sim)/\!\multimap$ and E_1, E_2 be in C. If E_1 is repetition-enforcing than E_2 as well.*

For eight of the ten classes of $(\mathcal{E}/\!\sim)/\!\multimap$ it is shown that they are repetition-enforcing. In the remaining cases, counter-examples will be given.

Lemma 29. *Classes 3 (represented by $(e,e,e \mid o,o)$) and 7 (represented by $(e,e,o \mid o,o)$) are not repetition-enforcing.*

Proof. Equations of these classes that have solutions which are not $[f]$-repetitions can be obtained by extending the examples in Remark 22.

Consider $w = \mathsf{aaba}$ and f the mirror image on $\Sigma = \{\mathsf{a}, \mathsf{b}\}$. Although w is not an $[f]$-repetition, the following holds: $www = \mathsf{aabaaabaaaba} = \mathsf{a}f(w)f(w)\mathsf{aba}$, so class 3 is not repetition enforcing.

Also, the following holds $wwf(w) = \mathsf{aabaaabaabaa} = \mathsf{a}f(w)f(w)\mathsf{baa}$, so class 7 is not repetition enforcing.

For some classes the repetition enforcement can be proven by Lemma 9 from the morphic case (Fig. 4). This is possible since the word $f^{b_1}(w)$ occurs inside $f^{a_1}(w)f^{a_2}(w)$ and a_1, a_2, b_1 are even (for short, e *occurs in* ee) in all equations contained in the classes 1, 2, 4, and 5. In class 10 we again have that e occurs in ee: $f^{a_2}(w)$ is a factor of $f^{b_1}(w)f^{b_2}(w)$, and a_2, b_1, b_2 are all even. Class 8 may appear to be different but in fact it contains a similar structure. The characteristic of the aforementioned pattern is, that - neglecting the permutations for a moment - a word is split into $w = xy$ and x occurs also as a suffix and y also as an infix. Having a deeper look into the representative of class 8 reveals that a prefix x of $f^{b_1}(w)$ is a suffix of $f^{a_1}(w)$ and a suffix y of $f^{b_2}(w)$ is a prefix of $f^{a_3}(w)$ with $|xy| = |w|$. So, $f^{a_1-a_3}(y)$ is a prefix of $f^{a_1}(w)$. Therefore, $f^{a_1}(w)$ is a factor of $f^{a_1-a_3}(w)f^{b_1}(w)$ and $a_1, a_1 - a_3, b_1$ are all even. Accordingly, the following lemma holds.

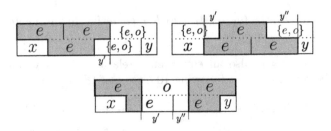

Fig. 4. With the aforementioned abbreviations e and o for an arbitrary even resp. odd permutation of f the classes 1, 2, 4, and 5 are given by the first picture. Class 10 is represented by the second picture and the picture below shows that the necessary 1-in-2 pattern (here visualised as a grey T) occurs if the middle part is ignored.

Lemma 30. *The classes 1, 2, 4, 5, 8 and 10 are repetition-enforcing.*

Before showing that class 9 is also repetition enforcing, several more definitions are needed. If $w = f^{i_1}(s)f^{i_2}(s)\cdots f^{i_k}(s)$, for some $s \in \Sigma^*$ and $k \geq 1$, and $i_j \not\equiv_2 i_{j+1}$ for all $1 \leq j \leq k-1$, then w is called an alternating $[f, s]$-repetition. If the word w is an alternating $[f, s]$-repetition for some s, but this word s is not important to us, then we just say that w is an alternating $[f]$-repetition.

It can be shown that if w is an alternating $[f, s]$-repetition then $f^{a_1}(w)f^{a_2}(w)f^{a_3}(w)$ is also an alternating $[f, s]$-repetition, where $a_1 \equiv_2 a_3 \not\equiv_2 a_2$. Indeed, assume a_1, a_3 are even, and a_2 is odd (the other case is similar). If $w = f^{i_1}(s)f^{i_2}(s)\cdots f^{i_k}(s)$, for some $s \in \Sigma^*$ and $k \geq 1$, and $i_j \not\equiv_2 i_{j+1}$ for all $1 \leq j \leq k-1$, then

$$f^{a_1}(w)f^{a_2}(w)f^{a_3}(w) = f^{a_1+i_1}(s)\cdots f^{a_1+i_k}(s)\cdot$$
$$f^{a_2+i_k}(s)\cdots f^{a_2+i_1}(s)\cdot$$
$$f^{a_3+i_1}(s)\cdots f^{a_3+i_k}(s).$$

As $a_1 + i_k$ has a different parity then $a_2 + i_k$, and $a_2 + i_1$ has a different parity then $a_3 + i_1$, the claim follows.

Next, it is shown that class 9 is repetition enforcing, and, moreover, that if w is a solution of an equation from the respective class, then w is an alternating $[f, s]$-repetition for some word s.

Lemma 31. *Class 9 is repetition-enforcing. More precisely, if*

$$f^{a_1}(w)f^{a_2}(w)f^{a_3}(w) = xf^{b_1}(w)f^{b_2}(w)y$$

with $x, y \neq \varepsilon$ and $(a_1, a_2, a_3 \mid b_1, b_2)$ in class 9, then there exists $s \in \Sigma^*$ such that $xf^{b_1}(w)$ and w are alternating $[f, s]$-repetitions.

The fact that class 6 is also repetition-enforcing follows now.

Lemma 32. *Class 6 is repetition-enforcing.*

Proof. Consider the equation $E : f^{a_1}(w)f^{a_2}(w)f^{a_3}(w) = xf^{b_1}(w)f^{b_2}(w)y$ corresponding to the representative of class 6. Then y is a suffix of $f^{a_3}(w)$. Thus $f^{b_2-a_3}(y)$ is a prefix of $f^{b_2}(w)$ (as $b_2 - a_3$ is odd and $f^{b_2}(w) = f^{b_2-a_3}(f^{a_3}(w))$). By the alignment of $f^{b_2}(w)$ inside $f^{a_2}(w)f^{a_3}(w)$, It follows that $f^{b_2-a_3}(y)$ is a suffix of $f^{a_2}(w)$. Therefore, y is a prefix of $f^{a_2+a_3-b_2}(w)$ and $a_2 + a_3 - b_2$ is odd. Therefore, we get that $f^{a_2}(w)f^{a_3}(w)$ occurs inside $f^{b_1}(w)f^{b_2}(w)f^{a_2+a_3-b_2}(w)$, which leads to an equation represented by $(o, e, o \mid e, o)$, so from class 9. Such equations are repetition enforcing, by Lemma 32.

To conclude this section we propose a series of applications of our repetition enforcing results. In the first one, a repetition enforcing result in the style of FWT is presented.

Theorem 33. *Let $u, v \in \Sigma^+$ such that $|u| < |v|$. Let f be an antimorphic permutation of Σ and $\alpha = f^{i_1}(u)f^{i_2}(u) \cdots f^{i_k}(u)$, $\beta = f^{j_1}(v)f^{j_2}(v) \cdots f^{j_p}(v)$ be two words such that: $k, p \geq 3$, $j_t \neq j_{t+1}$ for all $1 \leq t \leq p-1$, and the common prefix of α and β is longer than $2|v| + |u|$. Then there exists $\gamma \in \Sigma^+$ such that $v, u \in \{f^i(\gamma) \mid 0 \leq i \leq 2\text{ord}(f)\}^+$.*

The second application shows that an extension of the LSE is repetition enforcing.

Theorem 34. *Let f be an antimorphic permutation of an alphabet Σ. Consider the equation:*

$$f^{i_1}(u) \dots f^{i_r}(u)f^{j_1}(v) \dots f^{j_s}(v) = f^{k_1}(w) \dots f^{k_t}(w),$$

with $r, s \geq 3$, $t \geq 6$, and $i_p \neq_2 i_{p+1}$ for $1 \leq p \leq r-1$, $j_p \neq_2 j_{p+1}$ for $1 \leq p \leq s-1$, $k_p \neq_2 k_{p+1}$ for $1 \leq p \leq t-1$. Then there exists γ such that $u, v, w \in \{f^i(\gamma) \mid 0 \leq i \leq 2\text{ord}(f)\}^+$.

5 Further Directions

In this paper we presented a series of equations on words whose solutions are necessarily repetitions under anti-/morphic permutations. The main problem that still remains open is to characterise exactly the triples (r, s, t) for which the equation

$$f^{i_1}(u) \dots f^{i_r}(u)f^{j_1}(v) \dots f^{j_s}(v) = f^{k_1}(w) \dots f^{k_t}(w),$$

with f antimorphic permutation, has only solutions which are $[f]$-repetitions. While Theorem 21 shows that the classical result of Lyndon and Schützenberger is preserved in the generalised case of morphic permutations, we expect that in the case of antimorphic permutations the results obtained in [17] for restricted case of antimorphic involutions should still hold.

References

1. Chiniforooshan, E., Kari, L., Xu, Z.: Pseudopower avoidance. Fund. Inf. **114**(1), 55–72 (2012)
2. Currie, J., Manea, F., Nowotka, D.: Unary patterns with permutations. In: Potapov, I. (ed.) DLT 2015. LNCS, vol. 9168, pp. 191–202. Springer, Cham (2015). doi:10.1007/978-3-319-21500-6_15
3. Czeizler, E., Czeizler, E., Kari, L., Seki, S.: An extension of the Lyndon-Schützenberger result to pseudoperiodic words. Inf. Comput. **209**, 717–730 (2011)
4. Czeizler, E., Kari, L., Seki, S.: On a special class of primitive words. In: Ochmański, E., Tyszkiewicz, J. (eds.) MFCS 2008. LNCS, vol. 5162, pp. 265–277. Springer, Heidelberg (2008). doi:10.1007/978-3-540-85238-4_21
5. Czeizler, E., Kari, L., Seki, S.: On a special class of primitive words. Theoret. Comput. Sci. **411**(3), 617–630 (2010)
6. Fine, N.J., Wilf, H.S.: Uniqueness theorems for periodic functions. Proc. Am. Math. Soc. **16**, 109–114 (1965)
7. Gawrychowski, P., Manea, F., Mercaş, R., Nowotka, D., Tiseanu, C.: Finding pseudo-repetitions. In: Proceedings of STACS 2013, LIPIcs, vol. 20, pp. 257–268 (2013)
8. Gawrychowski, P., Manea, F., Nowotka, D.: Discovering hidden repetitions in words. In: Bonizzoni, P., Brattka, V., Löwe, B. (eds.) CiE 2013. LNCS, vol. 7921, pp. 210–219. Springer, Heidelberg (2013). doi:10.1007/978-3-642-39053-1_24
9. Gawrychowski, P., Manea, F., Nowotka, D.: Testing generalised freeness of words. In: Proceedings of STACS 2014, LIPIcs, vol. 25, pp. 337–349 (2014)
10. Gusfield, D.: Algorithms on Strings, Trees, and Sequences - Computer Science and Computational Biology. Cambridge University Press, Cambridge (1997)
11. Kari, L., Masson, B., Seki, S.: Properties of pseudo-primitive words and their applications. Internat. J. Found. Comput. Sci. **22**(2), 447–471 (2011)
12. Lothaire, M.: Combinatorics on Words. Cambridge University Press, Cambridge (1997)
13. Lothaire, M.: Applied Combinatorics on Words. Cambridge University Press, New York (2005)
14. Lyndon, R.C., Schützenberger, M.P.: The equation $a^m = b^n c^p$ in a free group. Mich. Math. J. **9**(4), 289–298 (1962)
15. Manea, F., Mercaş, R., Nowotka, D.: Fine and Wilf's theorem and pseudo-repetitions. In: Rovan, B., Sassone, V., Widmayer, P. (eds.) MFCS 2012. LNCS, vol. 7464, pp. 668–680. Springer, Heidelberg (2012). doi:10.1007/978-3-642-32589-2_58
16. Manea, F., Müller, M., Nowotka, D.: Cubic patterns with permutations. J. Comput. Syst. Sci. **81**(7), 1298–1310 (2015)
17. Manea, F., Müller, M., Nowotka, D., Seki, S.: The extended equation of lyndon and Schützenberger. J. Comput. Syst. Sci. **85**, 132–167 (2017)
18. Xu, Z.: A minimal periods algorithm with applications. In: Amir, A., Parida, L. (eds.) CPM 2010. LNCS, vol. 6129, pp. 51–62. Springer, Heidelberg (2010). doi:10.1007/978-3-642-13509-5_6

Matching Lexicographic and Conjugation Orders on the Conjugation Class of a Special Sturmian Morphism

David Clampitt[1(✉)] and Thomas Noll[2]

[1] The Ohio State University, Columbus, OH, USA
clampitt.4@osu.edu
[2] Escola Superior de Música de Catalunya, Barcelona, Spain
thomas.mamuth@gmail.com

Abstract. The conjugation class of a special Sturmian morphism carries a natural linear order by virtue of the two elementary conjugations $conj_a$ and $conj_b$ with the single letters a and b, with the standard morphism of the class as the smallest element in this order. We show that a lexicographic order on the morphisms of the given conjugation class can be defined that matches the conjugation order.

Keywords: Sturmian morphisms · Sturmian involution · Christoffel words · Standard words and their conjugates

1 Motivation

Conjugation classes of special Sturmian morphisms carry a natural linear order by virtue of the two elementary conjugations $conj_a$ and $conj_b$ with the single letters a and b (see [6]). For every morphism f in the class—except for the anti-standard morphism—either $conj_a \circ f$ or $conj_b \circ f$ belongs to the class and can be identified as the successor of f. Starting from the standard morphism in the class as the smallest element all the others can be iteratively reached in this way. The largest element in the order is the anti-standard morphism in the class. Figure 1 shows a directed graph, containing five such conjugation classes—including the trivial class of the identity morphism. The four non-trivial classes are aligned along concentric circular arcs around the identity morphism.

In addition to these linear graphs, whose counter-clockwise circular arrows are labeled with either $conj_a$ or $conj_b$, there are outward reaching arrows, which are labeled with the four generators $G, \tilde{G}, D, \tilde{D}$ of the special Sturmian monoid St_0. From every inner node of the graph depart two arrows, labeled either G and $\tilde{G}$ or D and $\tilde{D}$. Hence, there are $2^4 = 16$ paths leading from the central node to the nodes on the outermost arc. Intuitively these pathways can also be ordered in a counter-clockwise manner. The intuition will be made precise in Sect. 4.

Our initial consideration is the following: Each conjugation class is ordered counter-clockwise along the corresponding arc. Each node along this arc can also

© Springer International Publishing AG 2017
S. Brlek et al. (Eds.): WORDS 2017, LNCS 10432, pp. 85–96, 2017.
DOI: 10.1007/978-3-319-66396-8_9

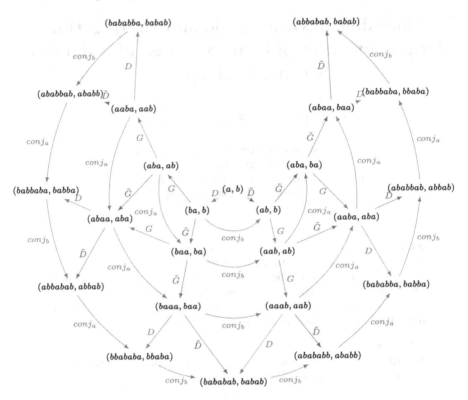

Fig. 1. The nodes along each of the concentric circular arcs form a complete conjugation class of special Sturmian morphisms. Each morphism f is represented by the pair $(f(a), f(b))$ of images of the letters a and b. The node (a, b) in the center represents the Identity map. Then from inside outwards the conjugation classes of D, GD, GGD and $DGGD$ are displayed. The graph forms a subgraph of the Cayley graph of the group $Aut(F_2)$ with respect to the generators $G, \tilde{G}, D, \tilde{D}, conj_a, conj_b$ (and E). Each single conjugation class forms a linear graph, whose arrows are all labeled with one of the conjugations $conj_a$ or $conj_b$. The outward reaching arrows, connecting nodes on successive arcs, are labeled with the generators $G, \tilde{G}, D, \tilde{D}$ of the special Sturmian monoid (= monoid of special positive automorphisms).

be reached along one or more pathways from the center. All the pathways from the center to the nodes on the same arc can also be ordered in a counterclockwise-outward right-to-left lexicographic manner as follows: For each node we postulate that $\overset{G}{\leftarrow}$ precedes $\overset{\tilde{G}}{\leftarrow}$ or that $\overset{D}{\leftarrow}$ precedes $\overset{\tilde{D}}{\leftarrow}$, in accordance with the counter-clockwise arrangement of these arrows. Paths can be ordered lexicographically from right to left (= from the center outward). $\overset{D}{\leftarrow}\overset{G}{\leftarrow}\overset{G}{\leftarrow}\overset{D}{\leftarrow}$ precedes $\overset{\tilde{D}}{\leftarrow}\overset{G}{\leftarrow}\overset{G}{\leftarrow}\overset{D}{\leftarrow}$ precedes $\overset{D}{\leftarrow}\overset{\tilde{G}}{\leftarrow}\overset{G}{\leftarrow}\overset{D}{\leftarrow}$ precedes $\overset{\tilde{D}}{\leftarrow}\overset{\tilde{G}}{\leftarrow}\overset{G}{\leftarrow}\overset{D}{\leftarrow}$, etc. The last path is $\overset{\tilde{D}}{\leftarrow}\overset{\tilde{G}}{\leftarrow}\overset{\tilde{G}}{\leftarrow}\overset{\tilde{D}}{\leftarrow}$. Hence the question arises, whether the two orders match.

In the strictest sense the orders do not match: The path $\xleftarrow{\tilde{D}}\xleftarrow{\tilde{G}}\xleftarrow{G}\xleftarrow{D}$ precedes the path $\xleftarrow{D}\xleftarrow{G}\xleftarrow{\tilde{G}}\xleftarrow{D}$. Yet the morphism $DG\tilde{G}D$ with the node label $(babbaba, babba)$ precedes the morphism $\tilde{D}\tilde{G}GD$ with the node label $(abbabab, abbab)$ in the conjugation order: $conj_b \circ DG\tilde{G}D = \tilde{D}\tilde{G}GD$. There is a weaker sense, though, according to which the two orders match. In the concrete example, one may bring to bear that the morphisms G and $\tilde{G}$ commute. Hence with $\xleftarrow{D}\xleftarrow{\tilde{G}}\xleftarrow{G}\xleftarrow{D}$ there is an equivalent path to the node $(babbaba, babba)$ which is lexicographically smaller than $\xleftarrow{\tilde{D}}\xleftarrow{\tilde{G}}\xleftarrow{G}\xleftarrow{D}$, which is the smallest path to the node $(abbabab, abbab)$. Section 4 establishes a general result to that effect, proving a conjecture made in [3].

The following section is of a preparatory nature and—among other things—it inspects various commutative triangles and squares in Fig. 1, such as:

$$
\begin{aligned}
conj_a \circ G &= G \circ conj_a = \tilde{G} & conj_b \circ D &= D \circ conj_b = \tilde{D} \\
conj_a \circ \tilde{G} &= \tilde{G} \circ conj_a & conj_b \circ \tilde{G} &= G \circ conj_b \\
conj_a \circ \tilde{D} &= D \circ conj_a & conj_b \circ \tilde{D} &= \tilde{D} \circ conj_b.
\end{aligned}
$$

2   Special Sturmian Morphisms and Conjugation

Let F_2 denote the free group generated by the two letters a and b. Following [4] we consider the special Sturmian monoid St_0 as a submonoid of the automorphism group $Aut(F_2)$. It is generated by the four positive automorphisms $G, \tilde{G}, D, \tilde{D}$. On the letters a and b they are defined as follows:

$$
\begin{array}{c|c|c|c}
G(a) = a & \tilde{G}(a) = a & D(a) = ba & \tilde{D}(a) = ab \\
G(b) = ab & \tilde{G}(b) = ba & D(b) = b & \tilde{D}(b) = b.
\end{array}
$$

Lemma 1. *On the inverted letters a^{-1} and b^{-1} the morphisms $G, \tilde{G}, D, \tilde{G}$ have the following images:*

$$
\begin{array}{c|c|c|c}
G(a^{-1}) = a^{-1} & \tilde{G}(a^{-1}) = a^{-1} & D(a^{-1}) = a^{-1}b^{-1} & \tilde{D}(a^{-1}) = b^{-1}a^{-1} \\
G(b^{-1}) = b^{-1}a^{-1} & \tilde{G}(b^{-1}) = a^{-1}b^{-1} & D(b^{-1}) = b^{-1} & \tilde{D}(b^{-1}) = b^{-1}
\end{array}
$$

Proof. All four morphisms f have to satisfy $f(a^{-1}) = f(a)^{-1}$ as $\epsilon = f(aa^{-1}) = f(a)f(a^{-1})$, analogously for b. Thus, we obtain $G(a^{-1}) = G(a)^{-1} = a^{-1}$, $G(b^{-1}) = G(b)^{-1} = (ab)^{-1} = b^{-1}a^{-1}$ etc.

For some purposes it is useful to know the inverses of $G, \tilde{G}, D, \tilde{D}$ within the group $Aut(F_2)$:

Lemma 2. *The inverses of the special Sturmian morphisms $G, \tilde{G}, D, \tilde{D}$ within the automorphism group $Aut(F_2)$ are given as follows on the letters and the inverted letters:*

$$
\begin{array}{c|c|c|c}
G^{-1}(a) = a & \tilde{G}^{-1}(a) = a & D^{-1}(a) = b^{-1}a & \tilde{D}^{-1}(a) = ab^{-1} \\
G^{-1}(b) = a^{-1}b & \tilde{G}^{-1}(b) = ba^{-1} & D^{-1}(b) = b & \tilde{D}^{-1}(b) = b \\
G^{-1}(a^{-1}) = a^{-1} & \tilde{G}^{-1}(a^{-1}) = a^{-1} & D^{-1}(a^{-1}) = a^{-1}b & \tilde{D}^{-1}(a^{-1}) = ba^{-1} \\
G^{-1}(b^{-1}) = b^{-1}a & \tilde{G}^{-1}(b^{-1}) = ab^{-1} & D^{-1}(b^{-1}) = b^{-1} & \tilde{D}^{-1}(b^{-1}) = b^{-1}
\end{array}
$$

Proof. All four inverses have to satisfy $f^{-1}(f(a)) = f(f^{-1}(a)) = a$ and $f^{-1}(f(b)) = f(f^{-1}(b)) = b$. So we check $G^{-1}(G(a)) = G^{-1}(a) = a$, $G^{-1}(G(b)) = G^{-1}(ab) = aa^{-1}b = b$, etc.

As indicated in Sect. 1 we verify now the equations behind the commutative triangles and squares in Fig. 1, which we need later in Sect. 4.

Proposition 1. *Let* $conj_w : F_2 \to F_2$ *denote the conjugation automorphism with the element* w, *i.e.,* $conj_w(u) = w^{-1}uw$. *Then the following equalities hold:* $\tilde{G}G^{-1} = G^{-1}\tilde{G} = conj_a$ *and* $\tilde{D}D^{-1} = D^{-1}\tilde{D} = conj_b$.

Proof. It suffices to verify $\tilde{G}G^{-1} = conj_a$ and $\tilde{D}D^{-1} = conj_b$ on the letters a and b and to take into consideration that $G^{-1}\tilde{G} = \tilde{G}G^{-1}$ and $\tilde{D}D^{-1} = D^{-1}\tilde{D}$. Thus we verify $\tilde{G}G^{-1}(a) = \tilde{G}(a) = a = a^{-1}aa = conj_a(a)$, $\tilde{G}G^{-1}(b) = \tilde{G}(a^{-1}b) = a^{-1}ba = conj_a(b)$ and $\tilde{D}D^{-1}(a) = \tilde{D}(b^{-1}a) = b^{-1}ab = conj_b(a)$, $\tilde{D}D^{-1}(b) = \tilde{D}(b) = b = b^{-1}bb = conj_b(b)$.

Corollary 1. $conj_a \circ \tilde{G} = \tilde{G} \circ conj_a$ *and* $conj_b \circ \tilde{D} = \tilde{D} \circ conj_b$.

Proof. Substituting $conj_a = \tilde{G}G^{-1}$ and $conj_b = \tilde{D}D^{-1}$ we obtain:

$$conj_a \circ \tilde{G} = (\tilde{G}G^{-1}) \circ \tilde{G} = \tilde{G} \circ (G^{-1}\tilde{G}) = \tilde{G} \circ conj_a$$
$$conj_b \circ \tilde{D} = (\tilde{D}D^{-1}) \circ \tilde{D} = \tilde{D} \circ (D^{-1}\tilde{D}) = \tilde{D} \circ conj_b$$

Proposition 2. $conj_b \circ \tilde{G} = G \circ conj_b$ *and* $conj_a \circ \tilde{D} = D \circ conj_a$.

Proof. We apply the morphisms to the letters a and b and compare both sides:

$$
\begin{aligned}
conj_b(\tilde{G}(a)) &= b^{-1}\tilde{G}(a)b = b^{-1}ab &&= G(b^{-1}aaa^{-1}b) = G(b^{-1}ab) &&= G(conj_b(a)) \\
conj_b(\tilde{G}(b)) &= b^{-1}\tilde{G}(b)b = b^{-1}bab = G(b) &&&&= G(b^{-1}bb) &&= G(conj_b(b)) \\
conj_a(\tilde{D}(a)) &= a^{-1}\tilde{D}(a)a = a^{-1}aba = D(a) &&&&= D(a^{-1}aa) &&= D(conj_a(a)) \\
conj_a(\tilde{D}(b)) &= a^{-1}\tilde{D}(b)a = a^{-1}ba &&= D(a^{-1}bbb^{-1}a) = D(a^{-1}ba) &&= D(conj_a(b)).
\end{aligned}
$$

To define the linear *conjugation order* on the conjugation class of a special Sturmian morphism, we recall the following known facts (e.g., see [2,4]).

1. The outer group $Out(F_2) = Aut(F_2)/Inn(F_2)$ of automorphisms of the free group F_2 modulo conjugations is isomorphic to the automorphism group $Aut(\mathbb{Z}^2) = GL_2(\mathbb{Z})$ of the commutative image $\mathbb{Z}^2$ of F_2. This implies that all conjugates of a given special Sturmian morphism f are characterized by the fact that they share the same incidence matrix $M_f = \begin{pmatrix} |f(a)|_a & |f(b)|_a \\ |f(a)|_b & |f(b)|_b \end{pmatrix}$.

2. The incidence matrices of special Sturmian morphisms belong to the monoid $SL_2(\mathbb{N})$, freely generated by the matrices $M_G = M_{\tilde{G}} = R = \begin{pmatrix} 1 & 1 \\ 0 & 1 \end{pmatrix}$ and $M_D = M_{\tilde{D}} = L = \begin{pmatrix} 1 & 0 \\ 1 & 1 \end{pmatrix}$. This implies that the representations of conjugate morphisms share the same sequence of basic letters G and D and differ at most in the distribution of diacritic $\sim$ marks attached to these letters.

3. With every special Sturmian morphism $f \in St_0$ we may associate the word $w = f(ab) = w_1 \ldots w_n \in \{a, b\}^*$. By conjugating the word w with its first letter w_1 we obtain the word $w_1^{-1} w w_1 = w_2 \ldots w_n w_1$. By iterating these conjugations with the respective first letter we obtain a full cycle of n conjugated words, namely $w_1 \ldots w_n$, $w_2 \ldots w_n w_1$, $\ldots$, $w_n w_1, \ldots w_{n-1}$.

4. All but one of these conjugated words (the bad conjugate) are images of the type $g(ab)$ of a special Sturmian morphism from the same conjugation class as f, and these $n-1$ morphisms also exhaust the conjugation class. Removing the bad conjugate from the full cycle of single-letter conjugations thereby induces a linear order $f_1 < f_2 < \cdots < f_{n-1}$ on the conjugation class of f. Its initial element f_1 is a special standard morphism, i.e., $f_1 \in \langle G, D \rangle$ and its terminal element f_{n-1} is a special anti-standard morphism, i.e., $f_{n-1} \in \langle \tilde{G}, \tilde{D} \rangle$.

Definition 1. *Consider a special standard morphism $f_1 \in \langle G, D \rangle \subset St_0$ and let $n = |f_1(ab)|$ denote the length of the image of the word ab. The linear order $f_1 < f_2 < \cdots < f_{n-1}$ on the conjugation class of f_1 (as described above) is called* conjugation order.

3 The Path Monoid and the Abacus Relations

In addition to the special Sturmian monoid $St_0 = \langle G, \tilde{G}, D, \tilde{D} \rangle \subset Aut(F_2)$ we consider the *path monoid* $\Sigma^* = \{\mathcal{G}, \tilde{\mathcal{G}}, \mathcal{D}, \tilde{\mathcal{D}}\}^*$, freely generated over the set $\Sigma = \{\mathcal{G}, \tilde{\mathcal{G}}, \mathcal{D}, \tilde{\mathcal{D}}\}$ of four formal symbols, which we distinguish from the four generating Sturmian morphisms $G, \tilde{G}, D, \tilde{D}$ themselves. The projection $\mu : \Sigma^* \to St_0$ with

$$\mu(\mathcal{G}) = G, \mu(\tilde{\mathcal{G}}) = \tilde{G}, \mu(\mathcal{D}) = D, \mu(\tilde{\mathcal{D}}) = \tilde{D},$$

mediating between the path monoid and the special Sturmian monoid, is well-understood by virtue of the following result from [4] (proposition 2.1):

Proposition 3 *(Kassel and Reutenauer). The special Surmian monoid has a presentation of the form*

$$St_0 \cong \left\langle \mathcal{G}, \tilde{\mathcal{G}}, \mathcal{D}, \tilde{\mathcal{D}} \mid \mathcal{G}\mathcal{D}^k\tilde{\mathcal{G}} = \tilde{\mathcal{G}}\tilde{\mathcal{D}}^k\mathcal{G}, \mathcal{D}\mathcal{G}^k\tilde{\mathcal{D}} = \tilde{\mathcal{D}}\tilde{\mathcal{G}}^k\mathcal{D} \text{ for all } k \in \mathbb{N} \right\rangle.$$

We will refer to these relations on paths as the *abacus relations*.

Definition 2. *On the set $\Sigma = \{\mathcal{G}, \tilde{\mathcal{G}}, \mathcal{D}, \tilde{\mathcal{D}}\}$ we introduce the total order $\mathcal{G} < \tilde{\mathcal{G}} < \mathcal{D} < \tilde{\mathcal{D}}$.[1] This order induces a natural right-to-left lexicographic order on the free monoid Σ^*. For any two words $U, V \in \Sigma^*$ consider the longest common suffix Y, such that $U = X_1 L_1 Y$ and $V = X_2 L_2 Y$ for $X_1, X_2 \in \Sigma^*$ and $L_1, L_2 \in \Sigma$. The two symbols L_1 and L_2 necessarily differ from each other. So we say $U < V$ iff $L_1 < L_2$.*

---

[1] The setting $\tilde{\mathcal{G}} < \mathcal{D}$ is arbitrary here. It could be likewise $\mathcal{D} < \tilde{\mathcal{G}}$ without consequences for the content of the article.

Lemma 3. *The following permutations* $\tau, \tau_{\mathcal{G}}, \tau_{\mathcal{D}} : \Sigma \to \Sigma$ *generate monoid automorphisms on* Σ^*:

$$
\begin{aligned}
\tau(\mathcal{G}) &:= \tilde{\mathcal{G}}, \quad \tau(\tilde{\mathcal{G}}) := \mathcal{G}, & \tau(\mathcal{D}) &:= \tilde{\mathcal{D}}, \quad \tau(\tilde{\mathcal{D}}) := \mathcal{D}, \\
\tau_{\mathcal{G}}(\mathcal{G}) &:= \tilde{\mathcal{G}}, \quad \tau_{\mathcal{G}}(\tilde{\mathcal{G}}) := \mathcal{G}, & \tau_{\mathcal{G}}(\mathcal{D}) &:= \mathcal{D}, \quad \tau_{\mathcal{G}}(\tilde{\mathcal{D}}) := \tilde{\mathcal{D}}, \\
\tau_{\mathcal{D}}(\mathcal{G}) &:= \mathcal{G}, \quad \tau_{\mathcal{D}}(\tilde{\mathcal{G}}) := \tilde{\mathcal{G}}, & \tau_{\mathcal{D}}(\mathcal{D}) &:= \tilde{\mathcal{D}}, \quad \tau_{\mathcal{D}}(\tilde{\mathcal{D}}) := \mathcal{D}.
\end{aligned}
$$

Proof. This is true for any permutation of Σ.

Proposition 4. *Consider a special standard morphism* $f \in \langle G, D \rangle$ *and its conjugation class* $\mathfrak{F} \subset St_0$. *Let* $\mathfrak{W} = \mu^{-1}(\mathfrak{F}) \subset \Sigma^*$ *denote the set of all words representing these Sturmian morphisms. With respect to lexicographic order for any two words* $U, V \in \mathfrak{W}$ *the following holds:*

$$ U < V \quad iff \quad \tau(V) < \tau(U). $$

Proof. Consider the evaluation $ev : \Sigma \to \{0,1\}$ with $ev(\mathcal{G}) = ev(\mathcal{D}) := 0$ and $ev(\tilde{\mathcal{G}}) = ev(\tilde{\mathcal{D}}) := 1$. Let m denote the common length of all words $W \in \mathfrak{W}$. We define $ev^* : \mathfrak{W} \to \{0, 1, \dots 2^m - 1\}$ with $ev^*(W_1, ..., W_m) = \sum_{k=1}^m ev(W_k) 2^{k-1}$. The map ev^* is an order-preserving bijection, i.e., we have $U < V$ in lexicographic order if and only if $ev^*(U) < ev^*(V)$ in the order of natural numbers. Furthermore, we have $ev^*(\tau(W)) = 2^m - ev^*(W)$ for any word W. Hence, for any two words $U, V \in \mathfrak{W}$ we have $U < V$ iff $ev^*(U) < ev^*(V)$ iff $2^m - ev^*(V) < 2^m - ev^*(U)$ iff $ev^*(\tau(V)) < ev^*(\tau(U))$ iff $\tau(V) < \tau(U)$.

4   Matching Lexicographic and Conjugation Order

Proposition 5. *Consider a word* $W \in \{\tilde{\mathcal{G}}, \tilde{\mathcal{D}}\}^*$ *and the associated anti-standard morphism* $f = \mu(W)$. *The following equations hold:*

$$ conj_a \circ f \circ G = \mu(\tau_{\mathcal{D}}(W)) \circ \tilde{G}, \quad conj_b \circ f \circ D = \mu(\tau_{\mathcal{G}}(W)) \circ \tilde{D}. $$

Proof. Any anti-standard morphism f can be expressed in the form:

$$ f = \tilde{G}^{n_k} \tilde{D}^{m_k} \dots \tilde{G}^{n_1} \tilde{D}^{m_1} $$

where $k \geq 0$, $n_k, m_1 \geq 0$ and $n_1, \dots, n_{k-1}, m_2, \dots, m_k > 0$. Iteratively applying equations from Proposition 2 we obtain:

$$
\begin{aligned}
conj_a \circ f \circ G &= conj_a \circ \tilde{G}^{n_k} \tilde{D}^{m_k} \dots \tilde{G}^{n_1} \tilde{D}^{m_1} G \\
&= \tilde{G}^{n_k} \circ conj_a \circ \tilde{D}^{m_k} \dots \tilde{G}^{n_1} \tilde{D}^{m_1} G \\
&= \tilde{G}^{n_k} \tilde{D}^{m_k} \circ conj_a \dots \tilde{G}^{n_1} \tilde{D}^{m_1} G \\
&= \tilde{G}^{n_k} \tilde{D}^{m_k} \dots \tilde{G}^{n_1} \tilde{D}^{m_1} \circ conj_a \circ G \\
&= \tilde{G}^{n_k} \tilde{D}^{m_k} \dots \tilde{G}^{n_1} \tilde{D}^{m_1} \tilde{G} \\
&= \mu(\tau_{\mathcal{D}}(W)) \circ \tilde{G}
\end{aligned}
$$

The proof for the second equation is analogous.

Proposition 6. *Consider a word* $W \in \{\mathcal{G}, \mathcal{D}\}^*$ *and the associated standard morphism* $f = \mu(W)$. *The following equations hold:*

$$f \circ \tilde{G} = conj_a \circ \mu(\tau_{\mathcal{D}}(W)) \circ G, \quad f \circ \tilde{D} = conj_b \circ \mu(\tau_{\mathcal{G}}(W)) \circ D.$$

Proof. Any standard morphism f can be expressed in the form:

$$f = G^{n_k} D^{m_k} \dots G^{n_1} D^{m_1}$$

where $k \geq 0$, $n_k, m_1 \geq 0$ and $n_1, \dots, n_{k-1}, m_2, \dots, m_k > 0$. Then applying the lemma once, and iteratively applying the abacus relation and commutativity of G and $\tilde{G}$:

$$
\begin{aligned}
conj_a \circ \mu(\tau_{\mathcal{D}}(W)) \circ G &= conj_a \circ G^{n_k} \tilde{D}^{m_k} \dots G^{n_1} \tilde{D}^{m_1} \circ G \\
&= \tilde{G} G^{n_k-1} \tilde{D}^{m_k} \dots G^{n_1} \tilde{D}^{m_1} G \\
&= G^{n_k-1} (\tilde{G} \tilde{D}^{m_k} G) G^{n_k-1-1} \dots G^{n_1} \tilde{D}^{m_1} G \\
&= G^{n_k-1} (G D^{m_k} \tilde{G}) G^{n_k-1-1} \dots G^{n_1} \tilde{D}^{m_1} G \\
&= G^{n_k} D^{m_k} G^{n_k-1} D^{m_k-1} \dots \tilde{G} G^{n_1-1} \tilde{D}^{m_1} G \\
&= G^{n_k} D^{m_k} G^{n_k-1} D^{m_k-1} \dots G^{n_1-1} (\tilde{G} \tilde{D}^{m_1} G) \\
&= G^{n_k} D^{m_k} G^{n_k-1} D^{m_k-1} G^{n_k-2} \dots G^{n_1} D^{m_1} \tilde{G} \\
&= f \circ \tilde{G}
\end{aligned}
$$

The proof for the second equation is analogous.

Proposition 7. *Consider a non-anti-standard special Sturmian morphism* $f \in St_0$ *and let* $\mathfrak{F} \subset St_0$ *denote the conjugation class of* f. *Consider the smallest representative* $U \in \mu^{-1}(f)$ *of* f *in lexicographic order. Let* $W \in \{\tilde{\mathcal{G}}, \tilde{\mathcal{D}}\}^*$ *denote the maximal anti-standard prefix of* U *such that* $U = W\mathcal{L}X$ *with a letter* $\mathcal{L} \in \{\mathcal{G}, \mathcal{D}\}$ *and some suffix* $X \in \Sigma^*$. *Then the word* $U' = \tau_{\mathcal{L}}(W)\tilde{\mathcal{L}}X$ *is the smallest representative of the successor of* f' *of* f *in conjugation order.*

Proof. For a moment we consider the special case where X is empty. We then have to show that every word $V\tilde{\mathcal{L}} \in \{\mathcal{G}, \tilde{\mathcal{G}}, \mathcal{D}, \tilde{\mathcal{D}}\}^*$, which is lexicographically larger then $W\mathcal{L}$ and smaller than $\tau_{\mathcal{L}}(W)\tilde{\mathcal{L}}$ represents a Sturmian morphism, which—in conjugation order—either precedes or coincides with f.

We look at the case where $\mathcal{L} = \mathcal{G}$. The proof for $\mathcal{L} = \mathcal{D}$ is completely analogous. For $n_k, m_1 \geq 0$ and $n_1, \dots, n_{k-1}, m_2, \dots, m_k > 0$ we obtain the following general form for $W\mathcal{L}$ and $\tau_{\mathcal{G}}(W)\tilde{\mathcal{L}}$:

$$W\mathcal{L} = \tilde{\mathcal{G}}^{n_k} \tilde{\mathcal{D}}^{m_k} \dots \tilde{\mathcal{G}}^{n_1} \tilde{\mathcal{D}}^{m_1} \mathcal{G}, \quad \tau_{\mathcal{G}}(W)\tilde{\mathcal{L}} = \tilde{\mathcal{G}}^{n_k} \mathcal{D}^{m_k} \dots \tilde{\mathcal{G}}^{n_1} \mathcal{D}^{m_1} \tilde{\mathcal{G}}$$

Now we consider a word V satisfying $W\mathcal{L} < V\tilde{\mathcal{L}} < \tau_{\mathcal{L}}(W)\tilde{\mathcal{L}}$. It is specified by the exponents $l_j > 0$ and $l_{j+1}, \dots, l_k, h_{j+1}, \dots, h_k \geq 0$ as follows:

$$
\begin{aligned}
V = &(\tilde{\mathcal{G}}^{n_k-l_k} \mathcal{G}^{l_k})(\tilde{\mathcal{D}}^{m_k-h_k} \mathcal{D}^{h_k}) \dots \\
&\dots (\tilde{\mathcal{G}}^{n_{j+1}-l_{j+1}} \mathcal{G}^{l_{j+1}})(\tilde{\mathcal{D}}^{m_{j+1}-h_{j+1}} \mathcal{D}^{h_{j+1}})(\tilde{\mathcal{G}}^{n_j-l_j} \mathcal{G}^{l_j}) \mathcal{D}^{m_j} \tilde{\mathcal{G}}^{n_{j-1}} \mathcal{D}^{m_{j-1}} \dots \tilde{\mathcal{G}}^{n_1} \mathcal{D}^{m_1}.
\end{aligned}
$$

The index j marks the right-most factor in V, where there is a nontrivial power of $\mathcal{G}$. Powers of $\tilde{\mathcal{D}}$ can only occur on the left side of the factor with index j, where they have no influence on the following calculation. If they were to be to the right of that index, the resulting word would be lexicographically larger than $\tau_{\mathcal{L}}(W)\tilde{\mathcal{L}}$, contrary to the assumption.

We substitute $V\tilde{\mathcal{L}}$ with the help of suitable abacus relations until the final $\tilde{\mathcal{L}}$ is replaced by $\mathcal{L}$. This implies that the substitution is smaller or equal to $W\mathcal{L}$.

$$
\begin{aligned}
V\tilde{\mathcal{L}} &= \ldots (\tilde{\mathcal{G}}^{n_j-l}\mathcal{G}^l)\mathcal{D}^{m_j}\tilde{\mathcal{G}}^{n_j-1}\mathcal{D}^{m_j-1}\ldots \tilde{\mathcal{G}}^{n_1}\mathcal{D}^{m_1}\tilde{\mathcal{G}} \\
&= \ldots \tilde{\mathcal{G}}^{n_j-l}\mathcal{G}^{l-1}(\mathcal{G}\mathcal{D}^{m_j}\tilde{\mathcal{G}})\tilde{\mathcal{G}}^{n_j-1-1}\mathcal{D}^{m_j-1}\ldots \tilde{\mathcal{G}}^{n_1}\mathcal{D}^{m_1}\tilde{\mathcal{G}} \\
&\cong \ldots \tilde{\mathcal{G}}^{n_j-l}\mathcal{G}^{l-1}(\tilde{\mathcal{G}}\tilde{\mathcal{D}}^{m_j}\mathcal{G})\tilde{\mathcal{G}}^{n_j-1-1}\mathcal{D}^{m_j-1}\ldots \tilde{\mathcal{G}}^{n_1}\mathcal{D}^{m_1}\tilde{\mathcal{G}} \\
&\cong \ldots \tilde{\mathcal{G}}^{n_j-l}\mathcal{G}^{l-1}\tilde{\mathcal{G}}\tilde{\mathcal{D}}^{m_j}\tilde{\mathcal{G}}^{n_j-1-1}(\mathcal{G}\mathcal{D}^{m_j-1}\tilde{\mathcal{G}})\tilde{\mathcal{G}}^{n_j-2-1}\ldots \tilde{\mathcal{G}}^{n_1}\mathcal{D}^{m_1}\tilde{\mathcal{G}} \\
&\cong \ldots \tilde{\mathcal{G}}^{n_j-l}\mathcal{G}^{l-1}\tilde{\mathcal{G}}\tilde{\mathcal{D}}^{m_j}\tilde{\mathcal{G}}^{n_j-1-1}(\tilde{\mathcal{G}}\tilde{\mathcal{D}}^{m_j-1}\mathcal{G})\tilde{\mathcal{G}}^{n_j-2-1}\ldots \tilde{\mathcal{G}}^{n_1}\mathcal{D}^{m_1}\tilde{\mathcal{G}} \\
&\quad\vdots \\
&\cong \ldots \tilde{\mathcal{G}}^{n_j-l}\mathcal{G}^{l-1}\tilde{\mathcal{G}}\tilde{\mathcal{D}}^{m_j}\tilde{\mathcal{G}}^{n_j-1}\tilde{\mathcal{D}}^{m_j-1}\tilde{\mathcal{G}}^{n_j-2}\ldots \tilde{\mathcal{G}}^{n_1-1}(\mathcal{G}\mathcal{D}^{m_1}\tilde{\mathcal{G}}) \\
&\cong \ldots \tilde{\mathcal{G}}^{n_j-l}\mathcal{G}^{l-1}\tilde{\mathcal{G}}\tilde{\mathcal{D}}^{m_j}\tilde{\mathcal{G}}^{n_j-1}\tilde{\mathcal{D}}^{m_j-1}\tilde{\mathcal{G}}^{n_j-2}\ldots \tilde{\mathcal{G}}^{n_1-1}(\tilde{\mathcal{G}}\tilde{\mathcal{D}}^{m_1}\mathcal{G})
\end{aligned}
$$

The calculation remains valid if the exponent m_1 of the rightmost power of $\tilde{\mathcal{D}}$ vanishes. In this case the abacus relation reduces to $\tilde{\mathcal{G}}\mathcal{G} \cong \mathcal{G}\tilde{\mathcal{G}}$. If the suffix X is not empty, nothing in the above argument changes.

Corollary 2. *Consider a non-standard special Sturmian morphism $f \in St_0$ and let $\mathfrak{F} \subset St_0$ denote the conjugation class of f. Consider the largest representative $U \in \mu^{-1}(f)$ of f in lexicographic order. Let $W \in \{\mathcal{G}, \mathcal{D}\}^*$ denote the maximal standard prefix of U such that $U = W\tilde{\mathcal{L}}X$ with a letter $\tilde{\mathcal{L}} \in \{\tilde{\mathcal{G}}, \tilde{\mathcal{D}}\}$ and some suffix $X \in \{\mathcal{G}, \tilde{\mathcal{G}}, \mathcal{D}, \tilde{\mathcal{D}}\}^*$. Then the word $U' = \tau_{\mathcal{L}}(W)\mathcal{L}X$ is the largest representative of the predecessor of f' of f in conjugation order.*

Proof. This follows from the application of Proposition 4 to Proposition 7.

Theorem 1. *Consider a conjugation class $\mathfrak{F} = \{f_1 < \cdots < f_{n-1}\} \subset St_0$ of special Sturmian morphisms in conjugation order. Let $\mathfrak{W} = \mu^{-1}(\mathfrak{F}) \subset \Sigma^*$ denote the set of all their representing words. $\mathfrak{W} = \mathfrak{W}_1 \sqcup \cdots \sqcup \mathfrak{W}_{n-1}$, where $\mathfrak{W}_k = \mu^{-1}(f_k)$, for $k = 1, \ldots, n-1$. Let $U_k, V_k \in \mathfrak{W}_k$ denote the lexicographically smallest and largest elements of $\mathfrak{W}_k$, respectively. Then the following holds:*

1. *$\tau(\mathfrak{W}_k) = \mathfrak{W}_{n-k}$ and $\tau(U_k) = V_{n-k}$ for $k = 1, \ldots, n-1$.*
2. *$U_1 < U_2 < \cdots < U_{n-1}$ and $V_1 < V_2 < \cdots < V_{n-1}$ in lexicographic order.*

Figure 2 illustrates the theorem with an example.

$GDG̃DG$	$(abaababa, abaababaabaab)$
$G̃DG̃DG$	$(baababaa, baababaabaaba)$
$GDD̃GDG$	$(aababaab, aababaabaabab)$
$G̃DD̃GDG \cong GDG̃D̃DG$	$(ababaaba, ababaabaababa)$
$G̃DG̃D̃DG$	$(babaabaa, babaabaababaa)$
$GDD̃G̃DG \cong GDG̃D̃G̃DG$	$(abaabaab, abaabaababaab)$
$G̃DD̃G̃DG \cong G̃DG̃D̃G̃DG$	$(baabaaba, baabaabaababa)$
$GDD̃GD̃DG$	$(aabaabab, aabaababaabab)$
$G̃DD̃GD̃DG \cong GDG̃DD̃DG \cong GDGDG̃$	$(abaababa, abaababaababa)$
$G̃DG̃DD̃DG \cong G̃DGDG̃$	$(baababaa, baababaababaa)$
$GDD̃G̃DD̃G \cong GDGDG̃$	$(aababaab, aababaabababab)$
$G̃DD̃G̃DD̃G \cong G̃DGDD̃G \cong GDGDG̃$	$(ababaaba, ababaabaababa)$
$G̃DD̃GD̃DG̃$	$(babaabaa, babaababaabaa)$
$GDD̃G̃DD̃G̃ \cong GDGDD̃G̃$	$(abaabaab, abaababaabaab)$
$G̃DD̃G̃DD̃G̃ \cong G̃DGDD̃G̃$	$(baabaaba, baabaabaabaaba)$
$GDD̃GD̃DG̃$	$(aabaabab, aababaabaabab)$
$G̃DD̃GD̃DG̃ \cong GDGDD̃G̃$	$(abaababa, ababaabaababa)$
$G̃DD̃GD̃DG$	$(baababaa, babaabaababaa)$
$GDD̃G̃D̃DG̃$	$(aababaab, abaabaababaab)$
$G̃DD̃G̃D̃DG̃$	$(ababaaba, baabaababaaba)$

Fig. 2. The complete 20-element conjugation class of the special standard morphism $GDGDG$ is listed row by row in terms of all representing words (left side) and in terms of the pairs $(f(a), f(b))$ of images of the letters a and b (right side). The rows are ordered according to the conjugation order. Within each row the equivalent words are lexicographically ordered. Smallest representatives are placed to the left, largest words are placed to the right. The thick polygon traverses the $32 = 2^5$ words in lexicographic order. On this trajectory the smallest words are traversed in conjugation order. The same is true for the largest words.

5 Dualizing the Network

Berthé et al. [2] introduce Sturmian involution, an anti-automorphism of the monoid St_0 that sends f in St_0 to f^* by fixing G and $G̃$ while exchanging D and $D̃$. They relate conjugation order on the morphisms f_i to the lexicographic order on words $f_i^*(ab)$, where f_1 is a standard morphism and f_1^* is a Christoffel morphism. They show that with $a < b$, $f_1^*(ab) < f_2^*(ab) < \cdots < f_{n-1}^*(ab)$. Another perspective on the lexicographic ordering, via the Burrows-Wheeler Transform, is available in [5]. In this section we revisit and illustrate the finding of [2] by constructing an isography between the diagram in Fig. 1 and a dualized diagram to be fed from the former by applying Sturmian involution to all its components.

The Sturmian monoid St_0 generates the subgroup $ST_0 = M^{-1}(SL_2(\mathbb{Z}))$ of index 2 within $Aut(F_2)$, of all group automorphisms with incidence matrices of determinant 1. ST_0 acts on itself from the left via $\lambda : ST_0 \times ST_0 \to ST_0$ with $\lambda_g(f) = g \circ f$ and from the right via $\rho : ST_0 \times ST_0 \to ST_0$ where $\rho_g(f) = f \circ g$.

In order to manage the right action ρ in terms of conventional function application we consider the generating transformations and the conjugations separately:

$$\rho_G = \Gamma, \rho_{\tilde{G}} = \tilde{\Gamma}, \rho_D = \Delta, \rho_{\tilde{D}} = \tilde{\Delta}, \rho_{conj_a} = \chi_a, \rho_{conj_b} = \chi_b : ST_0 \to ST_0 \text{ with:}$$

$$
\begin{array}{c|c|c}
\Gamma(f) = f \circ G & \Delta(f) = f \circ D & \chi_a(f) = conj_{f(a)}, \\
\tilde{\Gamma}(f) = f \circ \tilde{G} & \tilde{\Delta}(f) = f \circ \tilde{D} & \chi_b(f) = conj_{f(b)}.
\end{array}
$$

Lemma 4. *The transformations* $\Gamma, \tilde{\Gamma}, \Delta, \tilde{\Delta}$ *satisfy the equations:*

$$\tilde{\Gamma} \Delta^k \Gamma = \Gamma \tilde{\Delta}^k \tilde{\Gamma} \quad and \quad \tilde{\Delta} \Gamma^k \Delta = \Delta \tilde{\Gamma}^k \tilde{\Delta}$$

Proof. $\tilde{\Gamma} \Delta^k \Gamma = \rho_{GD^k \tilde{G}} \cong \rho_{\tilde{G} \tilde{D}^k G} = \Gamma \tilde{\Delta}^k \tilde{\Gamma}$. Analogously for $\tilde{\Delta} \Gamma^k \Delta$.

Here we regard Sturmian Involution as an anti-automorphism $* : ST_0 \to ST_0$ generated by $G^* = G$, $\tilde{G}^* = \tilde{G}$, $D^* = \tilde{D}$, $\tilde{D}^* = D$. Applying the anti-automorphism $*$ to all components of the left action λ naturally yields a transformation into the right action ρ (see diagram below):

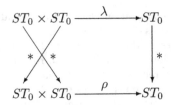

The diagram in Fig. 3 is the result of a thorough application of Sturmian involution to all components (nodes and arrows) of the diagram in Fig. 1. Thereby we may revisit the relations between the generators and the conjugations: from Sect. 2.

Proposition 8.

$$
\begin{array}{lll}
\chi_a \circ \Gamma = \Gamma \circ \chi_a = \tilde{\Gamma} & \chi_b \circ \Delta = \Delta \circ \chi_b = \tilde{\Delta} \Leftrightarrow \chi_{b^{-1}} \circ \tilde{\Delta} = \tilde{\Delta} \circ \chi_{b^{-1}} = \Delta \\
\chi_a \circ \tilde{\Gamma} = \tilde{\Gamma} \circ \chi_a & \chi_b \circ \Gamma = \tilde{\Gamma} \circ \chi_b & \Leftrightarrow \chi_{b^{-1}} \circ \tilde{\Gamma} = \Gamma \circ \chi_{b^{-1}} \\
\chi_a \circ \Delta = \tilde{\Delta} \circ \chi_a & \chi_b \circ \tilde{\Delta} = \Delta \circ \chi_b & \Leftrightarrow \chi_{b^{-1}} \circ \tilde{\Delta} = \tilde{\Delta} \circ \chi_{b^{-1}}
\end{array}
$$

Proof. These relations arise from translating the analogous relations (Proposition 1, Corollary 1, Proposition 2) from the left action λ to the right action ρ:

$$
\begin{array}{lll}
f \circ conj_a \circ G = [\Gamma \circ \chi_a](f) & \text{and } f \circ G \circ conj_a = [\chi_a \circ \Gamma](f) & \text{and } f \circ \tilde{G} = \tilde{\Gamma}(f) \\
f \circ conj_b \circ D = [\Delta \circ \chi_b](f) & \text{and } f \circ D \circ conj_b = [\chi_b \circ \Delta](f) & \text{and } f \circ \tilde{D} = \tilde{\Delta}(f) \\
f \circ conj_b \circ \tilde{G} = [\tilde{\Gamma} \circ \chi_b](f) & \text{and } f \circ G \circ conj_b = [\chi_b \circ \Gamma](f) & \\
f \circ conj_a \circ \tilde{G} = [\tilde{\Gamma} \circ \chi_a](f) & \text{and } f \circ \tilde{G} \circ conj_a = [\chi_a \circ \tilde{\Gamma}](f) & \\
f \circ conj_a \circ \tilde{D} = [\tilde{\Delta} \circ \chi_a](f) & \text{and } f \circ D \circ conj_a = [\chi_a \circ \Delta](f) & \\
f \circ conj_b \circ \tilde{D} = [\tilde{\Delta} \circ \chi_b](f) & \text{and } f \circ \tilde{D} \circ conj_b = [\chi_b \circ \tilde{\Delta}](f) &
\end{array}
$$

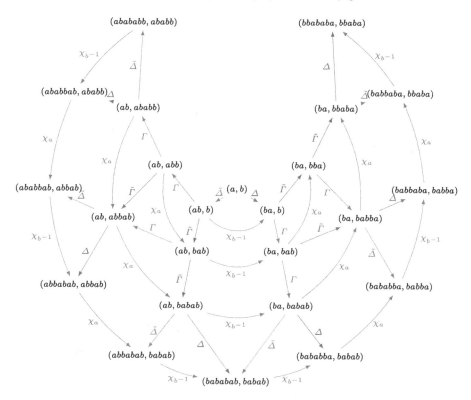

Fig. 3. We obtain the diagram in this Figure from the isographic diagram in Fig. 1 by replacing their node and arrow labels as follows: (1) Each node label $(f(a), f(b))$ is replaced by $(f^*(a), f^*(b))$; (2) each arrow label $conj_a$ or $conj_b$ is replaced by χ_a or $\chi_{b-1} = \chi_b^{-1}$, respectively; (3) each arrow label G or $\tilde{G}$ is replaced by Γ or $\tilde{\Gamma}$, respectively; (4) each arrow label D or $\tilde{D}$ is replaced by $\tilde{\Delta}$ or Δ, respectively.

Our insight about the incidence of the lexicographic order of the smallest (or largest) paths with the conjugation order for each conjugation class is faithfully transferred along the duality. The conjugation order has another meaning though, in so far as the conjugating elements are no longer the letters a and b. Also the lexicographic order of the paths has to be modified in view of the Sturmian involution. After involution it is induced by the order $\mathcal{G} < \tilde{\mathcal{G}} < \tilde{\mathcal{D}} < \mathcal{D}$ on Σ. In Fig. 3 we observe that the node labels of each conjugation class $(f^*(a), f^*(b))$ are lexicographically ordered. They are aligned along the corresponding arc in the left-to-right ordering of the images $f^*(ab)$ which is induced by the ordered alphabet $\{a < b\}$. With the final considerations we intend to relate this order to the right-to-left lexicographic order of the words encoding paths. To that end we need to define a suitable lexicographic order on a given conjugation class:

Definition 3. *Consider a conjugation class $\mathfrak{F} \subset St_0$ of special Sturmian morphisms. We say that its elements are in left-to-right lexicographic order: $\{f_1 \prec$*

$\cdots \prec f_{n-1}\}$ *iff their images of the word* $ab \in \{a < b\}^*$ *are in left-to-right lexicographic order:* $\{f_1(ab) < \cdots < f_{n-1}(ab)\}$.

Lemma 5. *For any special Sturmian morphism* $f \in St_0$ *one has* $f(ab) < f(ba)$ *with respect to the left-to-right lexicographic order in* $\{a < b\}^*$.

Proof. For special standard words one has $f(a)f(b) = cab$ and $f(b)f(a) = cba$, where c is the associated central word (see [1]). Hence $f(ab) < f(ba)$. Conjugation with the prefixes w of c preserves the order relation: $conj_w f(ab) < conj_w f(ba)$. And this exhausts the conjugation class of f.

Corollary 3. *For all* $f \in St_0$ *one has*

$$\Gamma(f(a), f(b)) \prec \tilde{\Gamma}(f(a), f(b)) \ and \ \tilde{\Delta}(f(a), f(b)) \prec \Delta(f(b), f(a))$$

Proof. For any $f \in St_0$ we have

$$\Gamma(f(a), f(b)) \prec \tilde{\Gamma}(f(a), f(b)) \text{ iff } (f(a), f(ab)) \prec (f(a), f(ba)) \text{ iff } f(ab) < f(ba)$$
$$\tilde{\Delta}(f(a), f(b)) \prec \Delta(f(a), f(b)) \text{ iff } (f(ab), f(b)) \prec (f(ba), f(b)) \text{ iff } f(ab) < f(ba)$$

The condition $f(ab) < f(ba)$ is always satisfied by virtue of Lemma 5.

From this result we may finally conclude, that the inherited conjugation order after the application of Sturmian involution coincides with the lexicographic order $\prec$ from Definition 3.

References

1. Berstel, J., Lauve, A., Reutenauer, C., Saliola, F.V.: Combinatorics on words: Christoffel Words and Repetitions in Words. CRM Monograph Series, vol. 27. American Mathematical Society, Providence, RI (2009)
2. Berthé, V., de Luca, A., Reutenauer, C.: On an involution of Christoffel words and Sturmian morphisms. European J. Combin. **29**(2), 535–553 (2008). http://dx.doi.org/10.1016/j.ejc.2007.03.001
3. Clampitt, D.: Lexicographic orderings of modes and morphisms. In: Pareyón, G., Pina-Romero, S., Augustín-Aquino, O.A., Emilio, L.P. (eds.) The Musical-Mathematical Mind: Patterns and Transformations. Computational Music Science, pp. 91–99. Springer, Berlin (2017). https://books.google.ca/books?id=9kAavgAACAAJ
4. Kassel, C., Reutenauer, C.: Sturmian morphisms, the braid group B_4, Christoffel words and bases of F_2. Ann. Mat. Pura Appl. **186**(2), 317–339 (2007). http://dx.doi.org/10.1007/s10231-006-0008-z
5. Mantaci, S., Restivo, A., Sciortino, M.: Burrows-Wheeler transform and Sturmian words. Inf. Process. Lett. **86**(5), 241–246 (2003). http://dx.doi.org/10.1016/S0020-0190(02)00512-4
6. Séébold, P.: On the conjugation of standard morphisms. Theoret. Comput. Sci. **195**(1), 91–109 (1998). http://dx.doi.org/10.1016/S0304-3975(97)00159-X. Mathematical foundations of computer science (Cracow, 1996)

More on the Dynamics of the Symbolic Square Root Map

(Extended Abstract)

Jarkko Peltomäki[1,2(✉)] and Markus Whiteland[2]

[1] Turku Centre for Computer Science TUCS, Turku, Finland
[2] Department of Mathematics and Statistics, University of Turku, Turku, Finland
{jspelt,mawhit}@utu.fi

Abstract. In our paper [A square root map on Sturmian words, Electron. J. Combin. 24.1 (2017)], we introduced a symbolic square root map. Every optimal squareful infinite word s contains exactly six minimal squares and can be written as a product of these squares: $s = X_1^2 X_2^2 \cdots$. The square root $\sqrt{s}$ of s is the infinite word $X_1 X_2 \cdots$ obtained by deleting half of each square. We proved that the square root map preserves the languages of Sturmian words (which are optimal squareful words). The dynamics of the square root map on a Sturmian subshift are well understood. In our earlier work, we introduced another type of subshift of optimal squareful words which together with the square root map form a dynamical system. In this paper, we study these dynamical systems in more detail and compare their properties to the Sturmian case. The main results are characterizations of periodic points and the limit set. The results show that while there is some similarity it is possible for the square root map to exhibit quite different behavior compared to the Sturmian case.

1 Introduction

Kalle Saari showed in [5,6] that every Sturmian word contains exactly six minimal squares (that is, squares having no proper square prefixes) and that each position of a Sturmian word begins with a minimal square. Thus a Sturmian word s can be expressed as a product of minimal squares: $s = X_1^2 X_2^2 X_3^2 \cdots$. In our earlier work [3], see also [2], we defined the square root $\sqrt{s}$ of the word s to be the infinite word $X_1 X_2 X_3 \cdots$ obtained by deleting half of each square X_i^2. We proved that the words s and $\sqrt{s}$ have the same language, that is, the square root map preserves the languages of Sturmian words. More precisely, we showed that if s has slope α and intercept ρ, then $\sqrt{s}$ has intercept $\psi(\rho)$, where $\psi(\rho) = \frac{1}{2}(\rho+1-\alpha)$. The simple form of the function ψ immediately describes the dynamics of the square root map in the subshift Ω_α of Sturmian words of slope α: all words in Ω_α are attracted to the set $\{01\mathbf{c}_\alpha, 10\mathbf{c}_\alpha\}$ of words of intercept $1 - \alpha$; here $\mathbf{c}_\alpha$ is the standard Sturmian word of slope α.

The square root map makes sense for any word expressible as a product of squares. Saari defines in [6] an intriguing class of such infinite words which he

© Springer International Publishing AG 2017
S. Brlek et al. (Eds.): WORDS 2017, LNCS 10432, pp. 97–108, 2017.
DOI: 10.1007/978-3-319-66396-8_10

calls optimal squareful words. Optimal squareful words are aperiodic infinite words containing the least number of minimal squares such that every position begins with a square. It turns out that such a word must be binary, and it must contain exactly six minimal squares; less than six minimal squares forces the word to be ultimately periodic. Moreover, the six minimal squares must be the minimal squares of some Sturmian language; the set of optimal squareful words is however larger than the set of Sturmian words. The six minimal squares of an optimal squareful word take the following form for some integers a and b such that $a \geq 1$ and $b \geq 0$:

$$0^2, \qquad (10^a)^2,$$
$$(010^{a-1})^2, \qquad (10^{a+1}(10^a)^b)^2,$$
$$(010^a)^2, \qquad (10^{a+1}(10^a)^{b+1})^2.$$

It is natural to ask if there are non-Sturmian optimal squareful words whose languages the square root map preserves. In [3], we proved by an explicit construction that such words indeed exist. The construction is as follows. The substitution

$$\tau: \begin{array}{c} S \mapsto LSS \\ L \mapsto SSS \end{array}$$

produces two infinite words $\Gamma_1^* = SSSLSSLSS\cdots$ and $\Gamma_2^* = LSSLSSLSS\cdots$ having the same language $\mathcal{L}$. Let $\tilde{s}$ be a (long enough) reversed standard word in some Sturmian language and $L(\tilde{s})$ be the word obtained from $\tilde{s}$ by exchanging its first two letters. By substituting the language $\mathcal{L}$ by the substitution σ mapping the letters S and L respectively to $\tilde{s}$ and $L(\tilde{s})$, we obtain a subshift Ω consisting of optimal squareful words. We proved that the words Γ_1 and Γ_2, the σ-images of Γ_1^* and Γ_2^*, are fixed by the square root map and, more generally, either $\sqrt{\mathbf{w}} \in \Omega$ or $\sqrt{\mathbf{w}}$ is periodic for all $\mathbf{w} \in \Omega$.

The aim of this paper is to study the dynamics of the square root map in the subshift Ω in the slightly generalized case where $\tau(S) = LS^{2c}$ and $\tau(L) = S^{2c+1}$ for some positive integer c and to see in which ways the dynamics differ from the Sturmian case. Our main results are the characterization of periodic and asymptotically periodic points and the limit set. We show that asymptotically periodic points must be ultimately periodic points and that periodic points must be fixed points; there are only two fixed points: Γ_1 and Γ_2. We prove that any word in Ω that is not an infinite product of the words $\sigma(S)$ and $\sigma(L)$ must eventually be mapped to a periodic word, thus having a finite orbit, while products of the words $\sigma(S)$ and $\sigma(L)$ are always mapped to aperiodic words. It follows from our results that the limit set of the square root map contains exactly the words that are products of $\sigma(S)$ and $\sigma(L)$. In addition, we study the injectivity of the square root map on Ω: only certain left extensions of the words Γ_1 and Γ_2 may have more than one preimage.

Let us make a brief comparison with the Sturmian case to see that the obtained results indicate that the square root map behaves somewhat differently on Ω.

The mapping ψ, defined above, is injective, so in the Sturmian case all words have at most one preimage. As ψ maps points strictly towards the point $1 - \alpha$ on the circle, all points are asymptotically periodic (see Definition 17) and all periodic points are fixed points. The fixed points are the two words $01\mathbf{c}_\alpha$ and $10\mathbf{c}_\alpha$ mentioned above, and the limit set consists only of these two fixed points.

The paper is organized as follows. The following section gives needed results on Sturmian words and standard words and it describes the construction of the subshift Ω in full detail. In Sect. 3, we proceed to characterize the limit set and to study injectivity. Section 4 contains results on periodic points.

2 Notation and Preliminary Results

Due to space constraints we refer the reader to [1] for basic notation, results on words, and for basic concepts such as *prefix*, *suffix*, *factor*, *language*, *primitive word*, *conjugate*, *ultimately periodic word*, *aperiodic word*, and *subshift*. We distinguish finite words from infinite words by writing the symbols referring to infinite words in boldface.

If w is a word such that $w = u^2$, then we call w a *square* with *square root* u. A square is *minimal* if it does not have a square as a proper prefix. If w is a word, then by $L(w)$ we denote the word obtained from w by exchanging its first two letters (we will not apply L to too short words). The language of a subshift Ω is denoted by $\mathcal{L}(\Omega)$, and the shift operator on infinite words is denoted by T. We index words from 0. We write $u \lhd v$ if the word u is lexicographically less than v. For binary words over $\{0, 1\}$, we set $0 \lhd 1$.

2.1 Sturmian Words and Standard Words

Several proofs in [3] regarding Sturmian words and the square root map require knowledge on continued fractions. In this paper, only some familiarity with continued fractions is required. We only recall that every irrational real number α has a unique infinite continued fraction expansion:

$$\alpha = [a_0; a_1, a_2, a_3, \ldots] = a_0 + \cfrac{1}{a_1 + \cfrac{1}{a_2 + \cfrac{1}{a_3 + \ldots}}} \tag{1}$$

with $a_0 \in \mathbb{Z}$ and $a_k \in \mathbb{Z}_+$ for $k \geq 1$. The numbers a_i are called the *partial quotients* of α. An introduction to continued fractions in relation to Sturmian words can be found in [2, Chap. 4].

We view here Sturmian words as the infinite words obtained as codings of orbits of points in an irrational circle rotation with two intervals. For alternative definitions and further details, see [1,4]. We identify the unit interval $[0, 1)$ with the unit circle $\mathbb{T}$. Let α in $(0, 1)$ be irrational. The map $R \colon \mathbb{T} \to \mathbb{T}$, $\rho \mapsto \{\rho + \alpha\}$, where $\{\rho\}$ stands for the fractional part of the number ρ, defines a rotation on $\mathbb{T}$. Divide the circle $\mathbb{T}$ into two intervals I_0 and I_1 defined by the points 0 and $1 - \alpha$.

Then define the coding function ν by setting $\nu(\rho) = 0$ if $\rho \in I_0$ and $\nu(\rho) = 1$ if $\rho \in I_1$. The coding of the orbit of a point ρ is the infinite word $\mathbf{s}_{\rho,\alpha}$ obtained by setting its $n^{\text{th}}, n \geq 0$, letter to equal $\nu(R^n(\rho))$. This word $\mathbf{s}_{\rho,\alpha}$ is defined to be the Sturmian word of slope α and intercept ρ. To make the definition proper, we need to define how ν behaves in the endpoints 0 and $1 - \alpha$. We have two options: either take $I_0 = [0, 1 - \alpha)$ and $I_1 = [1 - \alpha, 1)$ or $I_0 = (0, 1 - \alpha]$ and $I_1 = (1 - \alpha, 1]$. The difference is seen in the codings of the orbits of the points $\{-n\alpha\}$. This choice is largely irrelevant in this paper with the exception of the definition of the mapping ψ in the next subsection. The only difference between Sturmian words of slope $[0; 1, a_2, a_3, \dots]$ and Sturmian words of slope $[0; a_2 + 1, a_3, \dots]$ is that the roles of the letters 0 and 1 are reversed. We make the typical assumption that $a_1 \geq 2$ in (1). Since the sequence $(\{n\alpha\})_{n \geq 0}$ is dense in $[0, 1)$—as is well-known—Sturmian words of slope α have a common language (that is, the set of factors) denoted by $\mathcal{L}(\alpha)$. The Sturmian words of slope α form the Sturmian subshift Ω_α, which is minimal and aperiodic.

Let (d_k) be a sequence of positive integers. Corresponding to (d_k), we define a sequence (s_k) of *standard words* by the recurrence

$$s_k = s_{k-1}^{d_k} s_{k-2}$$

with initial values $s_{-1} = 1$, $s_0 = 0$. The sequence (s_k) converges to an infinite word $\mathbf{c}_\alpha$, which is a Sturmian word of intercept α and slope α, where α is an irrational with continued fraction expansion $[0; d_1 + 1, d_2, d_3, \dots]$. Thus standard words related to the sequence (d_k) are called standard words of slope α. The standard words are the basic building blocks of Sturmian words, and they have rich and surprising properties. For this paper, we only need to know that standard words are primitive and that the final two letters of a (long enough) standard word are different. Actually, in connection to the square root map, it is more natural to consider reversed standard words obtained by writing standard words from right to left. If s is a standard word in $\mathcal{L}(\alpha)$, then also the reversed standard word $\tilde{s}$ is in $\mathcal{L}(\alpha)$ because $\mathcal{L}(\alpha)$ is closed under reversal. For more on standard words, see [1, Chap. 2.2].

2.2   Optimal Squareful Words and the Square Root Map

An infinite word is *squareful* if its every position begins with a square. An infinite word is *optimal squareful* if it is aperiodic and squareful and it contains the least possible number of distinct minimal squares. In [6], Kalle Saari proves that optimal squareful words contain six distinct minimal squares; a squareful word containing at most five minimal squares is necessarily ultimately periodic. Moreover, Saari shows that optimal squareful words are binary and that the six minimal squares are of very restricted form. The square roots of the six minimal squares of an optimal squareful word are

$$
\begin{aligned}
S_1 &= 0, & S_4 &= 10^a, \\
S_2 &= 010^{a-1}, & S_5 &= 10^{a+1}(10^a)^b, \\
S_3 &= 010^a, & S_6 &= 10^{a+1}(10^a)^{b+1},
\end{aligned}
\tag{2}
$$

for some integers $\mathfrak{a}$ and $\mathfrak{b}$ such that $\mathfrak{a} \geq 1$ and $\mathfrak{b} \geq 0$. We call an optimal squareful word containing the minimal square roots of (2) an *optimal squareful word with parameters* $\mathfrak{a}$ *and* $\mathfrak{b}$. Throughout this paper, we reserve this meaning for the fraktur letters $\mathfrak{a}$ and $\mathfrak{b}$. Furthermore, we agree that the symbols S_i always refer to the minimal square roots of (2).

Let $\mathbf{s}$ be an optimal squareful word and write it as a product of minimal squares: $\mathbf{s} = X_1^2 X_2^2 \cdots$ (such a product is unique). The *square root* $\sqrt{\mathbf{s}}$ of $\mathbf{s}$ is the word $X_1 X_2 \cdots$ obtained by deleting half of each minimal square X_i^2. We reserve the notation $\sqrt[n]{\mathbf{s}}$ for the nth square root of $\mathbf{s}$. We chose this notation for its simplicity; the nth square root of a number x would typically be denoted by $\sqrt[2^n]{x}$. We often consider square roots of finite words. We let $\Pi(\mathfrak{a}, \mathfrak{b})$ to be the language of all nonempty words w such that w is a factor of some optimal squareful word with parameters $\mathfrak{a}$ and $\mathfrak{b}$ and w is factorizable as a product of minimal squares (2). Let $w \in \Pi(\mathfrak{a}, \mathfrak{b})$, that is, $w = X_1^2 \cdots X_n^2$ for minimal square roots X_i. Then we can define the square root $\sqrt{w}$ of w by setting $\sqrt{w} = X_1 \cdots X_n$. The square root map (on infinite words) is continuous with respect to the usual topology on infinite words (see [1, Sect. 1.2.2]). The following lemma, used later, sharpens this observation.

Lemma 1. *Let* $\mathbf{u}$ *and* $\mathbf{v}$ *be two optimal squareful words with the same parameters* $\mathfrak{a}$ *and* $\mathfrak{b}$. *If* $\mathbf{u}$ *and* $\mathbf{v}$ *have a common prefix of length* ℓ, *then* $\sqrt{\mathbf{u}}$ *and* $\sqrt{\mathbf{v}}$ *have a common prefix of length* $\lceil \ell/2 \rceil$.

Proof. Say $\mathbf{u}$ and $\mathbf{v}$ have a nonempty common prefix w. We may suppose that $w \notin \Pi(\mathfrak{a}, \mathfrak{b})$ as otherwise the claim is clear. Let z be the longest prefix of w that is in $\Pi(\mathfrak{a}, \mathfrak{b}) \cup \{\varepsilon\}$, and let X^2 and Y^2 respectively be the minimal square prefixes of the words $T^{|z|}(\mathbf{u})$ and $T^{|z|}(\mathbf{v})$. Hence $\sqrt{\mathbf{u}}$ begins with $\sqrt{z}X$ and $\sqrt{\mathbf{v}}$ begins with $\sqrt{z}Y$. Since X and Y begin with the same letter, it is easy to see that either X is a prefix of Y or Y is a prefix of X. By symmetry, we suppose that X is a prefix of Y. It follows that $\sqrt{\mathbf{u}}$ and $\sqrt{\mathbf{v}}$ have a common prefix of length $|zX^2|/2$. By the maximality of z, we have $|zX^2| > |w|$ proving that $\sqrt{\mathbf{u}}$ and $\sqrt{\mathbf{v}}$ have a common prefix of length $\lceil |w|/2 \rceil$. $\qquad \square$

Sturmian words are a proper subset of optimal squareful words. If $\mathbf{s}$ is a Sturmian word of slope α having continued fraction expansion as in (1), then it is an optimal squareful word with parameters $\mathfrak{a} = a_1 - 1$ and $\mathfrak{b} = a_2 - 1$. The square root map is especially interesting for Sturmian words because it preserves their languages. Define a function $\psi \colon \mathbb{T} \to \mathbb{T}$ as follows. For $\rho \in (0, 1)$, we set

$$\psi(\rho) = \frac{1}{2}(\rho + 1 - \alpha),$$

and we set

$$\psi(0) = \begin{cases} \frac{1}{2}(1 - \alpha), & \text{if } 0 \in I_0, \\ 1 - \frac{\alpha}{2}, & \text{if } 0 \notin I_0. \end{cases}$$

The mapping ψ moves a point ρ on $\mathbb{T}$ towards the point $1 - \alpha$ by halving the distance between the points ρ and $1 - \alpha$. The distance to $1 - \alpha$ is measured in the interval I_0 or I_1 depending on which of these intervals the point ρ belongs to. In [3], we proved the following result relating the intercepts of a Sturmian word and its square root.

Theorem 2. Let $\mathbf{s}_{\rho,\alpha}$ be a Sturmian word of slope α. Then $\sqrt{\mathbf{s}_{\rho,\alpha}} = \mathbf{s}_{\psi(\rho),\alpha}$.

Specific solutions to the word equation

$$X_1^2 X_2^2 \cdots X_n^2 = (X_1 X_2 \cdots X_n)^2 \qquad (3)$$

in the Sturmian language $\mathcal{L}(\alpha)$ play an important role. We are interested only in the solutions of (3) where all words X_i are *minimal square roots* (2), i.e., primitive roots of minimal squares. Thus we give the following definition.

Definition 3. A nonempty word w is a *solution to* (3) if w can be written as a product of minimal square roots $w = X_1 X_2 \cdots X_n$ which satisfy the word equation (3). The solution is *primitive* if w is primitive.

Consider for example the word $S_2 S_1 S_4$ for $\mathfrak{a} = 1$ and $\mathfrak{b} = 0$. We have

$$(S_2 S_1 S_4)^2 = (01 \cdot 0 \cdot 10)^2 = 01010 \cdot 01010 = (01)^2 \cdot 0^2 \cdot (10)^2 = S_2^2 S_1^2 S_4^2,$$

so the word $S_2 S_1 S_4$ is a solution to (3).

In [3, Theorem 5.2], the following result was proved.

Theorem 4. If $\widetilde{s}$ is a reversed standard word, then the words $\widetilde{s}$ and $L(\widetilde{s})$ are primitive solutions to (3).

Solutions to (3) are important as they can be used to build fixed points of the square root map. If (u_k) is a sequence of solutions to (3) with the property that u_k^2 is a proper prefix of u_{k+1} for $k \geq 1$, then the infinite word $\mathbf{w}$ obtained as the limit $\lim_{k \to \infty} u_k$ has arbitrarily long prefixes $X_1^2 \cdots X_n^2$ with the property that $X_1 \cdots X_n$ is a prefix of $\mathbf{w}$. In other words, the word $\mathbf{w}$ is a fixed point of the square root map. All known constructions of fixed points rely on this method. For example, the two Sturmian words $01\mathbf{c}_\alpha$ and $10\mathbf{c}_\alpha$ of slope α and intercept $1 - \alpha$ both have arbitrarily long squares u^2 as prefixes, where $u = L(\widetilde{s})$ for a reversed standard word $\widetilde{s}$ [3, Proposition 6.3]. In the next subsection, we see that the dynamical system studied in this paper is also fundamentally linked to fixed points obtained from solutions of (3).

The following lemma [3, Lemma 5.5] is of technical nature, but it conveys an important message: under the assumptions of the lemma, swapping two adjacent and distinct letters that do not occur as a prefix of a minimal square affects a product of minimal squares only locally and does not change its square root. This establishes the often-used fact that $\widetilde{s}\widetilde{s}$ and $\widetilde{s}L(\widetilde{s})$ are both in $\Pi(\mathfrak{a}, \mathfrak{b})$ and have the same square root for a reversed standard word $\widetilde{s}$. For example, if $\widetilde{s} = 1001001010010$, then

$$\widetilde{s}\widetilde{s} = 1001001010 \cdot 0101 \cdot 00 \cdot 1001010010 \quad \text{and}$$
$$\widetilde{s}L(\widetilde{s}) = 1001001010 \cdot 010010 \cdot 1001010010,$$

so the change is indeed local and does not affect the square root. Notice that every long enough standard word has S_6 as a proper suffix.

Lemma 5. *Let u and v be words such that*

- *u is a nonempty suffix of S_6,*
- *$|v| \geq |S_5 S_6|$,*
- *v begins with xy for distinct letters x and y,*
- *uv and $L(v)$ are factors of some optimal squareful words with the same parameters.*

Suppose there exists a minimal square X^2 such that $|X^2| > |u|$ and X^2 is a prefix of uv or $uL(v)$. Then there exist minimal squares $Y_1^2, \ldots, Y_n^2$ such that X^2 and $Y_1^2 \cdots Y_n^2$ are prefixes of uv and $uL(v)$ of the same length and $X = Y_1 \cdots Y_n$.

2.3 The Subshift Ω

In this subsection, we define the main object of study of this paper. The results presented were obtained in [3] in the case $\mathfrak{c} = 1$, the generalization being straightforward.

Let $\mathfrak{c}$ be a fixed positive integer. Repeated application of the substitution

$$\tau: \begin{array}{l} S \mapsto LS^{2\mathfrak{c}} \\ L \mapsto S^{2\mathfrak{c}+1} \end{array}$$

to the letter S produces two infinite words

$$\Gamma_1^* = SS^{2\mathfrak{c}}(LS^{2\mathfrak{c}})^{2\mathfrak{c}}(S^{2\mathfrak{c}+1}(LS^{2\mathfrak{c}})^{2\mathfrak{c}})^{2\mathfrak{c}} \cdots \text{ and}$$
$$\Gamma_2^* = LS^{2\mathfrak{c}}(LS^{2\mathfrak{c}})^{2\mathfrak{c}}(S^{2\mathfrak{c}+1}(LS^{2\mathfrak{c}})^{2\mathfrak{c}})^{2\mathfrak{c}} \cdots$$

with the same language $\mathcal{L}$. We set Ω^* to be the minimal and aperiodic subshift with language $\mathcal{L}$.

Fix integers $\mathfrak{a}$ and $\mathfrak{b}$ such that $\mathfrak{a} \geq 1$ and $\mathfrak{b} \geq 0$, and let α be an irrational with continued fraction expansion $[0; \mathfrak{a} + 1, \mathfrak{b} + 1, \ldots]$. Let w to be a word such that $w \in \{\tilde{s}_k, L(\tilde{s}_k)\}$ where $\tilde{s}_k$ is a reversed standard word of slope α such that $|\tilde{s}_k| > |S_6|$.[1] Let then σ be the substitution mapping S to w and L to $L(w)$. By substituting the letters S and L in words of Ω^*, we obtain a new minimal and aperiodic subshift $\sigma(\Omega^*)$, which we denote by Ω_A. We also set $\Gamma_1 = \sigma(\Gamma_1^*)$ and $\Gamma_2 = \sigma(\Gamma_2^*)$. The subshift Ω_A is generated by both of the words Γ_1 and Γ_2. The words Γ_1 and Γ_2 differ only by their first two letters. This difference is often irrelevant to us, so we let Γ to stand for either of these words. Further, we let the symbol γ_k to stand for the word $\sigma(\tau^k(S))$ and $\overline{\gamma}_k$ to stand for $\sigma(\tau^k(L))$.

It is easy to see that $\Gamma_1 = \lim_{k\to\infty} \gamma_{2k}$ and $\Gamma_2 = \lim_{k\to\infty} \overline{\gamma}_{2k}$. In what follows, we often consider infinite products of γ_k and $\overline{\gamma}_k$, and we wish to argue independently of the index k. Hence we make a convention that γ and $\overline{\gamma}$ respectively stand for γ_k and $\overline{\gamma}_k$ for some $k \geq 0$. The words γ and $\overline{\gamma}$ are primitive; see [3, Lemma 8.2]. For simplification, we abuse notation and write S for γ_0 and L

---

[1] Without this condition the subshift Ω, defined below, does not consist of optimal squareful words; see the remark after [3, Lemma 8.3].

for $\overline{\gamma}_0$. It will always be clear from context if letters S and L or words S and L are meant.

It can be shown that the words of Ω_A are optimal squareful words with parameters $\mathfrak{a}$ and $\mathfrak{b}$; see [3, Lemma 8.3]. Therefore the square root map is defined for words in Ω_A. The square root map on Ω_A has the following crucial properties.

Lemma 6. *The following properties hold:*

$$- \sqrt{\gamma\gamma} = \gamma,$$
$$- \sqrt{\gamma\overline{\gamma}} = \gamma,$$
$$- \sqrt{\overline{\gamma}\gamma} = \overline{\gamma}, \text{ and}$$
$$- \sqrt{\overline{\gamma}\overline{\gamma}} = \overline{\gamma}.$$

Proof. The proof of [3, Proposition 8.1] works essentially as it is. $\square$

Lemma 6 shows that the words Γ_1 and Γ_2 are fixed points of the square root map. Namely, the word γ_{k+2} has γ_k^2 as a prefix and $\overline{\gamma}_{k+2}$ has $\overline{\gamma}_k^2$ as a prefix. Thus by Lemma 6, we have, e.g.,

$$\sqrt{\Gamma_1} = \sqrt{\lim_{k\to\infty} \gamma_{2k}^2} = \lim_{k\to\infty} \gamma_{2k} = \Gamma_1.$$

The words in Ω_A can be (uniquely) written as a product of the words S and L up to a shift. Consider a word $\mathbf{w}$ in Ω_A and write $\mathbf{w} = T^\ell(\mathbf{w}')$ for some $\mathbf{w}' \in \Omega_A \cap \{S, L\}^\omega$ and ℓ such that $0 \leq \ell < |S|$. There are four distinct possibilities (types):

(A) $\ell = 0$,
(B) $\ell > 0$ and the prefix of $\mathbf{w}$ of length $|S| - \ell$ is in $\Pi(\mathfrak{a}, \mathfrak{b})$,
(C) $\ell > 0$ and the prefix of $\mathbf{w}$ of length $2|S| - \ell$ is in $\Pi(\mathfrak{a}, \mathfrak{b})$, or
(D) none of the above applies.

These possibilities are mutually exclusive: cases (B) and (C) cannot simultaneously apply because $S, L \notin \Pi(\mathfrak{a}, \mathfrak{b})$.[2] In our earlier paper, we proved the following theorem, see [3, Theorem 8.7].[3]

Theorem 7. *Let $\mathbf{w} \in \Omega_A$. If $\mathbf{w}$ is of type (A),(B), or (C), then $\sqrt{\mathbf{w}} \in \Omega_A$. If $\mathbf{w}$ is of type (D), then $\sqrt{\mathbf{w}}$ is periodic with minimal period conjugate to S.*

Thus to make Ω_A a proper dynamical system, we need to adjoin a periodic part to it. Let

$$\Omega_P = \{T^\ell(S^\omega) \colon 0 \leq \ell < |S|\},$$

and define $\Omega = \Omega_A \cup \Omega_P$. Clearly Ω is compact and $\sqrt{\Omega_A} \subseteq \Omega$ by Theorem 7. Further, as the proof of Theorem 7 in [3] applies to arbitrary products of S and L, it follows that $\sqrt{\mathbf{w}}$ is periodic with minimal period conjugate to S if $\mathbf{w} \in \Omega_P$.

---

[2] If S or L were in $\Pi(\mathfrak{a}, \mathfrak{b})$, then they would be nonprimitive as solutions to (3).
[3] In the proof of [3, Theorem 8.7] only the case $\mathfrak{c} = 1$ was considered. This is of no consequence as the proof given applies to arbitrary product ofs the words S and L.

Thus $\sqrt{\Omega_P} \subseteq \Omega_P$, and the pair $(\Omega, \sqrt{\cdot})$ is a valid dynamical system. Notice further that $L^\omega \in \Omega_P$; it is a special property of a reversed standard word $\tilde{s}$ that $\tilde{s}$ and $L(\tilde{s})$ are conjugates, see [3, Proposition 2.6].

Let us recall next what is known about the structure of the words in Ω. The word Γ is by definition an infinite product of the words γ_k and $\overline{\gamma}_k$ for all $k \geq 0$. Thus all words in Ω_A are (uniquely) factorizable as products of γ_k and $\overline{\gamma}_k$ up to a shift. Let us for convenience denote by Ω_γ the set $\Omega \cap \{\gamma, \overline{\gamma}\}^\omega$ consisting of words of Ω that are infinite products of γ and $\overline{\gamma}$. The following lemma describes two important properties of factorizations of words of Ω_A as products of γ and $\overline{\gamma}$. This result is an immediate property of the substitution τ that generates Ω^*.

Lemma 8. *Consider a factorization of a word in $\Omega_A \cap \Omega_\gamma$ as a product of γ and $\overline{\gamma}$. Such factorization has the following properties:*

- *Between two occurrences of $\overline{\gamma}$ there is always $\gamma^{2\mathfrak{c}}$ or $\gamma^{4\mathfrak{c}+1}$.*
- *Between two occurrences of $\overline{\gamma}\gamma^{4\mathfrak{c}+1}\overline{\gamma}$ there is always $\gamma^{2\mathfrak{c}}$ or $(\gamma^{2\mathfrak{c}}\overline{\gamma})^4 \cdot \overline{\gamma}^{-1}$.*

We also need to know how certain factors synchronize or align in a product of γ and $\overline{\gamma}$. The proof is a straightforward application of the elementary fact that a primitive word cannot occur nontrivially in its square.

Lemma 9 (Synchronizability Properties). *Let $\mathbf{w} \in \Omega_\gamma$. If z is a word in $\{\gamma\gamma, \gamma\overline{\gamma}, \overline{\gamma}\gamma\}$ occurring at position ℓ of $\mathbf{w}$, then the prefix of $\mathbf{w}$ of length ℓ is a product of γ and $\overline{\gamma}$.*[4]

The preceding lemma shows that if $\mathbf{w}$ is a word in Ω_A, then for each k there exists a unique ℓ such that $0 \leq \ell < |\gamma_k|$ and $T^\ell(\mathbf{w}) \in \Omega_{\gamma_k}$. We then say that the γ_k-factorization of $\mathbf{w}$ starts at the position ℓ of $\mathbf{w}$.

Let us conclude this subsection by making a remark regarding the subshift Ω^*. It is possible to define a counterpart for the square root map of Ω. Write a word $\mathbf{w}$ of Ω^* as a product of pairs of the letters S and L: $\mathbf{w} = X_1 X_1' \cdot X_2 X_2' \cdots$, where $X_i X_i' \in \{SS, SL, LS, LL\}$. We define the square root $\sqrt{\mathbf{w}}$ of $\mathbf{w}$ to be the word $X_1 X_2 \cdots$. Based on the above, it is not difficult to see that $\sigma(\sqrt{\mathbf{w}}) = \sqrt{\sigma(\mathbf{w})}$ for $\mathbf{w} \in \Omega^*$. In other words, the square root map for words in $\Omega_S \cap \Omega_A$ has the same dynamics as the square root map in Ω^*.

3 The Limit Set and Injectivity

In this section, we consider what happens for words of Ω when the square root map is iterated. We extend Theorem 7 and show that also the words of type (B) and type (C) are eventually mapped to a periodic word. In fact, we prove a stronger result: the number of steps required is bounded by a constant depending only on the word S. These results enable us to characterize the limit set as the set Ω_S. In other words, asymptotically the square root map on Ω has the same dynamics as the counterpart mapping on $\Omega^* \cup \{S^\omega, L^\omega\}$. We also show that the

---

[4] In general, e.g., the word γ^2 can be a factor of $\overline{\gamma}^3$.

square root map is mostly injective on Ω_A, only certain left extensions of Γ may have two preimages.

Let us first look at an example. Let $\mathfrak{a} = 1$, $\mathfrak{b} = 0$, and $S = 01010010$. Set $\mathbf{w} = T^4(S^2\mathbf{u})$ for some $S^2\mathbf{u} \in \Omega_S \cap \Omega_A$. The word $\mathbf{w}$ is of type (C) as the word $T^4(S^2)$, which equals $00 \cdot 1001010010$, is in $\Pi(\mathfrak{a}, \mathfrak{b})$. Now $\sqrt{\mathbf{w}} = 010010 \cdot \sqrt{\mathbf{u}}$ and $\sqrt{\mathbf{w}} \in \Omega_A$ by Theorem 7. So $\sqrt{\mathbf{w}}$ is of type (B), and $\sqrt[3]{\mathbf{w}} = 010 \cdot \sqrt[3]{\mathbf{u}}$. Still we have $\sqrt[3]{\mathbf{w}} \in \Omega_A$. It is clear now that $\sqrt[3]{\mathbf{w}}$ is not of type (A) or (B). The word $\sqrt[3]{\mathbf{u}}$ begins with S or L, and neither $010 \cdot S$ nor $010 \cdot L$ is in $\Pi(\mathfrak{a}, \mathfrak{b})$, so $\sqrt[3]{\mathbf{w}}$ is not of type (C) either. Thus it is of type (D), so $\sqrt[3]{\mathbf{w}}$ is periodic. The minimal period of $\sqrt[3]{\mathbf{w}}$ is readily checked to be 01010010, that is, $\sqrt[3]{\mathbf{w}} = S^\omega$. With some effort it can be verified that in this particular case $\sqrt[3]{\mathbf{v}}$ is periodic for all $\mathbf{v} \in \Omega \setminus \Omega_S$. Notice that the parameter $\mathfrak{c}$ is irrelevant to all of the preceding arguments.

Theorem 10. *There exists an integer n, depending only on the word S, such that $\sqrt[n]{\mathbf{w}} \in \{S^\omega, L^\omega\}$ for all $\mathbf{w} \in \Omega \setminus \Omega_S$.*

Theorem 10 can be proven using the following two lemmas, the first of which is the important Embedding Lemma.

Lemma 11 (Embedding Lemma). *Let $\mathbf{w} \in \Omega$ and u_1 and u_2 to respectively be the prefixes of $\mathbf{w}$ and $\sqrt{\mathbf{w}}$ of length $|S|$.*

(i) If $\mathbf{w}$ begins with 0 and $u_1 \neq u_2$, then $u_1 \lhd u_2$.
(ii) If $\mathbf{w}$ begins with 1 and $u_1 \neq u_2$, then $u_1 \rhd u_2$.

Lemma 12. *Let w be any of the words SS, SL, LS, or LL. If ℓ is an odd integer such that $0 < \ell < |S|$, then $T^\ell(w) \notin \Pi(\mathfrak{a}, \mathfrak{b})$.*

Proof. Let ℓ be an odd integer such that $0 < \ell < |S|$. Since $|T^\ell(w)| = |S^2| - \ell$, we see that $|T^\ell(w)|$ is odd. Thus it is impossible that $T^\ell(w) \in \Pi(\mathfrak{a}, \mathfrak{b})$. $\quad\square$

Next we turn our attention to injectivity. The results provided next give sufficient information to characterize the limit set. There is a slight imperfection in the following results. Namely, we are unable to characterize the preimage of the periodic part Ω_P, and we believe no nice characterization exists. First of all, the words S^ω and L^ω must have several preimages, periodic and aperiodic, by Theorem 10. Secondly, if $\mathbf{w}$ in Ω_A is of type (D), then not only is $\sqrt{\mathbf{w}}$ periodic with minimal period conjugate to S but the square root of any word in Ω_A that shares a prefix of length $3|S|$ with $\mathbf{w}$ is periodic with the same minimal period.[5] Therefore here we only focus on characterizing preimages of words in the aperiodic part Ω_A.

The next theorem says that the square root map is not injective on Ω_A but is almost injective: only words of restricted form may have more than one preimage and even then there is at most two preimages. In the Sturmian case, all words have at most one preimage.

---

[5] See the proof of [3, Theorem 8.7] for precise details.

Theorem 13. *If $\mathbf{w}$ is a word in Ω_A having two preimages $\mathbf{u}$ and $\mathbf{v}$ in Ω under the square root map, then $\mathbf{u} = zS\Gamma_1$ and $\mathbf{v} = zS\Gamma_2$ where zS is a suffix of some γ_k such that $z \in \Pi(\mathfrak{a}, \mathfrak{b})$.*

Theorem 13 can be proven using the following lemma.

Lemma 14. *Suppose that $\mathbf{u}$ and $\mathbf{v}$ are words in Ω_γ such that $\sqrt{\mathbf{u}} = \sqrt{\mathbf{v}}$. If $\mathbf{u} = \gamma\gamma\cdots$ and $\mathbf{v} = \gamma\overline{\gamma}\cdots$, then $\mathbf{u} = \gamma\gamma\gamma^{2c}\overline{\gamma}\cdots$ and $\mathbf{v} = \gamma\overline{\gamma}\gamma^{2c}\overline{\gamma}\cdots$ and both $\mathbf{u}$ and $\mathbf{v}$ must be preceded by $\overline{\gamma}\gamma^{2c-1}$ in Ω.*

The *limit set* Λ is the set of words that have arbitrarily long chains of preimages, that is,

$$\Lambda = \bigcap_{n=0}^{\infty} \sqrt[n]{\Omega}.$$

In the Sturmian case, the limit set contains only the two fixed points of the square root map. For the subshift Ω, the limit set is much larger. In fact, the limit set contains all words that are products of the words S and L.

Theorem 15. *We have $\Lambda = \Omega_S$.*

4   Periodic Points

In this section, we characterize the periodic points of the square root map in Ω. The result is that the only periodic points are fixed points. We further characterize asymptotically periodic points and show that all asymptotically periodic points are ultimately periodic points.

Recall that a word $\mathbf{w}$ is a *periodic point* of the square root map with period n if $\sqrt[n]{\mathbf{w}} = \mathbf{w}$.

Theorem 16. *If $\mathbf{w}$ is a periodic point in Ω, then $\mathbf{w} \in \{\Gamma_1, \Gamma_2, S^\omega, L^\omega\}$.*

The case with the Sturmian periodic points is similar: periodic points are fixed points and the fixed points are obtained as limits from solutions of (3).

Next we consider the dynamical notion of an asymptotically periodic point and characterize asymptotically periodic points in Ω.

Definition 17. *Let (X, f) be a dynamical system. A point x in X is asymptotically periodic if there exists a periodic point y in X such that*

$$\lim_{n\to\infty} d(f^n(x), f^n(y)) = 0.$$

If this is the case, then we say that the point x is *asymptotically periodic to y.*

The following proposition essentially says that if a word in Ω is asymptotically periodic, then it is an ultimately periodic point. The situation is opposite to the Sturmian case where all words are asymptotically periodic and only periodic points are ultimately periodic points.

Proposition 18. *If* $\mathbf{w} \in \Omega_S$, *then* $\mathbf{w}$ *is asymptotically periodic if and only if* $\mathbf{w} \in \{\Gamma_1, \Gamma_2, S^\omega, L^\omega\}$, *that is, if and only if* $\mathbf{w}$ *is a periodic point. If* $\mathbf{w} \in \Omega \setminus \Omega_S$, *then* $\mathbf{w}$ *is asymptotically periodic to* S^ω *or* L^ω.

Acknowledgments. The work of the first author was supported by the Finnish Cultural Foundation by a personal grant. He also thanks the Department of Computer Science at Åbo Akademi for its hospitality. The second author was partially supported by the Vilho, Yrjö and Kalle Väisälä Foundation. Jyrki Lahtonen deserves our thanks for fruitful discussions.

References

1. Lothaire, M.: Algebraic Combinatorics on Words. No. 90 in Encyclopedia of Mathematics and Its Applications. Cambridge University Press, Cambridge (2002)
2. Peltomäki, J.: Privileged Words and Sturmian Words. Ph.D. dissertation, Turku Centre for Computer Science, University of Turku (2016). http://www.doria.fi/handle/10024/124473
3. Peltomäki, J., Whiteland, M.: A square root map on Sturmian words. Electron. J. Comb. **24**(1) (2017). Article No. P1.54
4. Pytheas Fogg, N.: Sturmian Sequences. In: Berthé, V., Ferenczi, S., Mauduit, C., Siegel, A. (eds.) Substitutions in Dynamics, Arithmetics and Combinatorics. Lecture Notes in Mathematics, vol 1794. Springer, Berlin, Heidelberg (2002)
5. Saari, K.: On the frequency and periodicity of infinite words. Ph.D. dissertation, Turku Centre for Computer Science, University of Turku (2008). http://users.utu.fi/kasaar/pubs/phdth.pdf
6. Saari, K.: Everywhere α-repetitive sequences and Sturmian words. Eur. J. Comb. **31**, 177–192 (2010)

Study of Christoffel Classes: Normal Form and Periodicity

Mélodie Lapointe[(✉)]

Laboratoire de Combinatoire et d'Informatique Mathématique,
Université du Québec à Montréal, Montréal, Canada
lapointe.melodie@courrier.uqam.ca

Abstract. We characterize the left normal forms of conjugates of Christoffel words and compute their minimal period. This answers two open questions in Reutenauer (2015).

Keywords: Christoffel words · Sturmian factors · Normal form · Minimal period

1 Introduction

Christoffel words appeared at the end of the 19th century in the works of Markoff and Christoffel. The terminology "Christoffel word" was coined in 1990 by Berstel [1]. Some characterizations of Sturmian sequences have been extended to Christoffel words by inspecting their conjugates [5,6]. Variants of Christoffel words, such as central words and standard words [2,9], have been extensively studied [10–13]. They also have several applications in various field such as number theory, discrete geometry, symbolic dynamics and combinatorics on words.

The present article is motivated by the article [14] on periods of conjugates of Christoffel word. The (left) normal form of sturmian factors was introduced in [14]; this factorization is based on the fact that the minimal period of a sturmian factor is the length of some Christoffel factor, as shown by de Luca and De Luca [11]. The normal form of a sturmian factor w is $w = sc^n p$, such that c is a Christoffel word with palindromic factorization $v'u'$, s is a proper suffix of u' and p is a proper prefix of c; in this case, the minimal period of w is $|c|$. Some periodic properties of conjugates of Christoffel words can be deduced from this factorization. The left normal form exist and is unique for all sturmian factor which are not a power of a letter [14]. In the same article, some classes of Christoffel words are characterized by their left normal form e.g. left special sturmian factor, central word, right special Sturmain word, bispecial sturmian factor and power of Christoffel word.

In a related work by Currie and Saari [7], the set of minimal periods of the factors of any Sturmian sequence is determined. They show that this set is essentially the set of lengths of standard words (equivalently Christoffel words) which are factors of the sequence. For this, they establish a result [7, Lemma 6]

© Springer International Publishing AG 2017
S. Brlek et al. (Eds.): WORDS 2017, LNCS 10432, pp. 109–120, 2017.
DOI: 10.1007/978-3-319-66396-8_11

which gives a period for each conjugate of a Christoffel word; however they don't give the minimal period of each conjugate, as it is done here in Corollary 7. Also, Hegedus and Nagy [8] defines explicitly the set of all non-trivial weak period of circular Christoffel words $P(w)$ i.e. if $q \in P(w)$, then there exist at least one conjugate c of w such that q is a period of c.

The main result is a description of the normal forms of the conjugates of a Christoffel words (Theorem 6) and a characterization of these words by their normal forms (Theorem 10). This answers two problems in [14, p.55]; namely, in Proposition 11.2 of this article, several classes of sturmian factors (special, palindromes) are characterized through property of their normal form; however, the characterization of conjugate's of Christoffel words is missing.

In Sect. 2, we recall some basic definitions and results concerning combinatorics of words. In Sect. 3, we compute some periods that appears in the conjugates of a Christoffel words. In Sect. 4, we explicit the left normal form of conjugates of Christoffel words and give a formula for their minimal period. In Sect. 5, we characterize the left normal form of conjugates of Christoffel word.

2   Preliminaries

An *alphabet* is a nonempty finite set A whose elements are called *letters*. A finite sequence $a_1 a_2 \ldots a_n$ of elements of A is called a *word* over A. The set of all words over A is denoted by A^*. The *concatenation* is the binary operation on two words $u, v \in A^*$ such that $u \cdot v = uv$. The empty sequence, called the *empty word*, is denoted by ε. Let $w = a_1 a_2 \ldots a_n$ be a word in A^*. The length of w, denoted by $|w|$, is the number of letters in w, that is $|w| = n$. Recall that, the length of ε is 0. The number of occurrences of the letter x in w is denoted by $|w|_x$ ($x \in A$). The *reversal* of w, denoted by $R(w)$, is $R(w) = a_n a_{n-1} \ldots a_2 a_1$. A word such that $w = R(w)$ is called a *palindrome*. The set of all palindrome in A^* is denoted by PAL_A.

A word u is a *factor* of w if there exist $x, y \in A^*$ such that $w = xuy$. The set of all factors of w is denoted by $\mathrm{Fact}(w)$. If $x = \varepsilon$, then u is a *prefix* of w and if $y = \varepsilon$, then u is a *suffix* of w. A prefix (resp. suffix) is called *proper* if $y \neq \varepsilon$ (resp. $x \neq \varepsilon$). The prefix (resp. suffix) of w of length i is denoted by $\mathrm{Pref}_i(w)$ (resp. $\mathrm{Suf}_i(w)$). A word w is *balanced* if for all $x, y \in \mathrm{Fact}(w)$ and $|x| = |y|$, $||x|_a - |y|_a| \leq 1$. We call *factorization* of a word w a pair (u, v) such that $w = uv$; sometimes, we simply write a factorization as uv.

Two words w and v are *conjugates* if there exist two words $x, y \in A^*$ such that $w = xy$ and $v = yx$. The mapping $C : A^* \to A^*$ is defined by $C(a_1 \ldots a_n) = a_2 \ldots a_n a_1$. Observe that $C^i(w) = yx$ if $w = xy$ and $x \in \mathrm{Pref}_{i \bmod |w|}(w), \forall i \in \mathbb{Z}$.

We say that a word $w = a_1 a_2 \ldots a_n$ has a *period* p if $a_i = a_{i+p}$ for all $i \in \{1, \ldots, n - p\}$. A period is nontrivial if $0 < p < n$. A word w is periodic if it has a *nontrivial* period p. The next lemma is useful.

Lemma 1 *[10]. A palindrome w has a nontrivial period p if and only if w has a palindrome suffix or prefix of length $|w| - p$.*

Consider a *lattice path*, which is a sequence of elementary steps in the plane; each elementary step is $[(x, y), (x + 1, y)]$ (horizontal step) or $[(x, y), (x, y + 1)]$ (vertical step). Let p and q be relatively prime integers. Consider the segment from $(0, 0)$ to (p, q) and the lattice path l between them located below the segment such that the polygon delimited by the segment and the lattice path has no interior integer point.

Let $A = \{a, b\}$ be a binary alphabet. The *lower Christoffel word* of slope q/p is the word in A^* encoding l, where a (resp. b) codes a horizontal (resp. vertical) step. The *upper Christoffel word* of slope q/p is defined in the same way by taking the lattice path above the segment instead of the one below the segment. Figure 1, show the upper and the lower Christoffel word of slope $7/13$. We say that a word is a *Christoffel word* if it is a lower or an upper Christoffel word. A Christoffel word is *proper* if its length is at least 2. The only non proper Christoffel words are a and b. Let us denote by $\mathcal{C}$ the set of all Christoffel words.

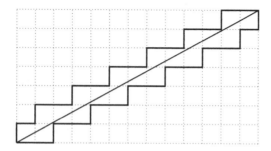

Fig. 1. The lower (resp. upper) Christoffel word of slope $7/13$ is *aabaabaabaabaabaabab* (resp. *babaabaabaabaabaabaa*)

Now, let us recall some useful properties of Christoffel words. Let w be a Christoffel word. The word $R(w)$ is also in $\mathcal{C}$. Moreover, w and $R(w)$ are conjugates. All proper Christoffel words can be written as xmy with m a palindrome, $x, y \in A$ and $x \neq y$. Each proper Christoffel words has a unique *standard factorization* (u, v) such that $u, v \in \mathcal{C}$ [4]. They have another factorization called palindromic factorization (v', u') such that $u', v' \in \mathrm{PAL}_A$ and $|u| = |u'|$ and $|v| = |v'|$. For example, the standard factorization of the lower Christoffel words of slopes $7/13$ (see Fig. 1) is $(aab, aabaabaabaabaabab)$ and its palindromic factorization is $(aabaabaabaabaabaa, bab)$.

The *Christoffel tree* [3] is the complete binary tree where the root is labelled by (a, ab, b) and each vertex (u, uv, v) has two children given by the following rules: the *left child* is (u, uuv, uv) and the *right child* is (uv, uvv, v). The middle term of each node is the Christoffel word given by this node and the two other words describe its standard factorization. Each proper lower Christoffel word appears exactly once as middle element in the Christoffel tree. Let w be a Christoffel word and (u, v) its standard factorization. The construction rules of the Christoffel tree implies that for every node in the tree except the root, u is a proper prefix of v or v is a proper suffix of u.

If w is a proper Christoffel word such that $|w| > 2$, one has either $|u| < |v|$ or $|u| > |v|$. In the first case, one has $w = u^t \alpha$ with t maximum (α is shorter than u and even a proper suffix of u) and in the second $w = \beta v^t$ with t maximum (β is shorter than v and even a proper prefix of u). We denote this by $t(w)$. One has $t = \lceil |v|/|u| \rceil$ if $|u| < |v|$ and $t = \lceil |u|/|v| \rceil$ if $|v| < |u|$. Moreover, α is the second element of the standard factorization of u and β is the first element in the standard factorization of v.

If $|w| > 2$, then $t(w) \geq 2$. Moreover, $\forall i \leq t$, the words u, v, $u^i \alpha$ and βv^i are Christoffel words. For example, if w is the lower Christoffel word in Fig. 1, then u is a prefix of v, $w = (aab)^6 ab$ and aab, ab, $(aab)^5 ab$, $(aab)^4 ab$, $(aab)^3 ab$, $(aab)^2 ab$ and $aabab$ are Christoffel words.

If $|w| = 2$, then $|u| = |v|$ and neither u nor v is a prefix or a suffix of the other one. Furthermore, w does not have conjugates which are not Christoffel words, since $w = ab$ or $w = ba$ in this case.

A word over $\{a, b\}$ is called a Sturmian factor if it is balanced. The following statements were proved in [14]. Let m be a sturmian factor which is not a power of a letter. The *left normal form* of m is (s, c^n, p), where $c \in C$ with palindromic factorization $c = (v', u')$, $n \geq 1$, s is a proper suffix of u' and p is a proper prefix of c. Equivalently, if $|c|$ is a period of m and s is shorter than u', then s is a proper suffix of u'. Also, if $|c|$ is a period of m and p is shorter than c, then p is a proper prefix of c. The left normal form of sturmian factor is unique. For example, if $m = ba^2 ba^2 a^2 ba^2 ba^2 baba^2 ba^2$, then the left normal form of w is $(b, (a^2 ba^2 ba^2 ba^2 bab)^1, a^2 ba^2)$. There is also the *right normal form* (s, c^n, p) of m where s is a proper prefix of c and p is a proper prefix of v'. In the left normal form, if p is a proper prefix of v', then the left and the right normal form coincide. Otherwise, the right normal form of m is $(s', R(c)^n, p')$ with $s' = sv'$ and $p = v'p'$. Moreover, the reversal of the left (resp. right) normal form (s, c^n, p) of m is $(R(p), R(c)^n, R(s))$: it is the right (resp. left) normal of $R(m)$, since $R(c)$ is a Christoffel word with palindromic factorization (u', v').

The normal form of m gives information about its minimal period: since c is an unbordered factor of m, the minimal period of m is exactly $|c|$ and c is a periodic pattern of m. Note that conjugates of Christoffel words are sturmian factors, so that they have a unique left normal form (See [14] Sect. 11 for more information.).

3   Period in Conjugates of Christoffel Words

The conjugacy relation is an equivalence relation. To represent each class in the relation, let us take the Christoffel word such that $|u| < |v|$ in its standard factorization (u, v); this is possible since each conjugacy class has a lower Christoffel word w and an upper Christoffel word $R(w)$ conjugate to it.

The function R is an involution, so that $R = R^{-1}$.

Lemma 2. $C = RC^{-1}R$

Proof. Let $w = a_1 \ldots a_n$ be any word.

$$RC^{-1}R(w) = RC^{-1}R(a_1 \ldots a_n) = RC^{-1}(a_n \ldots a_1) = R(a_1 a_n \ldots a_2) = C(w).$$

$\square$

Corollary 3. *If w is a proper Christoffel word and (u, v) is its standard factorization, then*

$$\forall i \in \mathbb{Z}, \; C^i(w) = RC^{|v|-i}(w).$$

Proof. Let $j = |v|$. We first prove that $w = RC^j(w)$. Suppose that u and v are both proper Christoffel words. Then $w = xw_1 yxw_2 y = xw_2 xyw_1 y$ with $w_1, w_2, w_1 xyw_2 \in \mathrm{PAL}_A$ and $v = xw_2 y$. Thus,

$$RC^j(w) = RC^j(xw_2 xyw_1 y) = R(yw_1 yxw_2 x) = xw_2 xyw_1 y = w$$

since $|xw_2 x| = |v| = j$. If $|u|$ or $|v|$ is not a proper Christoffel word, then clearly $RC^j(w) = w$. By Lemma 2, we have $C = RC^{-1}R = RC^{-1}R^{-1}$, thus

$$C^i(w) = RC^{-i}R(w) = RC^{-i}RRC^j(w) = RC^{j-i}(w)$$

since $w = RC^j(w)$. Hence, $C^i(w) = RC^{j-i}(w)$. $\square$

In particular, the conjugacy class of Christoffel words is closed under reversal.

Lemma 4. *Let w be a Christoffel word of length at least 3 and (u, v) its standard factorization with $|u| < |v|$. If $w = xy$ with $|x| = k|u| + 1$ and $k \in \{0, 1, \ldots, \lfloor |v|/|u| \rfloor\}$ and if p is the prefix of w of length $|v| - (k|u| + 1)$, then the word yp has a nontrivial period of length $|v| - k|u|$.*

Note that $k|u| + 1 < |v| + 1 \le |w|$, since $\gcd(|u|, |v|) = 1$. So, $|v| - (k|u| + 1) \ge 0$.

Proof. Suppose first that u is proper so that v is proper too. Note that the word w can be written as $w = a_1 \ldots a_{|v|} a_{|v|+1} \ldots a_{|w|}$. Recall that $w = xw_1 yxw_2 y = xw_2 xyw_1 y$ with $w_1, w_2, w_1 xyw_2 \in \mathrm{PAL}_A$ and $x \ne y$ such that $x, y \in A$.

Next, we prove that $yp = a_{k|u|+2} \ldots a_{|w|} a_1 \ldots a_{|v|-(k|u|+1)}$ is a palindrome for all $0 \le k \le \lfloor |v|/|u| \rfloor$. If $k = 0$, then $yp = a_2 \ldots a_{|w|} a_0 \ldots a_{|v|-1}$ and yp is equal to $w_2 xyw_1 yxw_2$. Thus, yp is a palindrome since w_1 and w_2 are palindromes. If $0 < k \le \lfloor |v|/|u| \rfloor$, the word $a_{k|u|+1} \ldots a_{|w|} a_0 \ldots a_{|v|-(k|u|+2)}$ is a factor of the palindrome yp where a prefix and a suffix of length $k|u|$ is removed, then it is a palindrome.

Finally, we check that the suffix $w' = a_{|v|+2} \ldots a_{|w|} a_1 \ldots a_{|v|-(k|u|+1)}$ of yp is also a palindrome. If $k = 0$, then $w' = a_{|v|+2} \ldots a_{|w|} a_1 \ldots a_{|v|-1}$ and we have $w' = w_1 xyw_2$ which is a palindrome since it is the central word of w. Also, we have $|u|$ and $|v|$ are periods of w', then by Lemma 1, the prefixes of length $|w'| - k|u|$ for all $0 < k \le \lfloor |v|/|u| \rfloor$ are palindromes.

We have yp a palindrome with a palindrome suffix w', hence, by Lemma 1

$$|yp| - |w'| = 2|v| - 2k|u| + |u| - 2 - (|v| - k|u| + |u| - 2) = |v| - k|u|$$

is a period of yp. Therefore, the period is nontrivial since $|yp| > |v| - k|u|$.

We leave the case when $|u| = 1$ to the reader. $\square$

Lemma 5. *Let w be a Christoffel word of length at least 3 and (u,v) its standard factorization with $|u| < |v|$. Let $k \in \{0,1,\ldots,\lfloor|v|/|u|\rfloor\}$. For all $i \in \{k|u| + 1,\ldots,|v| - k|u| - 1\}$, $C^i(w)$ has the nontrivial period $|v| - k|u|$.*

Proof. The words $C^i(w)$ are exactly the factors of length $|w|$ of the word yp defined in Lemma 4. Since the word yp has a period $|v| - k|u|$, then its factors of length $|w|$ also have the same period which is nontrivial since $|w| = |u| + |v| > |v| - k|u|$. □

The previous result is useful for identifying many periods of conjugates of Christoffel words, but these words may have other periods. For example, the word $abaaaba$ is conjugate to the Christoffel word $(baa, baaa)$, but $abaaaba$ has a period of length 6, which is not given by Lemma 5. In Sect. 4, we prove that the smallest period given for each word by Lemma 5 is its minimal period.

4   Left Normal Form and Minimal Period

Before stating the results, observe that if w is a Christoffel word, then its normal form is $(\varepsilon, w, \varepsilon)$, so that we disregard this case in the sequel. Every conjugates of Christoffel word of length 1 and 2 are also Christoffel word, so we ignore those words. Recall that a Christoffel word and it reversal have the same conjugates so we suppose that its standard factorization satisfied $|u| < |v|$: indeed, if this is not true for w, then it is true for the Christoffel word $R(w)$.

Theorem 6. *Let w be a Christoffel word of length at least 3 and (u,v) its standard factorization. We assume that $|u| < |v|$ and that $w = u^t\alpha$ where $t = t(w)$. The left normal form of the conjugates of w which are not Christoffel words is (s, c^n, p) with one of the following conditions:*

n	c	ps
1	$u^k\alpha$ or $R(u^k\alpha)$	u^j or $R(u^j)$
t	u or $R(u)$	α or $R(\alpha)$
$t-1$	u	$u\alpha$

with $0 \leq j \leq \lfloor\frac{t}{2}\rfloor$ and $k = t - j$.

Proof. Let $C^i(w)$ be a conjugate of w which is not a Christoffel word, then $i \neq 0$ and $i \neq |v|$ by Corollary 3. We consider 5 cases depending on the value of i.

1. If $1 \leq i \leq |v|/2$, then $C^i(w) = s(u^{t-j}\alpha)p$ with $j - 1 = \lceil i/|u|\rceil$ as in Fig. 2(a): $u = u_1u_2$, $p = u^{j-1}u_1$ and $s = u_2$. Observe that if $u_2 = \varepsilon$, then $u_1 = u$. Moreover, $u^{t-j}\alpha$ is a Christoffel word and its palindromic factorization is (v', u') such that $|u'| = |u|$ and $|v'| = |u^{t-j-1}\alpha|$. The word s is a proper factor of u, hence s is a proper suffix of u'.
   Also, $j \leq \lceil t/2\rceil$, since

$$j = \left\lceil\frac{i}{|u|}\right\rceil \leq \left\lceil\frac{|v|}{2|u|}\right\rceil \leq \left\lceil\frac{t}{2}\right\rceil.$$

Thus, p is a proper prefix of $u^{t-j}\alpha$, since $\gcd(|u|, |v|) = 1$ which means that $j-1 < t-j$ and $u_2 \neq \varepsilon$. Finally, we conclude that the left normal form is (s, c, p) with $c = u^{t-j}\alpha$ and $ps = u^j$.

2. If $|v|/2 < i \leq |v|-1$, we have that $C^i(w) = RC^{|v|-i}(w)$ by Corollary 3. We have $1 \leq |v| - i < |v|/2$, therefore the left normal form of $C^{|v|-i}(w)$ is $(s_1, u^{t-j}\alpha, p_1)$ as in the previous case and $j = \lceil (|v| - i)/|u| \rceil \leq \lceil t/2 \rceil$. Hence, the right normal form of $C^i(w)$ is $(R(p_1), R(u^{t-j}\alpha), R(s_1))$. If $R(p_1)$ is a proper suffix of v', then the left normal form of $C^j(w)$ is $(s, c, p) = (R(p_1), R(u^{t-j}\alpha), R(s_1))$ and $ps = R(s_1)R(p_1) = R(u^j)$. Otherwise, $R(p_1) = p_1'v'$ since $|R(u^{t-j}\alpha)|$ is a period of $C^j(w)$ by Lemma 5. Hence, the left normal form of $C^j(w)$ is $(s, c, p) = (p_1', v'R(u^{t-j-1}\alpha)u', v'R(s_1)) = (p_1', u^{t-j}\alpha, v'R(s_1))$ with $p_1' = R(u_1)R(u^{j-2})u'$ and

$$ps = v'R(s_1)p_1' = v'R(s_1)R(u_1)R(u)^{j-2}u' = u^j.$$

From now on, let (u_2', u_1') be the palindromic factorization of u and (u_1, u_2) be the standard factorization of u; $u_2 = \alpha$ (see Sect. 2). Also, recall that u is a Christoffel word and $|u|$ is a period of $C^i(w)$ by Lemma 5.

3. If $|w| - |\alpha| \leq i \leq |w| - 1$, then $\alpha = p_1 s_1$ (p_1 may be empty, but not s_1) as in Fig. 2(b). Observe that p_1 is a proper prefix of u, since $|p_1| < |\alpha| < |u|$. There is two cases to consider:

- If $|s_1| < |u_1'|$, then s_1 is a proper suffix of u_1', hence the left normal form of $C^i(w)$ is $(s, c^n, p) = (s_1, u^t, p_1)$ and $n = t$, $c = u$, $ps = \alpha$.
- Otherwise, $|s_1| \geq |u_1'|$, s_1 is a proper suffix of u, since $|s_1| \leq |\alpha| < |u|$. Also, $|p_1| < |u_2'|$, since $|s_1| + |p_1| < |u| = |u_1'| + |u_2'|$. Thus, p_1 is a proper prefix of u_2' and (s_1, u^t, p_1) is the right normal form of $C^i(w)$. Now, we have that $s_1 = s_1'u_1'$ and the left normal form of $C^i(w)$ is $(s, c^t, p) = (s_1', R(u)^t, u_1'p_1)$.

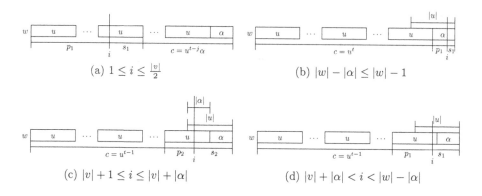

(a) $1 \leq i \leq \frac{|v|}{2}$

(b) $|w| - |\alpha| \leq |w| - 1$

(c) $|v| + 1 \leq i \leq |v| + |\alpha|$

(d) $|v| + |\alpha| < i < |w| - |\alpha|$

Fig. 2. Represent a conjugates $C^i(w)$ of w in different cases of Theorem 6

Recall that $\alpha = u_2$, thus $|\alpha| \geq |s_1| \geq |u_1'| = |u_1|$. Suppose that $|u_1| < |\alpha|$, the standard factorization of α is (u_1, δ) by Sect. 2 and is palindromic factorization is (δ', u_1') with $|\delta'| = |\delta|$ and $|u_1'| = |u_1|$. Hence, $\alpha = \delta' u_1' = p_1 s_1$. Now, $|s_1| \geq |u_1'|$, then p_1 is a prefix of δ' and $\delta' = p_1 s_1'$. Thus,

$$ps = u_1' p_1 s_1' = u_1' \delta' = R(\alpha).$$

Suppose now, that $|u_1| = |\alpha|$. Then, $u_1 = x$, $\alpha = y$ and $\{x, y\} = \{a, b\}$. Hence, the right normal form of $C^i(w)$ is $(y, (xy)^t, \varepsilon)$ and the left normal form of $C^i(w)$ is $(\varepsilon, (yx)^t, y)$ and $ps = y = R(\alpha)$.

4. If $|v| + 1 \leq i \leq |v| + |\alpha|$, then $|u| - |\alpha| \leq |s_2| \leq |u| - 1$, since $|s_2| = |w| - i$ ($|s_2| \geq |w| - (|v| + |\alpha|) = |u| - |\alpha|$ and $|s_2| \leq |w| - (|v| + 1) = |u| - 1$) as in Fig. 2(c). Moreover, $|s_2| \geq |u_1'|$, since $|u| = |u_1'| + |\alpha|$. The reversal of $C^i(w)$ is $C^{|v|-i}(w) = C^{|w|+|v|-i}(w)$ by Corollary 3. Hence, $|w| - |\alpha| \leq |w| + |v| - i \leq |w| - 1$. Thus, the left normal form of $C^{|v|-i}(w)$ is given by the case 3. Let (s_1, u^t, p_1) be the left or right normal form of $C^{|v|-i}(w)$. If $|u_1'| > |\alpha|$, then s_1 is always a proper suffix of u_1', since $|s_1| \leq |\alpha|$. Otherwise, $|u_1'| \leq |\alpha|$ and $|s_1|$ can be greater than $|u_1'|$. Suppose that $|s_1| \geq |u_1'|$, then

$$|w| + |v| - i \leq |w| - |u_1'| \Rightarrow i \leq |u_1'| + |v| = |w| - |\alpha|,$$

since $|w| = |u| + |v| = |u_1'| + |\alpha| + |v| \Rightarrow |u_1'| + |v| = |w| - |\alpha|$. Thus, $C^i(w)$ is already given by case 3. Therefore, the only unsolve case is when s_1 is a proper prefix of u_1'. Hence, the left normal form of $C^{|v|-i}(w)$ is (s_1, u^t, p_1) with $p_1 s_1 = \alpha$. Therefore, the right normal form of $C^i(w)$ is $(R(p_1), R(u)^t, R(s_1)))$. Observe that $|p_1| + |s_1| = |\alpha| = |u_2| = |u_2'|$ and $s_1 \neq \varepsilon$, thus p_1 is a proper prefix of u_2' and $R(p_1)$ is a proper suffix of $R(u_2')$. Thus, the left and the right normal form coincide. The left normal form of $C^i(w)$ is $(s, c^n, p) = (R(p_1), R(u)^t, R(s_1))$ and $ps = R(s_1)R(p_1) = R(p_1 s_1) = R(\alpha)$.

5. If $|v| + |\alpha| < i < |w| - |\alpha|$, we have $u = \gamma_1 \gamma_2$, $p_1 = \gamma_1$ and $s_1 = \gamma_2 \alpha$ as in Fig. 2(d) (γ_1 and γ_2 are not empty). Observe that $|\alpha| < |s_1| < |u| - |\alpha|$, since $|s_1| = |w| - i < |w| - |v| - |\alpha| = |u| - |\alpha|$. Moreover, $|u_1| = |u| - |\alpha|$, thus $|s_1| < |u_1| = |u_1'|$. Also, p_1 is a proper prefix of u, since γ_2 is not empty. Thus, the left normal form of $C^i(w)$ is $(s, c^n, p) = (s_1, u^{t-1}, p_1)$ and $n = t - 1$, $c = u$ and $ps = u\alpha$. $\qquad\square$

Knowing the left normal form, we also know the minimal period (see Sect. 2). Moreover, the right normal form can be deduced from Theorem 6.

Corollary 7. *Let w be a Christoffel word of length at least 3 and (u, v) its standard factorization. If $|u| < |v|$, then the minimal period of $C^i(w)$ is*

$$p_{min}(C^i(w)) = \begin{cases} |w| & \text{if } i \in \{0, |v|\} \\ |v| + \left(1 - \left\lceil \frac{i}{|u|} \right\rceil\right) |u| & \text{if } 1 \leq i \leq |v|/2 \\ |v| + \left(1 - \left\lceil \frac{|v|-i}{|u|} \right\rceil\right) |u| & \text{if } |v|/2 \leq i \leq |v| - 1 \\ |u| & \text{if } |v| + 1 \leq i \leq |w|. \end{cases}$$

Proof. This follows from the proof of Theorem 6. Recall that $v = u^{t-1}\alpha$ with $t = t(w)$. Hence, $|v| = |(t-1)|u| + |\alpha|$ and $|\alpha| = |v| - (t-1)|u|$. For $1 \le i \le |v|/2$, the minimal period is $|c|$ where $c = u^{t-j}\alpha$ and $j = \lceil i/|u| \rceil$. Therefore,

$$|c| = (t-j)|u| + |\alpha|$$
$$= (t-j)|u| + |v| - (t-1)|u|$$
$$= |v| + (1-j)|u|.$$

For $|v|/2 < i \le |v| - 1$, the minimal period is $|c|$, where $c = u^{t-j}\alpha$ or $c = R(u^{t-j}\alpha)$ and $j = \lceil (|v|-i)/|u| \rceil$. The length of $u^{t-j}\alpha$ or $R(u^{t-j}\alpha)$ is equal, thus as in the previous case $c = |v| + (1-j)|u|$. For $|v| < i < |w|$, the minimal period is $|c| = |u|$. $\square$

The minimal period in the interval $[0, |v|]$ and $[|v|+1, |\overset{\circ}{w}|]$ are symmetric (see Fig. 3). We deduce this symmetry from Corollary 3, since the minimal period of w and $R(w)$ is equal. Also, the conjugates $C^i(w)$ for $|v| < i < |w|$ have always the same minimal period.

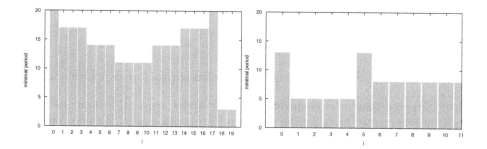

Fig. 3. Minimal period of conjugates of *aabaabaabaabaabaabab* and *aababaabababab*

Corollary 8. *Let w be a Christoffel word of length at least 3 and (u, v) its standard factorization. If u and v are proper Christoffel words, then the number of nontrivial minimal period of the conjugacy class of w is $\lceil t(w)/2 \rceil$. Otherwise, it is $\lceil t(w)/2 \rceil - 1$.*

Proof. Suppose that $|u| < |v|$. If $|u| = 1$, then there is no conjugate such that the left normal form is (s, u^n, p) or $(s, R(u)^n, p)$ which means that no conjugates has $|u|$ as period. Otherwise, there is $|u| - 1$ conjugates that have a minimal period of length $|u|$. Corollary 7 give us $\lceil t(w)/2 \rceil$ distinct minimal periods for the $|v| - 1$ other conjugates. Thus, if u and v are proper Christoffel word, there is $\lceil t(w)/2 \rceil$ distinct minimal period and $\lceil t(w)/2 \rceil + 1$ otherwise. $\square$

Standard words are particular conjugates of Christoffel words. A word s is *standard* if $s = gyx$ for some Christoffel word xgy with $x, y \in \{a, b\}$ and $x \ne y$. Hence, the standard word associate to w is $C(w)$.

Corollary 9. *Let s be a standard word conjugate to the Christoffel word w and* (u, v) *its standard factorization. If* $|u| < |v|$, *then the minimal period of s is* $|v|$. *Otherwise, the minimal period of s is* $|u|$.

Proof. This is straightforward by Corollary 7, since $s = C(w)$. $\square$

Corollary 9 is a generalization of a result in [10], where de Luca gives the minimal period of Standard words, when the standard factorization of $w = C^{-1}(s)$ satisfy $|v| > |u|$.

5   Normal Form of Conjugates of Christoffel Words

Now we want to characterize, among all left normal forms, those which represent conjugates of Christoffel words.

Theorem 10. *Let* (s, c^n, p) *be the left normal form of a sturmian factor* m, *which is not the power of a letter. Assume that the standard factorization of* c *is* (u, v). *Then* m *is conjugate to a Christoffel word, without being a Christoffel word, if and only if one of the following is satisfied:*

a. $n = 1$, $ps = u^j$ *or* v^j *for some* $j > 1$;
b. $n \geq 1$, $ps \in \{u, v, uv^2\}$.

The cases $ps = u^2v$ don't happens, since the left normal form is not symmetric.

Proof. Suppose that m is conjugate to a Christoffel word w, without being a Christoffel word. Let (x, y) be the standard factorization of w. We may assume that $|x| < |y|$, replacing w by $R(w)$ if necessary. Write $w = x^t\alpha$, $t = t(w)$. Recall that the word α is a Christoffel word and it is a proper prefix of the word x. Moreover, the standard factorization of x is (u_1, α). Then by Theorem 6, the left normal form of m is one of the following form:

i. $n = 1$ and either $c = x^i\alpha$, $ps = x^j$ or $c = R(x^i\alpha)$, $ps = R(x^j)$. In the first case, the standard factorization of c is $(u, v) = (x, x^{i-1}\alpha)$, then $ps = x^j = u^j$. In the second case, the standard factorization of c is $(u, v) = (R(x^{i-1}\alpha), R(x))$ and $ps = R(x^j) = v^j$. Also, $j > 0$ otherwise $ps = \varepsilon$ and m is a Christoffel word. We may assume that $j > 1$ since the case $j = 1$ is cover by the case (b).

ii. $n = t$ and either $c = x$, $ps = \alpha$ or $c = R(x)$, $ps = R(\alpha)$. In the first case, the standard factorization of $x = c$ is (u_1, α), thus $\alpha = v$ since the standard factorization is unique. Hence, $ps = v$. In the second case, the standard factorization of $c = R(x)$ is $(R(\alpha), R(u_1))$, thus $R(\alpha) = u$, since the standard factorization is unique. Hence, $ps = u$.

iii. $n = t - 1$ and $c = x$ and $ps = x\alpha$. The standard factorization of $c = x$ is $(u_1, \alpha) = (u, v)$, since the standard factorization is unique. Hence, $ps = x\alpha = uvv$.

Conversely, suppose that (s, c^n, p) is the left normal form of m with $c = (u, v)$ its standard factorization. The words u, v and c are Christoffel words and they form the Christoffel triplet (u, c, v). We want to prove that (s, c^n, p) is a conjugates to a Christoffel word without being a Christoffel word. First, (s, c^n, p) cannot be a Christoffel word since $ps \neq \varepsilon$.

If $n = 1$ and $ps = u^j$, the word $psc = u^j uv$ is a Christoffel word since the jth right child of (u, uv, v) is the triplet $(u, u^{j+1}v, u^j v)$. The word psc is conjugate to scp, thus scp is a conjugate to a Christoffel word. If $ps = v^j$, the word cps is a Christoffel word by similar arguments, then scp is a conjugate to a Christoffel word.

If $n \geq 1$ and $ps = v$. We want to prove that the word $c^n ps = (uv)^n v$ is a Christoffel word. The triplet (u, c, v) is in the Christoffel tree. His left child is the triplet (uv, uvv, v) and applied n times the right child rules give the triplet $(uv, (uv)^n v, (uv)^{n-1} v)$. Therefore, the word $c^n ps$ is a Christoffel word and $sc^n p$ is a Christoffel word conjugates. Similarly, one proves that $sc^n p$ is a Christoffel word, if $ps = u$.

If $ps = uv^2$, apply the right child rules on the triplet (u, uv, v) and we obtain the triplet (uv, uv^2, v) and applying n time the left child rule gives the triplet $(uv, (uv)^n uv^2, uv^2)$. Hence, the word $c^n ps$ is a Christoffel word and $sc^n p = (uv)^n uv^2$ is a conjugates to a Christoffel word. $\square$

If (s, c, p) is left normal form of a conjugate of a Christoffel word with $ps = u^j$ or $ps = v^j$, then $j \leq t(w)$ otherwise $|p| > |c|$ and p is not a proper prefix of c. For example, the left normal form of $abaabaabaabaabaabaab$ is $(a, (ba.a)^6, b)$ with $ps = ba$.

Corollary 11. *Let (s, c^n, p) be the left normal form of some conjugate of the Christoffel word w. Then c or $R(c)$ is a factor of w.*

Proof. This is a consequence of Theorem 10. $\square$

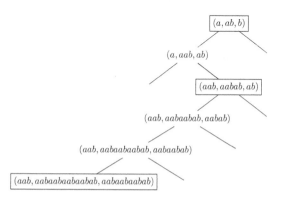

Fig. 4. The Christoffel tree before the word $w = aabaabaabaabaabab$. The framed node are the Christoffel words that appear in the left normal form of a conjugates of w

The word c or $R(c)$ are some ancestor of w or $R(w)$ in the Christoffel tree as in Fig. 4. Let w be a Christoffel word and (u, v) its standard factorization with $|u| < |v|$. Recall that $w = u^t \alpha$ with $t = t(w)$. The word c or its reversal are the last $\lceil t(w)/2 \rceil$ Christoffel words between $(u, u\alpha, \alpha)$ and (u, w, v) or the Christoffel word (u', u, α).

Acknowledgments. I would like to thank Christophe Reutenauer, Srecko Brlek, Alexandre Blondin Massé and Sébastien Labbé for their suggestions and their helpful comments. I was supported by NSERC (Canada).

References

1. Berstel, J.: Tracé de droites, fractions continues et morphismes itérés. Mots, pp. 298–309 (1990)
2. Berstel, J., Lauve, A., Reutenauer, C., Saliola, F.: Combinatorics on words: Christoffel words and repetitions in words. CRM Monograph Series, 27. American Mathematical Society, Providence (2009)
3. Berstel, J., de Luca, A.: Sturmian words, Lyndon words and trees. Theoret. Comput. Sci. **178**(1–2), 171–203 (1997)
4. Borel, J.P., Laubie, F.: Quelques mots sur la droite projective réelle. J. Théor. Nombres Bordeaux **5**, 15–27 (1993)
5. Borel, J.P., Reutenauer, C.: On Christoffel classes. RAIRO Theor. Inf. Appl. **40**(1), 15–27 (2006)
6. Chuan, W.F.: α-words and factors of characteristic sequences. Discrete Math. **177**, 33–50 (1997)
7. Currie, J.D., Saari, K.: Least periods of factors of infinite words. Theor. Inform. Appl. **43**(1), 165–178 (2009)
8. Hegedüs, L., Nagy, B.: On periodic properties of circular words. Discrete Math. **339**(3), 1189–1197 (2016)
9. Lothaire, M.: Combinatorics on Words. Addison Wesley, Boston (1983)
10. de Luca, A.: Sturmian words: structure, combinatorics, and their arithmetics. Theor. Comput. Sci. **183**, 45–82 (1997)
11. de Luca, A., Luca, A.D.: Some caracterizations of finite Sturmian words. Theor. Comput. Sci. **356**, 557–5573 (2006)
12. de Luca, A., Mignosi, F.: Some combinatorial properties of Sturmian palindromes. Theor. Comput. Sci. **136**(2), 541–546 (1994)
13. Pirillo, G.: A curious characteristic property of standard Sturmian words. In: Crapo, H., Senato, D. (eds.) Algebraic Combinatorics and Computer Science, pp. 541–546. Springer, Milano (2001)
14. Reutenauer, C.: Studies on finite Sturmian words. Theor. Comput. Sci. **591**, 106–133 (2015)

On Arithmetic Index in the Generalized Thue-Morse Word

Olga G. Parshina[1,2(✉)]

[1] Sobolev Institute of Mathematic SB RAS, 4 Acad. Koptyug Avenue,
630090 Novosibirsk, Russia
parolja@gmail.com

[2] Institut Camille Jordan, Université de Lyon, Université Claude Bernard Lyon 1,
43 Boulevard du 11 Novembre 1918, 69622 Villeurbanne Cedex, France

Abstract. Let q be a positive integer. Consider an infinite word $\omega = w_0 w_1 w_2 \cdots$ over an alphabet of cardinality q. A finite word u is called an arithmetic factor of ω if $u = w_c w_{c+d} w_{c+2d} \cdots w_{c+(|u|-1)d}$ for some choice of positive integers c and d. We call c the initial number and d the difference of u. For each such u we define its arithmetic index by $\lceil \log_q d \rceil$ where d is the least positive integer such that u occurs in ω as an arithmetic factor with difference d. In this paper we study the rate of growth of the arithmetic index of arithmetic factors of a generalization of the Thue-Morse word defined over an alphabet of prime cardinality. More precisely, we obtain upper and lower bounds for the maximum value of the arithmetic index in ω among all its arithmetic factors of length n.

Keywords: Arithmetic index · Arithmetic progression · Thue-Morse word

1 Introduction

One of the main characteristics of a given word is the subword complexity which counts the number of its distinct factors of each fixed length. We are interested in studying of so-called arithmetic factors. In other words, for a given infinite word $\omega = w_0 w_1 w_2 \cdots$ over a finite alphabet Σ we are studying the structure of its *arithmetic closure* – the set $A_\omega = \{w_c w_{c+d} w_{c+2d} \cdots w_{c+(n-1)d} | c \geq 0, d, n \geq 1\}$. Elements of A_ω are arithmetic subsequences or arithmetic factors with initial number c and difference d of the word ω. Of special interest are arithmetic factors having period 1, which are called *arithmetic progressions*. According to the classical Van der Waerden theorem [13], the arithmetic closure of each infinite word ω over an alphabet of cardinality q for every positive integer n contains an arithmetic progression of length n. A point of interest is to determine an upper bound on the minimal difference, with which the arithmetic progression of length n

This work was performed within the framework of the LABEX MILYON (ANR-10-LABX-0070) of Universite de Lyon, within the program "Investissements d'Avenir" (ANR-11-IDEX-0007) operated by the French National Research Agency (ANR).

© Springer International Publishing AG 2017
S. Brlek et al. (Eds.): WORDS 2017, LNCS 10432, pp. 121–131, 2017.
DOI: 10.1007/978-3-319-66396-8_12

appears in the arithmetic closure of a given word. The first result of the paper (see Theorem 1) concerns the distribution of words with period 1 in case ω is an infinite word over an alphabet of prime cardinality generalising the classical Thue-Morse word originally introduced by Thue in [11] (see also [1]). For a prime q and for every positive integer n this theorem provides the maximal length of an arithmetic progression with difference $d < q^n$ in the generalized Thue-Morse word over the alphabet of cardinality q and extends the earlier result on the generalized Thue-Morse word over the alphabet of cardinality 3 obtained by the author in [10].

The next question appearing in this context concerns the distribution of arithmetic subsequences with period 2. In the case of binary alphabet such subsequences have period 01, and we call them alternating subsequences. There is a conjecture, that in the Thue-Morse word alternating sequences are "the hardest to find", i.e. if d is the difference of the first occurrence of the alternating sequence of length n in the Thue-Morse word as an arithmetic factor, and h is the difference of the first arithmetic occurrence of some binary word of length n in the Thue-Morse word, then d is great or equal to h.

To carry out computer experiments and to check the conjecture we introduce the notion of arithmetic index. More precisely, given a positive integer q and an infinite q-automatic word ω, for every finite word u from its arithmetic closure we seek to determine the least positive integer d such that u occurs in ω with difference d. We call the q-ary representation of the difference d the *arithmetic index* of u in ω.

Computer experiments show that the set of words of the maximal arithmetic index in the Thue-Morse word contains alternating sequences, but they are not the sole members of this set. Describing this set even for a particular word did not appear to be an easy task. In this paper we try to determine upper and lower bounds on the rate of growth of the arithmetic index in case when ω is the generalized Thue-Morse word over the alphabet of prime cardinality. An upper bound is determined using the result on lengths of arithmetic progressions formulated in Theorem 1; a lower bound is obtained using the subword and arithmetical complexities of the word.

2   Preliminaries

Let q be a positive integer and Σ a finite alphabet of cardinality q. An infinite word over Σ is an infinite sequence $\omega = w_0 w_1 w_2 \cdots$ with $w_i \in \Sigma$ for every $i \in \mathbb{N}$. A finite word u over Σ is said to be a factor of ω if $u = w_j w_{j+1} \cdots w_{|u|+j-1}$ for some $j \in \mathbb{N}$.

For each positive integer d, let $A_\omega(d) = \{w_c w_{c+d} w_{c+2d} \cdots w_{c+(k-1)d} | c, k \in \mathbb{N}\}$ be the set of all arithmetic subsequences in the word ω of difference d. Elements of $A_\omega(d)$ are called arithmetic factors or arithmetic subwords of ω.

The arithmetic closure of ω is the set $A_\omega = \bigcup_{d=1}^{\infty} A_\omega(d)$ consisting of all its arithmetic factors, and the function $a_\omega(n) = |A_\omega \cap \Sigma^n|$ counting the number of distinct arithmetic factors of each fixed length n occurring in ω is called the *arithmetical complexity* of ω. The notion of arithmetical complexity was introduced by Avgustinovich, Fon-der-Flaass and Frid in [2]. Since $A_\omega(1)$ coincides with the set of factors of ω, it follows trivially that $a_\omega(n) \geq p_\omega(n)$. But aside from this basic inequality, there is no general relationship between the rates of growth of these two complexity functions. For instance, there exist infinite words of linear factor complexity and whose arithmetical complexity grows linearly or exponentially, as seen in [2]; arithmetical complexity of Sturmian words, which have subword complexity equals $n + 1$, grows as $O(n^3)$ (see [5]). A characterization of uniformly recurrent words having linear arithmetical complexity one can see in [7]. The question about lowest possible complexity among uniformly recurrent words was studied in [3]. A family of words with various sub-polynomial growths of arithmetical complexity was constructed in [8].

For a given infinite word ω and a finite word $u \in A_\omega$ we are interested in the least positive integer d such that u belongs to $A_\omega(d)$. We denote the length of the q-ary representation of such a minimal difference as $i_\omega(u)$ and call this quantity the *arithmetic index* of u in ω. For each positive integer n, we consider the function $I_\omega(n) = \max\limits_{u \in A_\omega \cap \Sigma^n} i_\omega(u)$. Let us note, that this function is defined over the set of arithmetic factors of ω.

In this we study the growth rate of the arithmetic index for a generalization of the Thue-Morse word defined over an alphabet $\Sigma_q = \{0, 1, ..., q-1\}$, where q is a prime number. Let $S_q : \mathbb{N} \to \Sigma_q^+$ be the function which assigns to each natural number x its base-q expansion. The length of this word is denoted by $|S_q(x)|$. Also let $s_q(x)$ be the sum modulo q of the digits in q-ary expansion of x. In other words, if $x = \sum\limits_{i=0}^{n-1} x_i \cdot q^i$, then $S_q(x) = x_{n-1} \cdots x_1 x_0$ and $s_q(x) = \sum\limits_{i=0}^{n-1} x_i$ mod q. We define the generalized Thue-Morse word $\omega_q = w_0 w_1 w_2 w_3 \cdots$ over the alphabet Σ_q by $w_i = s_q(i) \in \Sigma_q$. We note that this generalization differs from the one given in [12]. In case $q = 2$, we recover the classical Thue-Morse word which is known to be arithmetic universal, i.e. $a_{\omega_2}(n) = 2^n$, as it is shown in [2], moreover, using results of the paper it is easy to deduce that $a_{\omega_q}(n) = q^n$. In case $q = 3$, the generalized Thue-Morse word over ternary alphabet is given by:

$$\omega_3 = 012120201120201012201012120 \cdots$$

A lower and an upper bounds on the rate of growth of the function $I_{\omega_q}(n) = \max\limits_{u \in A_{\omega_q} \cap \Sigma_q^n} i_{\omega_q}(u)$ are obtained in the paper. The upper bound grows as $O(n \log n)$, the lower one grows linearly.

3   Upper Bound on Arithmetic Index in ω_q

An upper bound is based on the distribution of arithmetic progressions – arithmetic subsequences consisting of the same symbols – in the generalized Thue-Morse word formulated below.

3.1   Theorem on Arithmetic Progressions in ω_q

Let $L_\omega(c, d)$ be the function which outputs the length of an arithmetic progression with initial number c and difference d for positive integers c and d in an infinite word ω. The function $L_\omega(d) = \max\limits_{c} L_\omega(c, d)$ gives the length of the maximal arithmetic progression with the difference d in ω. Let us note, that for us the symbol of the alphabet on which the function $L_\omega(c, d)$ reaches its maxima is of no importance, since the set of arithmetic factors of the generalized Thue-Morse word is closed under adding a constant to each symbol.

Theorem 1. *Let q be a prime number and ω_q be the generalized Thue-Morse word over the alphabet Σ_q. For all integers $n \geq 1$ the following holds:*

$$\max_{d < q^n} L_{\omega_q}(d) = \begin{cases} q^n + 2q, & n \equiv 0 \bmod q, \\ q^n, & otherwise. \end{cases}$$

Moreover, the maximum is reached with the difference $d = q^n - 1$ in both cases.

Proof (of Theorem 1). Since the theorem is a generalization of the main result of [10], the technique of proving is similar to one presented there.

As the first step it should be proved that for a fixed n the inequality $d \neq q^n - 1$ implies $L_{\omega_q}(d) \leq q^n$. During the proof we have to manipulate with values $q - 1$ and $q - 2$, thus let us use the notations $\dot{q} := q - 1, \ddot{q} := q - 2$.

Case of $d \neq q^n - 1$. Let us note that subsequences of the ω_q which are composed of letters with indices having the same remainder of the division by q are equivalent to the word itself, so we do not need to consider differences which are divisible by q.

Lemma 1. *Let q be a prime number and ω_q be the generalized Thue-Morse word over Σ_q. For any positive integer n and $d \leq q^n - 1$ the length of the longest arithmetic progression with difference d in ω_q is not greater than q^n.*

Proof. Every number can be represented in the following way: $c = y \cdot q^n + x$, where x, y are arbitrary positive integers, $x < q^n$. Let us call x the suffix of c.

Consider a set $X = \{0, 1, 2, ..., q^n - 1\}$, its cardinality is $|X| = q^n$. As far as each difference d and suffix x belong to X and d is prime to $|X|$, the set X is an additive cyclic group, and d is a generator of X, thus for every $x \in X$ the set $\{x + i \cdot d\}_{i=0}^{q^n - 1}$ is precisely X. To proof the statement for this case, it is enough to provide for each $d \neq q^n - 1$ an element $x \in X$ with the following properties:

(a) $x + d < q^n$;
(b) $s_q(x + d) \neq s_q(x)$.

Indeed, consider the initial number of the form $c = y \cdot q^n + x$ with x satisfying (a) and (b) and y being an arbitrary positive integer. Because of (a), $c + d = y \cdot q^n + (x + d)$. Hence, $s_q(c) = s_q(y) + s_q(x) \bmod q$, $s_q(c + d) = s_q(y) + s_q(x + d)$ mod q, and because of (b), $s_q(c + d) \neq s_q(c)$. That means, if we consider an arithmetic subsequence with difference d starting with any symbol of generalizing Thue-Morse word and having length $q^n + 1$, then it will contain a symbol with the index of the form c mentioned above and thus at least two different symbols of the alphabet Σ_q. This implies that the arithmetic progression in this case has length less or equal to q^n.

If $s_q(d) \neq 0$, then $x = 0$ fits. In other case we use the inequation $d \neq q^n - 1$ which means that $S_q(d) = d_{n-1} \cdots d_1 d_0$ has at least one letter $d_j, j \in \{0, 1, ..., n-1\}$: $d_j \neq \dot{q}$. There are two possibilities:

1. There exists at least one index j such that $d_j < \dot{q}$ and $d_{j-1} = \dot{q}$.
   In this case $x = q^{j-1}$ fits. Indeed, c has the q-ary representation $S_q(y)S_q(x)$, where all symbols x_i are zeros except $x_{j-1} = 1$, thus $s_q(c) = s_q(y) + 1$. The difference d has the representation $d_{n-1} \cdots d_j \dot{q} d_{j-2} \cdots d_0$, and the sum of its digits equals zero modulo q. More precisely, $\sum_{i=0, i\neq j-1}^{n-1} d_i + \dot{q} \equiv 0 \bmod q$, or $\sum_{i=0, i\neq j-1}^{n-1} d_i - 1 \equiv 0 \bmod q$. When we add d to c, we obtain the number $c + d$ with representation $S_q(y)d_{n-1} \cdots (d_j + 1)0 d_{j-2} \cdots d_0$, where $d_j + 1 \leq \dot{q}$. Then $s_q(c) = s_q(y) + s_q(d) + 1 - \dot{q} = s_q(y) + 2$, which differs from $s_q(c) = s_q(y) + 1$.

2. For every j the fact $d_j \neq \dot{q}$ implies that all symbols having indices less than j are not equal to $\dot{q}$.
   If $j > 0$, then $d_1, d_0 \neq \dot{q}$, and $d_0 \neq 0$ since d is not divisible by q. In this case a suitable x is $q - d_0$, because $s_q(x + d) = s_q(1 - d_0) \neq s_q(q - d_0) = s_q(x)$. But there is no x satisfying (a) and (b) in the case $j = 0$, i.e. then $S_q(d) = \underbrace{\dot{q} \cdots \dot{q}}_{n-1} d_0$. However, we can take x with q-ary representation of the form $S_q(x) = x_{n-1} \cdots x_1 \dot{q}$, where $x_i \in \Sigma_q$ are arbitrary, and claim that for arbitrary value of y we obtain the number with the sum of digits different from $s_q(c) = s_q(y) + s_q(x)$ after at most two additions of the difference. Consider these two steps. After adding to c with $S_q(c) = S_q(y) x_{n-1} \cdots x_1 \dot{q}$ and $s_q(c) = s_q(y) + \sum_{i=1}^{n-1} x_i + \dot{q}$ the difference of the form $\dot{q} \cdots \dot{q} d_0$ we obtain the number $c + d$ with $S_q(c + d) = S_q(y + 1) x_{n-1} \cdots x_1 \dot{d_0}$. Its sum of digits is $s_q(y+1) + \sum_{i=1}^{n-1} x_i + \dot{d_0}$, and if it differs from $s_q(c)$, then this x fits. If the values $s_q(c)$ and $s_q(c+d)$ are equal, then the following holds: $s_q(y+1) + d_0 \equiv s_q(y) \bmod q$. This implies that q-ary representation of y ends with $0 \underbrace{\dot{q} \cdots \dot{q}}_{\dot{q} - d_0}$, and thus q-ary representation of $y + 1$ ends with zero. After the next addition of the difference there are two cases. If $2d_0 \geq q$, we obtain $S_q(y + 2) x_{n-1} \cdots x_1 (2\dot{d_0})$ with $s(y + 2) = s(y + 1) + 1$ and $s_q(c + 2d) = s_q(y + 1) + \sum_{i=1}^{n-1} x_i + 2d_0 \bmod q$, which implies $d_0 = \dot{q}$, but this is not the case. If $2d_0 < q$, we obtain the number of the form $S_q(c + 2d) = S_q(y + 2) x_{n-1} \cdots x_2 \dot{x_1} (2\dot{d_0})$ with $s_q(c + 2d) = s_q(y + 1) + \sum_{i=1}^{n-1} x_i + 2d_0 - 1$, which implies $d_0 \equiv 0 \bmod q$ and contradicts the fact $d_0 \neq 0$.

Thus there are $q(n-1)$ different values for x, such that for every positive integer c with the suffix x either $s_q(c+d) \neq s_q(c)$, or $s_q(c+2d) \neq s_q(c+d)$. That means that every arithmetic progression with the difference d with $S_q(d) = \underbrace{\dot{q} \cdots \dot{q}}_{n-1} d_0, d_0 \neq 0, d_0 \neq \dot{q}$ is not be longer than q^n.

Since all possible values of difference $d \neq q^n - 1$ are considered, the lemma is proved.

Case of $d = q^n - 1$. We start with the following lemma.

Lemma 2. *Let q be a prime number and w_q be the generalized Thue-Morse word over Σ_q. Let $d = q^n - 1$, $c = z \cdot q^{2n} + y \cdot q^n + x$, where $x + y = q^n - 1$, z is a non-negative integer, then*

$$\max_z L_{w_q}(c, d) = \begin{cases} x + q + 1, & n \equiv 0 \bmod q, \\ x + 1, & otherwise. \end{cases}$$

Proof. For descriptive reasons let us introduce a scheme where one can see q-ary representations of $c + id$ and values of $s_q(c + id)$ for each value of i, and let us give some comments on that.

i	$S_q(c + id)$									$s_q(c + id)$
0	$S_q(z)$	y_{n-1}	$\cdots$	$y_1 y_0$	x_{n-1}	$\cdots$	$x_1 x_0$			$n\dot{q} + s_q(z)$
$\vdots$		$\vdots$		$\vdots$			$\vdots$			$\vdots$
$\vdots$		$\vdots$	$+x \cdot d$	$\vdots$			$\vdots$			$\vdots$
$\vdots$		$\vdots$		$\vdots$			$\vdots$			$\vdots$
x	$S_q(z)$	$\dot{q}$	$\cdots$	$\dot{q}\,\dot{q}$	0	$\cdots$	$0\,0$			$n\dot{q} + s_q(z)$
$x+1$	$S_q(z)$	$\dot{q}$	$\cdots$	$\dot{q}\,\dot{q}$	$\dot{q}$	$\cdots$	$\dot{q}\,\dot{q}$			$2n\dot{q} + s_q(z)$
$x+2$	$S_q(z+1)$	0	$\cdots$	$0\,0$	$\dot{q}$	$\cdots$	$\dot{q}\,\dot{q}$			$n\dot{q} - 1 + s_q(z+1)$
$\vdots$		$\vdots$		$\vdots$			$\vdots$			$\vdots$
$\vdots$		$\vdots$	$+(q-2) \cdot d$	$\vdots$			$\vdots$			$\vdots$
$\vdots$		$\vdots$		$\vdots$			$\vdots$			$\vdots$
$x+q$	$S_q(z+1)$	0	$\cdots$	$0\,\ddot{q}$	$\dot{q}$	$\cdots$	$\dot{q}\,0$			$n\dot{q} - 1 + s_q(z+1)$
$x+q+1$	$S_q(z+1)$	0	$\cdots$	$0\,\dot{q}$	$\dot{q}$	$\cdots$	$\ddot{q}\,\dot{q}$			$n\dot{q} + \ddot{q} + s_q(z+1)$

Values in the third column are sums modulo q.

Since $d = q^n - 1$, we can regard the action $c + d$ as two simultaneous actions: $x - 1$ and $y + 1$. Thus, while the suffix of $c + id$ is greater then zero, the sum of digits in $S_q(c + id)$ equals $\dot{q}n$. This value holds during the first x additions of d (when $i = 0, 1, .., x$), and on the step number x the length of the arithmetic progression is $x + 1$.

On the next step ($i = x + 1$) the sum of digits in result's q-ary representation becomes $2\dot{q}n + s_q(z)$. To preserve the required property of progression members we need $\dot{q}n \equiv 2\dot{q}n \bmod q$, i.e., $n \equiv 0 \bmod q$.

After the next addition of the difference, z increases to $z + 1$, y becomes 0, $x = q^n - 2$ and the sum modulo q of digits in this number q-ary representation becomes $s_q(z+1) + s_q(x) = s_q(z+1) + n\dot{q} - 1$. We may choose a suitable z to hold the homogeneity of the progression, e.g. if $s_q(z) = 1$ we need $s_q(z + 1) = 2$, and z may be equal to 1. The value of the sum modulo q holds during the following $q - 2$ editions of d, and we get into the situation of $y = q - 2$, $x = q^n - \dot{q}$.

After the next addition y becomes $\dot{q}$, and $x = q^n - q - 1$. Now $s_q(y) + s_q(x) = n\dot{q} + \ddot{q} \bmod q$ and is not equal to its previous value $n\dot{q} - 1$.

Hence, in the case of $q|n$ the length of an arithmetic progression is $x + q + 1$ and it is $x + 1$ otherwise. The lemma is proved.

Lemma 3. *Let q be a prime number, and w_q be the generalized Thue-Morse word over Σ_q. Let $n \equiv 0 \bmod q$, $d = q^n - 1$, $c = z \cdot q^{2n} + y \cdot q^n + x$, $y = q^n - \dot{q}$, $x = \dot{q}$, and z be an arbitrary non-negative integer, then $\max_z L_{w_q}(c, d) = q^n + 2q$.*

Proof. The sum modulo q of digits in $S_q(c)$ is equal to $s_q(z) + n\dot{q} + 1$, and by arguments similar to the ones used in Lemma 2, this value is not changing while the suffix of $c + id$ is greater or equal to zero, i.e. during $\dot{q}$ steps; then we get into a situation when $y = x = 0$ and z is increased by 1. We may set z to be a zero to hold the homogeneity of a progression on this step.

After the next addition we get into conditions of Lemma 2 with $x = q^n - 1$, which provides us with an arithmetical progression of length $q^n + q$.

Now we subtract d from the initial c to make sure that $s_q(c - d) \neq s_q(c)$ and we cannot obtain longer arithmetical progression. Indeed, $c - d = z \cdot q^{2n} + (q^n - \dot{q}) \cdot q^n + q$ and the sum of digits in its q-ary representation is $n\dot{q} - \dot{q} + 1$, while in c it is $n\dot{q} + 1$. Hence the length of this progression is $1 + \dot{q} + q^n + q = q^n + 2q$, and the lemma is proved.

Now let us prove that we can not construct an arithmetical progression with the difference $d = q^n - 1$ longer than $q^n + 2q$.

Here we represent the initial number c of the progression this way: $c = y \cdot q^n + x$, $x < q^n$.

The case of initial number c with $x_j + y_j = \dot{q}$, $j = 0, 1, ..., n - 1$ is described in Lemma 2. In other case there is at least one index j such that $x_j + y_j \neq \dot{q}$. We choose j which is the minimal. There are $q \cdot \dot{q}$ possibilities of values (y_j, x_j): $(0, 0), (0, 1), ..., (0, \ddot{q}), (1, 0), ..., (\dot{q}, \dot{q})$.

Integers y and x have q-ary representations $S_q(y) = y_{s-1} \cdots y_{j+l+1} \dot{q} \cdots \dot{q} y_j \cdots y_0$ and $S_q(x) = x_{n-1} \cdots x_{j+m+1} 0 \cdots 0 x_j \cdots x_0$, where $0 \leq l \leq s - j$, $0 \leq m \leq n - j$, $y_{j+l+1} \neq \dot{q}$, and $x_{j+m+1} \neq 0$.

We add $q^{j+1} \cdot d$ to c. If $l \neq 0$, the block $\dot{q} \cdots \dot{q}$ in $S_q(y)$ transforms to the block of zeros; if $m \neq 0$, the block of zeros in $S_q(x)$ transforms to the block $\dot{q} \cdots \dot{q}$, y_{j+l+1} increases by one, and x_{j+m+1} decreases by one. To hold the homogeneity we need l and m to be equal modulo q.

There are two different cases.

1. If $x_j < \dot{q} - y_j$, then after $(x_j + 1) \cdot q^j$ additions of the difference we obtain the number with q-ary representation $y_{s-1} \cdots y_{j+l+2}(y_{j+l+1} + 1)0 \cdots 0(y_j + x_j + 1)y_{j-1} \cdots y_0 \, x_{n-1} \cdots x_{j+m+2}$

$x_{j+m+1}\dot{q}\cdots\dot{q}\dot{q}\dot{q}x_{j-1}\cdots x_0$ with the sum of digits $s_q(y) + s_q(x) + \dot{q} \neq s_q(y) + s_q(x) = s_q(c)$. Thus the length of an arithmetic progression is not greater than $q^j(q + x_j + 1) \leq q^n$ if $j < n - 1$.

2. If $x_j > \dot{q} - y_j$, we add $(q - y_j) \cdot q^j \cdot d$ and obtain a number of the form
$y_{s-1}\cdots y_{j+l+2}(y_{j+l+1} \qquad\qquad + \qquad\qquad 1)0\cdots 010 y_{j-1}\cdots y_0$
$x_{n-1}\cdots x_{j+m+2}x_{j+m+1}\dot{q}\cdots\dot{q}(x_j - q + y_j)x_{j-1}\cdots x_0$ with the sum of digits $s_q(y) + 1 - l\dot{q} + 1 - y_j + s_q(x) + m\dot{q} - 1 - q + y_j = s_q(y) + s_q(x) + 1$. The length of an arithmetic progression in this case is not greater than $q^j(2q - y_j) \leq q^n$, if $j < n - 1$.

Example 1. Let us consider an example for $q = 5$, $m = l = 3$, $n = 6$, $j = 1$, $(y_j, x_j) = (1, 1)$, $d = 15624$, $S_5(d) = 444444$.

number	S_5	s_5
$c = 97396881$	144413200011	1
	$+\quad 44444400$	
$c + 25 \cdot d$	200013144411	1
	$+\quad 4444440$	
$c + 30 \cdot d$	200023144401	1
	$+\quad 4444440$	
$c + 35 \cdot d$	200033144341	0

The case $j = n - 1$ needs a special consideration.

The way of acting is the same: we add $x \cdot d$ to c and nullify x by that, then add d necessary number of times. Thus the worst case is then $S_q(x) = x_{n-1}\dot{q}\cdots\dot{q}$ and $S_q(y) = y_{s-1}\cdots y_n y_{n-1}0\cdots 0$. The length of an arithmetic progression is $x + 2 < q^n + 2$ if $n \equiv 0 \bmod q$ and $x + 1 \leq q^n$ otherwise. But since $y_{n-1} + x_{n-1} \neq \dot{q}$, there are two cases to consider, let us introduce schemes for both.

1. $r = y_{n-1} + x_{n-1} < \dot{q}$.

i	$S_q(c + id)$				$s_q(c + id)$
0	$y_{s-1}\cdots y_n$	y_{n-1} $\ 0\cdots 0$	$x_{n-1}\dot{q}\cdots\dot{q}\dot{q}$		$n\dot{q} - \dot{q} + r + y_n + \ldots + y_{s-1}$
$\vdots$	$\vdots$	$\vdots$	$\vdots$	$\vdots$	$\vdots$
		$+x \cdot d$			
$\vdots$	$\vdots$	$\vdots$	$\vdots$	$\vdots$	$\vdots$
x	$y_{s-1}\cdots y_n$	r $\quad \dot{q}\cdots\dot{q}$	$0\ \ 0\cdots 0\ 0$		$n\dot{q} - \dot{q} + r + y_n + \ldots + y_{s-1}$
$x + 1$	$y_{s-1}\cdots y_n$	r $\quad \dot{q}\cdots\dot{q}$	$\dot{q}\ \ \dot{q}\cdots\dot{q}\dot{q}$		$2n\dot{q} - \dot{q} + r + y_n + \ldots + y_{s-1}$
$x + 2$	$y_{s-1}\cdots y_n(r + 1)$	$0\cdots 0$	$\dot{q}\ \ \dot{q}\cdots\dot{q}\dot{q}$		$n\dot{q} + r + y_n + \ldots + y_{s-1}$

2. $r = y_{n-1} + x_{n-1} > \dot{q}$. Here we denote by $y'_{s-1}\cdots y'_n$ symbols $y_{s-1}\cdots y_n$ transformed after increasing y_n by 1 on the step x; to hold the homogeneity their sum should be equal to $y_{s-1} + \ldots + y_n \bmod q$.

i	$S_q(c + id)$				$s_q(c + id)$
0	$y_{s-1}\cdots y_n$	y_{n-1}	$0\cdots 0\ x_{n-1}\ \dot q \cdots \dot q\ \dot q$		$n\dot q - \dot q + r + y_n + \dots + y_{s-1}$
$\vdots$	$\vdots$	$\vdots$	$\vdots$	$\vdots$	$\vdots$
$\vdots$	$\vdots$	$+x \cdot d$	$\vdots$	$\vdots$	$\vdots$
$\vdots$	$\vdots$	$\vdots$	$\vdots$	$\vdots$	$\vdots$
x	$y'_{s-1}\cdots y'_n\ (r-q)$	$\dot q \cdots \dot q$	0	$0\cdots 0\ 0$	$n\dot q - \dot q + r + y'_n + \dots + y'_{s-1}$
$x+1$	$y'_{s-1}\cdots y'_n\ (r-q)$	$\dot q \cdots \dot q$	$\dot q$	$\dot q \cdots \dot q\ \dot q$	$2n\dot q - \dot q + r + y'_n + \dots + y'_{s-1}$
$x+2$	$y'_{s-1}\cdots y'_n\ (r-q+1)\,0\cdots 0$	$\dot q$	$\dot q$	$\dot q \cdots \dot q\ \ddot q$	$n\dot q + r + y'_n + \dots + y'_{s-1}$

Hence all possible cases have been considered and the theorem is proved.

3.2 Upper Bound on the Arithmetic Index for ω_q

Let q be a prime, n be a positive integer, and u be a finite word over the alphabet Σ_q of length m, where $q^{n-1} \le m < q^n$. The goal is to find its occurrence in ω_q as an arithmetic factor. To reach the goal we use an arithmetic factor of ω_q of the form $0^{m-1}\beta_1 \cdots \beta_m$, where each $\beta_i \in \Sigma_q$ and $\beta_1 \neq 0$. Lemma 3 provides us with its initial symbol c and difference $d = q^n - 1$.

Then we define a basis $\{b_i\}_{i=1}^m$

$$
\begin{aligned}
b_1 &= 0\ 0\ 0\ \cdots\ 0\ \ \beta_1; \\
b_2 &= 0\ 0\ 0\ \cdots\ \beta_1\ \beta_2; \\
&\ \ \vdots \qquad\quad \vdots \qquad \vdots \\
b_m &= \beta_1\beta_2\beta_3\ \cdots\ \beta_{m-1}\ \beta_m,
\end{aligned}
$$

where all basis elements are arithmetic factors of ω_q with the difference d, and their initial numbers are $c_1 = c$, $c_{i+1} = c + i \cdot d$, $i = 1, ..., m-1$. The word u can be represented in the following form: $u = \bigoplus_{i=1}^m \alpha_i \cdot b_i$ with $\alpha_i \in \Sigma_q$ for every $i = 1, 2, ..., m$, where $\bigoplus$ means symbol-to-symbol addition modulo q.

Then let us construct the initial number c_u of the arithmetic factor u with q-ary representation $S_q(c_u) = \underbrace{S_q(c_1) \cdots S_q(c_1)}_{\alpha_1} \cdots \underbrace{S_q(c_m) \cdots S_q(c_m)}_{\alpha_m}$. According to Lemma 3 the length of each $S_q(c_i)$ is not greater than $2n + q$. To simplify the process let us put zeros left to the nonzero symbol with the maximal index in every $S_q(c_i)$ when it is necessary. Hence the length of $S_q(c_u)$ is not greater than $(2n + q)m\dot q$. Then we construct the difference d_u with $S_q(d_u) = \underbrace{0\cdots 0}_{n+q}\underbrace{\dot q\cdots \dot q}_{n}\cdots \underbrace{0\cdots 0}_{n+q}\underbrace{\dot q\cdots \dot q}_{n}$ of the same length, and the word u is guaranteed to appear in ω_q as an arithmetic factor with difference d_u and initial number c_u.

The worst case is when all basis elements in the representation of u are taken with coefficients $\dot{q}$. In this case the length of $S_q(d_u) = (2n + q)\dot{q} \cdot m = (2\lceil \log_q m \rceil + q)\dot{q} \cdot m$, which is the upper bound on the function of arithmetic index in ω_q. Thus we proved the following inequality:

$$I_{\omega_q}(m) \leq (2\lceil \log_q m \rceil + q)\dot{q} \cdot m$$

4   Lower Bound on Arithmetic Index

Consider the set $A_\omega(d)$ of all arithmetic subsequences with the difference d in the word ω over the alphabet Σ_q. Define a function $A_\omega(d, m) = |A_\omega(d) \cap \Sigma_q^m|$ counting the number of different arithmetic factors with difference d and of length m in ω. Clearly, $|A_\omega(1) \cap \Sigma_q^m| = p_\omega(m)$ and, more general, $A_\omega(d, m) = |A_\omega(d) \cap \Sigma_q^m| \leq d \cdot p_\omega(m)$.

Obtaining a lower bound on the function $I_\omega(m)$ is equivalent to obtaining the lower bound on x in the following inequality:

$$a_\omega(m) \leq \sum_{d=1}^{x} A_\omega(d, m) \leq \sum_{d=1}^{x} d \cdot p_\omega(m). \text{ Thus } x + \tfrac{1}{2} \geq \sqrt{\tfrac{1}{4} + \tfrac{2a_\omega(m)}{p_\omega(m)}}$$

We are interested in the integer part of $\log_q x$. The value $\lceil \log_q(x + 0.5) \rceil$ is either $\lceil \log_q x \rceil$ or $\lceil \log_q x \rceil + 1$, and $\lceil \log_q x \rceil + 1 \geq \lceil \log_q(x + \tfrac{1}{2}) \rceil \geq \lceil \log_q \sqrt{\tfrac{1}{4} + \tfrac{2a_\omega(m)}{p_\omega(m)}} \rceil$.

Thus $I_\omega(m) \geq \lceil \log_q x \rceil \geq \lceil \tfrac{1}{2} \log_q \tfrac{2a_\omega(m)}{p_\omega(m)} - 1 \rceil$

The Thue-Morse word and its generalization are fixed points of uniform morphisms, their subword complexity is known to grow linearly [6], i.e. $p_{\omega_q}(m) \leq C \cdot m$ for a positive integer C. As mentioned in Sect. 2, $a_{\omega_q}(m) = q^m$, thus the lower bound on the function of arithmetic index in ω_q is

$$I_{\omega_q}(m) \geq 0.5(m - C \log_q m) .$$

The subword complexity of ω_2 was computed in 1989 by Brlek [4] and de Luca and Varricino [9]. Using their result we can set the constant C to be 4 in this case, and the lower bound can be written in the following way:

$$I_{\omega_2}(m) \geq \frac{1}{2}m - 2 \log_q m .$$

5   Conclusion

To sum up, the upper bound on the function of arithmetic index grows as $O(m \log m)$, the lower bound grows as $O(m)$. The formula for computing the lower bound can be applied to words with known subword and arithmetic complexities; the upper bound is more difficult to compute and requires deeper knowledge of the word structure.

According to computer experiments, which were carried out for the Thue-Morse sequence, the real growth of the function $I_{\omega_2}(m)$ is closer to the lower bound. Moreover, both theoretical reasoning and computer data show, that alternating arithmetic subsequences have the maximal arithmetic index. But they are not the only subsequences contained in the set of words with extremal arithmetic index; this set is going to be described for ω_q and then for other automatic words.

References

1. Allouche, J.P., Shallit, J.: The ubiquitous Prouhet-Thue-Morse sequence. In: Ding, C., Helleseth, T., Niederreiter, H. (eds.) Sequences and their Applications. Discrete Mathematics and Theoretical Computer Science, pp. 1–16. Springer, London (1999)
2. Avgustinovich, S.V., Fon-Der-Flaass, D.G., Frid, A.E.: Arithmetical complexity of infinite words. In: Words, Languages and Combinatorics III, Kyoto 2000, pp. 51–62. World Science Publisher, River Edge (2003). doi:10.1142/9789812704979_0004
3. Avgustinovich, S.V., Cassaigne, J., Frid, A.E.: Sequences of low arithmetical complexity. Theor. Inform. Appl. 40(4), 569–582 (2006). doi:10.1051/ita:2006041
4. Brlek, S.: Enumeration of factors in the Thue-Morse word. Discrete Appl. Math. 24(1–3), 83–96 (1989). doi:10.1016/0166-218X(92)90274-E. First Montreal Conference on Combinatorics and Computer Science (1987)
5. Cassaigne, J., Frid, A.E.: On the arithmetical complexity of Sturmian words. Theoret. Comput. Sci. 380(3), 304–316 (2007). doi:10.1016/j.tcs.2007.03.022
6. Ehrenfeucht, A., Lee, K.P., Rozenberg, G.: Subword complexities of various classes of deterministic developmental languages without interactions. Theor. Comput. Sci. 1(1), 59–75 (1975)
7. Frid, A.E.: Sequences of linear arithmetical complexity. Theor. Comput. Sci. 339(1), 68–87 (2005). doi:10.1016/j.tcs.2005.01.009
8. Frid, A.E.: On possible growths of arithmetical complexity. Theor. Inform. Appl. 40(3), 443–458 (2006). doi:10.1051/ita:2006021
9. de Luca, A., Varricchio, S.: Some combinatorial properties of the Thue-Morse sequence and a problem in semigroups. Theor. Comput. Sci. 63(3), 333–348 (1989). doi:10.1016/0304-3975(89)90013-3
10. Parshina, O.G.: On arithmetic progressions in the generalized Thue-Morse word. In: Manea, F., Nowotka, D. (eds.) WORDS 2015. LNCS, vol. 9304, pp. 191–196. Springer, Cham (2015). doi:10.1007/978-3-319-23660-5_16
11. Thue, A.: Über die gegenseitige lage gleicher teile gewisser zeichenreichen. Skr. Vid.-Kristiana I. Mat. Naturv. Klasse 1, 1–67 (1912)
12. Tromp, J., Shallit, J.: Subword complexity of a generalized Thue-Morse word. Inform. Process. Lett. 54(6), 313–316 (1995). doi:10.1016/0020-0190(95)00074-M
13. Van der Waerden, B.: Beweis einer baudetschen vermutung. Nieuw Arch. Wisk. 15, 212–216 (1927)

Abelian Complexity of Thue-Morse Word over a Ternary Alphabet

Idrissa Kaboré[(⊠)] and Boucaré Kientéga

UFR–Sciences et Techniques, Université NAZI BONI,
01 BP 1091 Bobo-Dioulasso 01, Burkina Faso
ikaborei@yahoo.fr, boucarekientaga@yahoo.fr

Abstract. In this paper, we study the Thue-Morse word on a ternary alphabet. We establish some properties on special factors of this word and prove that it is 2-balanced. Moreover, we determine its Abelian complexity function.

Keywords: Infinite word · Factor · Morphism · Abelian complexity

Mathematics Subject Classification: 68R15 · 11B85

1 Introduction

Abelian complexity is a combinatorial notion used in the study of infinite words. It counts the number of Parikh vectors of given length in a word. The study of Abelian complexity was developed recently [6,7,9,16,17]. In particular, the Abelian complexity of some words and some classes of words have been studied [4,8,10,12,13,21,23].

The Thue-Morse word t_2 on the binary alphabet $\{0, 1\}$ is the infinite word generated by the morphism μ_2 defined by $\mu(0) = 01$, $\mu(1) = 10$. The study of this word goes back to the beginning of the twentieth century with the works of Thue [19,20]. It was extensively studied during the last three decades [1–3,14]. In [17] the authors have determined its Abelian complexity: for all $n \geq 1$, $\rho_{t_2}^{ab}(n) = 2$ if n is odd and $\rho_{t_2}^{ab}(n) = 3$ otherwise. The Thue-Morse word can be naturally generalized over an alphabet $\mathcal{A}_q$ of size $q \geq 3$. More precisely, it is, on the alphabet $\mathcal{A}_q = \{0, 1, ..., q-1\}$, the infinite word t_q generated by the morphism μ_q defined by: $\mu_q(k) = k(k+1)...(k+q-1)$, where the letters are expressed modulo q. A study of this word has been done in [18]. In this paper, we are interested in the study of the Abelian complexity of the Thue-Morse word over the alphabet $\mathcal{A}_3 = \{0, 1, 2\}$. More exactly, it is the word t_3 generated by the morphism μ_3 defined by $\mu_3(0) = 012$, $\mu_3(1) = 120$ and $\mu_3(2) = 201$.

The paper is organized as follows. After some definitions and notations, we recall in Sect. 2 some useful results. In Sect. 3, we establish some combinatorial properties of the word t_3. We determine, in particular, its triprolongable factors, then we show that it is 2-balanced. Lastly, in Sect. 4, we determine the Abelian complexity function of t_3.

© Springer International Publishing AG 2017
S. Brlek et al. (Eds.): WORDS 2017, LNCS 10432, pp. 132–143, 2017.
DOI: 10.1007/978-3-319-66396-8_13

2   Definitions and Notations

Let $\mathcal{A}$ be a finite alphabet. The set of finite words over $\mathcal{A}$ is noted $\mathcal{A}^*$ and ε represents the empty word. The set of non-empty finite words over $\mathcal{A}$ is denoted by $\mathcal{A}^+$. For all $u \in \mathcal{A}^*$, $|u|$ designates the length of u and the number of occurrences of a letter a in u is denoted $|u|_a$. A word u of length n formed by repeating a single letter x is denoted x^n.

An infinite word is a sequence of letters of $\mathcal{A}$, indexed by $\mathbb{N}$. We denote by $\mathcal{A}^\omega$ the set of infinite words on $\mathcal{A}$. The set of finite or infinite words on $\mathcal{A}$ is denoted $\mathcal{A}^\infty$.

Let u be a finite or infinite word and v a finite word on $\mathcal{A}$. The word v is called factor of u if there exists $u_1 \in \mathcal{A}^*$ and $u_2 \in \mathcal{A}^\infty$ such that $u = u_1 v u_2$. The factor v is called prefix (resp. suffix) if u_1 (resp. u_2) is the empty word. The set of the prefixes (resp. the suffixes) of u is denoted $pref(u)$ (resp. $suff(u)$).

Let u be an infinite word. The set of factors of length n of u is denoted $F_n(u)$. The set of all the factors of u is denoted $F(u)$.

Let v be a factor of u and a be a letter of $\mathcal{A}$. We say that v is right (resp. left) prolongable by a, if va (resp. av) is also a factor of u. The word va (resp. av) is called a right (resp. left) extension of v in u. The factor v is said to be right (resp. left) special it admits at least two right (resp. left) extensions. If v is both right special and left special, it is called bispecial.

An infinite word u is said to be recurrent if any factor of u appears infinitely often. It is said to be uniformly recurrent if for any natural n, it exists a natural n_0 such that any factor of length n_0 contains all the factors length n of u.

A morphism on $\mathcal{A}^*$ is a map $f : \mathcal{A}^* \to \mathcal{A}^*$ such that $f(uv) = f(u)f(v)$, for all $u, v \in \mathcal{A}^*$. A morphism f is said to be primitive if it exists a positive integer n such that, for all letter a in $\mathcal{A}$, $f^n(a)$ contains all the letters of $\mathcal{A}$. It is k-uniform, if $|f(a)| = k$ for all a in $\mathcal{A}$. A morphism f is said to be prolongable on a letter a if $f(a) = aw$ where $w \in \mathcal{A}^+$, and $f^n(w)$ is non empty for any natural n. A morphism f defined on an alphabet $A = \{a_1, a_2, ..., a_d\}$ is said to be left (resp. right) marked, if the first (resp. last) letters of $f(a_i)$ and $f(a_j)$ are different, for all $i \neq j$. If f is both left marked and right marked, it is said marked. An infinite word u is generated by a morphism f if there exists a letter a such that the words a, $f(a)$, ..., $f^n(a)$, ... are longer and longer prefixes of u. We note $u = f^\omega(a)$. An infinite word generated by a morphism is called purely morphic word. Let u be an infinite purely morphic word and w, a factor of u verifying

$$|w| \geq max\{|f(a)| : a \in \mathcal{A}\}.$$

Then w can be decomposed in the form

$$p_0 f(a_1) f(a_2)...f(a_n) s_{n+1},$$

where

- $n \geq 0$, $a_0, a_1, ..., a_{n+1} \in \mathcal{A}$;
- p_0 is a suffix of $f(a_0)$ and s_{n+1} is a prefix of $f(a_{n+1})$.

This decomposition is called *synchronization* [5].

Let u be an infinite word on an alphabet $\mathcal{A}_q = \{a_0, a_1, ..., a_{q-1}\}$ and v, a factor of u. The Parikh vector of v is the q-uplet $\psi(w) = (|v|_{a_0}, |v|_{a_1}, ..., |v|_{a_{q-1}})$. We denote by $\Psi_n(u)$, the set of the Parikh vectors of the factors of length n of u:

$$\Psi_n(u) = \{\psi(v) : v \in F_n(u)\}.$$

The Abelian complexity of u is the application of $\mathbb{N}$ to $\mathbb{N}$ defined by: $\rho^{ab}(n) = card(\Psi_n(u))$. Let θ be a natural. An infinite word u is said to be θ-balanced if for any letter a of $\mathcal{A}$ and any couple (v, w) of factors of u with the same length, one has $||v|_a - |w|_a| \leq \theta$.

Let u be an infinite word and v, a factor of u of length n. We denote by $u_{[n]}$ the prefix of u of length n. The relative Parikh vector [22] of v is:

$$\psi^{rel}(v) = \psi(v) - \psi(u_{[n]}).$$

The set of the relative Parikh vectors of the factors of u of length n will be simply denoted:

$$\Psi_n^{rel}(u) = \{\psi^{rel}(v) : v \in F_n(u)\}.$$

This set has the same cardinal as $\Psi_n(u)$. So,

$$\rho^{ab}(n) = card(\Psi_n^{rel}(u)).$$

If u is θ-balanced, then all the components of relative Parikh vector are bounded by θ [23].

Let us consider the alphabet $\mathcal{A}_3 = \{0, 1, 2\}$. The Thue-Morse word over $\mathcal{A}_3$ is the infinite word $\mathbf{t}_3$ generated by the morphism μ_3 defined by $\mu_3(0) = 012$, $\mu_3(1) = 120$, $\mu_3(2) = 201$:

$$\mathbf{t}_3 = \lim_{n \longrightarrow +\infty} \mu_3^{(n)}(0) = 0121202011202010122010121201202010122010121200012...$$

Theorem 2.1 *[11]. Let f be a primitive morphism, prolongable on a letter a. Then, the infinite word $f^{\omega}(a)$, generated by f on a, is uniformly recurrent.*

The morphism μ_3 being primitive and prolongable on 0, the word $\mathbf{t}_3 = \mu_3^{\omega}(0)$ is uniformly recurrent.

In the following, we consider the alphabet $\mathcal{A}_3 = \{0, 1, 2\}$.

3   Triprolongable Factors and Balance

In this section, we establish some combinatorial properties of $\mathbf{t}_3$, then we show that it is 2-balanced.

Recall the following useful lemma called synchronization lemma applied to the morphism μ_3.

Lemma 3.1 *Let u be a factor of $\mathbf{t}_3$. Then, there exist some factors v, δ_1 and δ_2 of $\mathbf{t}_3$ such that $u = \delta_1\mu_3(v)\delta_2$ with $|\delta_1|$, $|\delta_2| \leq 2$. This decomposition is unique if $|u| \geq 7$.*

Proposition 3.1 *Let u be a factor of $\mathbf{t}_3$. Then, u is right (resp. left) triprolongable if and only if $\mu_3(u)$ is right (resp. left) triprolongable.*

Proof: Let u be a factor of $\mathbf{t}_3$, right triprolongable. Then, for any $i \in \mathcal{A}_3$, ui is in $\mathbf{t}_3$. Therefore, $\mu_3(u)i$ is in $\mathbf{t}_3$, since $\mu_3(i)$ begins with i.

Conversely, let u be a factor of $\mathbf{t}_3$ such that $\mu_3(u)$ is right triprolongable with $|u| \geq 2$ (the case $|u| \leq 1$ is evident). Then, $\mu_3(u)i$ is in $\mathbf{t}_3$, for all $i \in \mathcal{A}_3$. So, the factor $\mu_3(u)i$ ends by the first letter of the image of $\mu_3(i)$, for all $i \in \mathcal{A}_3$; we use here the unicity in the Lemma 3.1 since $|\mu_3(u)i| \geq 7$. So, the factors $\mu_3(u)012$, $\mu_3(u)120$ and $\mu_3(u)201$ are in $\mathbf{t}_3$. These three factors can be written respectively $\mu_3(u0)$, $\mu_3(u1)$ and $\mu_3(u2)$. This proves that u is right triprolongable in $\mathbf{t}_3$. We proceed in the same way for the factors which are left triprolongable.   ∎

For the following, we denote by $BST(\mathbf{t}_3)$, the set of the factors of $\mathbf{t}_3$ which are both right triprolongable left triprolongable.

As a consequence of Proposition 3.1, a factor u is in $BST(\mathbf{t}_3)$ if and only if $\mu_3(u)$ is in $BST(\mathbf{t}_3)$.

Proposition 3.2. *Let u be an element of $BST(\mathbf{t}_3)$. If $|u| \geq 3$, it exists u' in $BST(\mathbf{t}_3)$ such that $u = \mu_3(u')$.*

Proof: Let u in $BST(\mathbf{t}_3)$ such that $|u| \geq 3$. One verifies manually the proposition for the case $3 \leq |u| \leq 6$. Now suppose $|u| \geq 7$. Then, the factor u can be written in a unique way in the form $u = \delta_1 \mu_3(u') \delta_2$, where u', δ_1 and δ_2 are factors of $\mathbf{t}_3$. Let us verify that factors δ_1 and δ_2 are empty. As u is right triprolongable, the factors $\delta_2 0$, $\delta_2 1$ and $\delta_2 2$ are in $\mathbf{t}_3$. So, one of the words $\delta_2 i$, contains the square of a letter. This is impossible because the image of no letter does contain a square. In the same way, we show that δ_1 is empty. Hence, $u = \mu_3(u')$. By Proposition 3.1, u' is in $BST(\mathbf{t}_3)$.   ∎

Theorem 3.1. *The set $BST(\mathbf{t}_3)$ is given by:*

$$BST(\mathbf{t}_3) = \bigcup_{n \geq 0} \{\mu_3^n(0), \mu_3^n(1), \mu_3^n(2), \mu_3^n(01), \mu_3^n(12), \mu_3^n(20)\} \cup \{\varepsilon\}.$$

Proof: Let u be an element of $BST(\mathbf{t}_3)$ with length at least 3. By Proposition 3.3, it exists u' in $BST(\mathbf{t}_3)$ such that $u = \mu_3(u')$. Hence, to obtain the set $BST(\mathbf{t}_3)$, it suffices to find its elements of length at most 2, since the others can be obtained by applying successively μ_3. These factors are 0, 1, 2, 01, 12 and 20.   ∎

Corollary 3.1 *Let u be a factor of $\mathbf{t}_3$ which is right triprolongable. If $|u| = 3^k$ or $|u| = 2 \times 3^k$, $k \geq 0$, then u is left triprolongable.*

Proof: Let u be a factor of $\mathbf{t}_3$, right triprolongable and verifying $|u| = 3^k$, $k \geq 1$. Then, u can be decomposed in the form $\delta_1 \mu_3(v) \delta_2$ where v, $\delta_1, \delta_2 \in F(\mathbf{t}_3)$. The factor u being right triprolongable, δ_2 is the empty word. So, $u = \delta_1 \mu_3(v)$. We know that $|\delta_1| \leq 2$ and $|\mu_3(v)|$ is multiple of 3. The factor u being of length 3^k then δ_1 is the empty word. Hence, $u = \mu_3(v)$ where v is a right triprolongable factor of length 3^{k-1}. By the same process, the factor v can be written $v = \mu_3(v')$, where v' is a right triprolongable factor of length 3^{k-2}. In a successive way, we succeed in $u = \mu_3^k(i)$, $i \in \mathcal{A}_3$. With Theorem 3.1, we conclude that u is left triprolongable. We proceed in the same way for the factors of length 2×3^k. ∎

Proposition 3.3. *Let u be in $BST(\mathbf{t}_3)$. Then, it exists a unique letter i in $\mathcal{A}_3$ such that iu (resp. ui) is right (resp. left) triprolongable.*

Proof: Let us construct the set $F(\mathbf{t}_3) \cap (\mathcal{A}_3 v \mathcal{A}_3)$ where $v \in BST(\mathbf{t}_3)$. We give those for which $|v| \leq 3$, and by induction we show that for those of upper length, the extensions respect the unicity. We have, for $i \in \mathcal{A}_3$:

$F(\mathbf{t}_3) \cap (\mathcal{A}_3 i \mathcal{A}_3) = \{0i1, 1i1, 2i0, 2i1, 2i2\}$;
$F(\mathbf{t}_3) \cap (\mathcal{A}_3 i(i+1)\mathcal{A}_3) = \{0i(i+1)2, 1i(i+1)2, 2i(i+1)0, 2i(i+1)1, 2i(i+1)2\}$;
$F(\mathbf{t}_3) \cap (\mathcal{A}_3 \mu_3(i)\mathcal{A}_3) = \{0\mu_3(i)1, 1\mu_3(i)0, 1\mu_3(i)1, 1\mu_3(i)2, 2\mu_3(i)0, 2\mu_3(i)1\}$;

where $i + 1$ is taken modulo 3.

Let us take a factor $v = \mu_3^n(0)$ and suppose that the set $F(\mathbf{t}_3) \cap \mathcal{A}_3 \mu_3^n(0)$ contains a single right triprolongable factor. Even if it means changing letter, let us take $0\mu_3^n(0)$ this factor. So, By Proposition 3.1, $2\mu_3^{n+1}(0)$ is a right triprolongable factor of $\mathbf{t}_3$. Let us verify that it is the only one. Suppose $0\mu_3^{n+1}(0)$ is right triprolongable. Then, $1\mu_3^n(0)$ is right triprolongable. This contradicts the recursion hypothesis. We proceed in the same way for the other factors. ∎

Proposition 3.4. *Let u be a factor of $\mathbf{t}_3$, right triprolongable. If u is left special, then it is left triprolongable.*

Proof: Let u be a factor of $\mathbf{t}_3$, right triprolongable and left special. Then, u can be written in the form $\delta_1 \mu_3(v_1)\delta_2$. As the factor u is right triprolongable, δ_2 is empty by Proposition 3.2. Furthermore, as u is left special, δ_1 is empty; because otherwise, δ_1 would be proper suffix of the image of some letter and by this fact u would be extended on left in a unique way. So, u can be synchronized in the form $u = \mu_3(v_1)$, where v_1 is a factor of $\mathbf{t}_3$. Since the morphism μ_3 is marked, then v_1 is left special. Moreover, it is right triprolongable by Proposition 3.1. Thus, v_1 can be synchronized in the form $v_1 = \mu_3(v_2)$, $v_2 \in F(\mathbf{t}_3)$. In a successive way, we succeed in $u = \mu_3^k(v_k)$ with $k \geq 0$ and v_k a right triprolongable factor, left special and of length at most 2. Therefore, v_k is left triprolongable by Theorem 3.1. ∎

Proposition 3.5. *For all positive natural n, $\mathbf{t}_3$ admits exactly three right (resp. left) triprolongable factors of length n.*

Proof: It is known that 0, 1 and 2 are the right triprolongable factors of length 1. Let us show that any right triprolongable factor of length n is suffix of a unique right triprolongable factor of length $n + 1$.

Let w be a right triprolongable factor of length n. If w admits a unique extension a on left, then aw is a right triprolongable factor since $\mathbf{t}_3$ is recurrent. If it admits at least two left extensions, then w is in $BST(\mathbf{t})$ by Proposition 3.4 and only one of its left extensions is right triprolongable by Proposition 3.3.

Thus, the number of right triprolongable factors of length $n + 1$ is equal to the number of right triprolongable factors of length n in t_3. In the same way, we treat the case of the left triprolongable factors. ∎

The following remark is a consequence of Proposition 3.5.

Remark 3.1. *Let u be a right (resp. left) triprolongable factor of t_3. If the length of u is $3k$, $k \geq 1$, then it exists a right (resp. left) triprolongable factor v of t_3 such that $u = \mu_3(v)$.*

Proposition 3.6. *For all positive natural n, the right (resp. left) triprolongable factors of length n begin (resp. end) with different letters.*

Proof: We proceed by induction on n. Suppose all the right triprolongable factors of t_3 of length at most n begin with different letters. Let u_1 and u_2 be two factors of t_3, right triprolongable of length n. We distinguish the two following cases.

Case 1: n is multiple of 3. Then, it exists some factors v_1 and v_2 of t_3 such that $u_1 = \mu_3(v_1)$ and $u_2 = \mu_3(v_2)$. Suppose there exists a letter a of $\mathcal{A}_3$ such that au_1 and au_2 are right triprolongable. Even if it means changing letter, let us take $a = 0$. Thus, the factors $120u_1$ and $120u_2$ are right triprolongable in t_3. These factors can be written respectively $\mu_3(1v_1)$ and $\mu_3(1v_2)$. By Proposition 3.1, $1v_1$ and $1v_2$ are right triprolongable factors. This fact contradicts the hypothesis of induction since $1v_1$ and $1v_2$ are of length lower than n.

Case 2: $n - 1$ is multiple of 3. Then, there exist some factors v_1 and v_2 of t_3, right triprolongable such that $u_1 = i\mu_3(v_1)$ and $u_2 = j\mu_3(v_2)$. As i and j are suffix of images of letters, they have each a unique left extension. Since they are different by hypothesis, the extensions are different.

Case 3: $n - 2$ is multiple of 3. We proceed similarly like previous cases. ∎

Theorem 3.2. *The word t_3 is 2-balanced.*

Proof: Let u_1 and u_2 be two factors of length $n \geq 7$ of t_3. Then, u_1 and u_2 can be synchronized in a unique way in the forms $u_1 = \delta_1 \mu_3(v_1)\delta_2$ and $u_2 = \delta_1' \mu_3(v_2)\delta_2'$, $v_1, v_2, \delta_1, \delta_2, \delta_1', \delta_2' \in F(t_3)$. Let us put $\alpha_i = |\delta_1|_i + |\delta_2|_i$, $\beta_i = |\delta_1'|_i + |\delta_2'|_i$, for all $i \in \mathcal{A}_3$. Consider the following cases.

Case 1: n is multiple of 3. Then, u_1 (resp. u_2) can be written uniquely in the form $\mu_3(v)$, $ij\mu_3(v)k$ or $i\mu_3(v)jk$, with i, j, $k \in \mathcal{A}_3$ and $v \in F(t_3)$. Consider the different forms taken by u_1 and u_2.

- Suppose $u_1 = \mu_3(v_1)$ and $u_2 = \mu_3(v_2)$. Then, we have $\psi(u_1) = \psi(u_2)$ and we have:

$$||u_1|_i - |u_2|_i| = 0,$$

  for any letter i.

- Suppose $u_1 = \mu_3(v_1)$ and $u_2 = i\mu_3(v_2)jk$. Write u_1 in the form $\mu_3(v_1')\mu_3(a)$, $a \in \mathcal{A}_3$. Thus, we have $|v_1'| = |v_2|$. As the letters have the same number of occurrences in image of each letter, we have $|\mu_3(a)|_i = 1$, for all $i \in \mathcal{A}_3$. Moreover, $\beta_i \leq 2$, for all $i \in \mathcal{A}_3$ since jk is the prefix of the image of some letter. Thus,

$$||u_1|_i - |u_2|_i| = ||\mu_3(a)|_i - \beta_i| \leq 1.$$

- Suppose $u_1 = ij\mu_3(v_1)k$ and $u_2 = l\mu_3(v_2)mn$, i, j, k, l, m, $n \in \mathcal{A}_3$. As previously, one verifies that $\alpha_i, \beta_i \leq 2$, for all $i \in \mathcal{A}_3$. Thus,

$$||u_1|_i - |u_2|_i| = |\alpha_i - \beta_i| \leq 2.$$

By taking $u_1 = 101212$ and $u_2 = 010120$, we observe that the bound 2 is reached.

Case 2: $n - 1$ is multiple of 3. Then, u_1 (resp. u_2) is of the form $i\mu_3(v)$, $\mu_3(v)k$ or $ij\mu_3(v)kl$, i, j, k, $l \in \mathcal{A}_3$, $v \in F(\mathbf{t}_3)$.

- Suppose $u_1 = i\mu_3(v_1)$ and $u_2 = \mu_3(v_2)j$. Then, we have $|v_1| = |v_2|$. So $|\alpha_i - \beta_i| \leq 1$, for all $i \in \mathcal{A}_3$.

- Suppose $u_1 = ij\mu_3(v_1)kl$ and $u_2 = i'j'\mu_3(v_2)k'l'$, where ij and $i'j'$ (resp. kl and $k'l'$) are suffix (resp. prefix) of images of letters. Then, note that $(i, k) \neq (j, l)$ and $(i', k') \neq (j', l')$. Thus, $|v_1| = |v_2|$. By analogy with the previous case, one verifies that $\alpha_i, \beta_i \leq 2$. Therefore,

$$||u_1|_i - |u_2|_i| = |\alpha_i - \beta_i| \leq 2,$$

for all $i \in \mathcal{A}_3$. By taking $u_1 = 01\mu_3(12)01$ and $u_2 = 20\mu_3(01)20$ we observe that the bound 2 is reached.

- Suppose $u_1 = i'\mu_3(v_1)$ and $u_2 = ij\mu_3(v_2)kl$. Then, we write u_1 in the form $i'\mu_3(v_1')\mu_3(a)$, $a \in \mathcal{A}_3$ and $v' \in F(\mathbf{t}_3)$. Thus, $|v_1'| = |v_2|$ and $\alpha_i, \beta_i \leq 2$. So

$$||u_1|_i - |u_2|_i| = |\alpha_i - \beta_i| \leq 2.$$

Case 3: $n - 2$ is multiple of 3. Suppose u_1 (resp. u_2) can be written in the form $ij\mu_3(v_1)$, $i\mu_3(v_1)k$ or $\mu_3(v_1)kl$ (resp. $ij\mu_3(v_2)$, $i\mu_3(v_2)k$ or $\mu_3(v_2)kl$). Then, we have $|v_1| = |v_2|$. In a similar way as previous cases, one verifies that $|\alpha_i - \beta_i| \leq 2$, for $i \in \mathcal{A}_3$.   ∎

4   Abelian Complexity

In this section we give an explicit formula of the Abelian complexity function ρ^{ab} of $\mathbf{t}_3$. We show that the sequence $(\rho^{ab}(n))_{n \geq 2}$ of the word $\mathbf{t}_3$ is 3-periodic.

Proposition 4.1. *For all $k \geq 1$, $\rho^{ab}(3k) = 7$.*

Proof: Let u be a factor of $\mathbf{t}_3$ of length $3k$, $k \geq 1$. Then, u synchronizes in the form $\mu_3(v)$, $i\mu_3(v)jk$ or $ij\mu_3(v)k$ with $i, j, k \in \mathcal{A}_3$, $ij, jk \in \{01, 12, 20\}$ and $v \in F(\mathbf{t}_3)$. As u is chosen arbitrary one verifies that these three forms are taken by u. As the prefix $\mathbf{t}_{3[3k]}$ begins with the image of some letter, it is in the form $\mu_3(v)$. For the sequel, we note $\mathbf{t}_{3[3k]} = \mu_3(v_1)$. We have three cases to discuss.

Case 1: The factor u is in the form $\mu_3(v_2)$. Then, $|v_1| = |v_2|$ and so $\psi^{rel}(u) = (0, 0, 0)$.

Case 2: The factor u is in the form $i\mu_3(v_2)jk$. Then, we have:

$$\psi(u) = (|v_2| + |ijk|_0, |v_2| + |ijk|_1, |v_2| + |ijk|_2).$$

Let us show that the set of the values taken by ijk is

$$\{001, 012, 020, 101, 112, 120, 201, 212, 220\}.$$

By Proposition 3.5, for any integer $k \geq 1$, $\mathbf{t}_3$ possesses exactly 3 right triprolongeable factors of length $3k$. Let us denote by R_1, R_2 and R_3 the right triprolongeable factors of length $3k - 3$. As these factors begin with different letters, we can suppose, even if it means changing the indexes, that $0R_1$, $1R_2$ and $2R_3$ are the right triprolongeable factors of $\mathbf{t}_3$ of length $3k - 2$. Therefore, the words $0R_101$, $0R_112$, $0R_120$, $1R_201$, $1R_212$, $1R_220$, $2R_301$, $2R_312$ and $2R_320$ are factors of $\mathbf{t}_3$ of length $3k$. Hence, ijk browses the announced set. So, $\psi(ijk)$ takes all the values of the following set

$$\{(2, 1, 0), (1, 1, 1), (2, 0, 1), (1, 2, 0), (0, 2, 1), (0, 1, 2), (1, 0, 2)\}.$$

Write the prefix $\mathbf{t}_{3[n]}$ in the form $\mu_3(v_1')\mu_3(l)$, $l \in \mathcal{A}_3$. Then, $|v_1'| = |v_2|$ and $\psi(\mu_3(l)) = (1, 1, 1)$. Thus, for all the factors u of length $3k$, $\psi^{rel}(u) = \psi(ijk) - \psi(\mu_3(l))$ takes all the values of the set

$$\{(1, 0, -1), (0, 0, 0), (1, -1, 0), (0, 1, -1), (-1, 1, 0), (-1, 0, 1), (0, -1, 1)\}.$$

Case 3: The factor u is in the form $ij\mu_3(v_2)k$. Then

$$\psi(u) = (|v_2| + |ijk|_0, |v_2| + |ijk|_1, |v_2| + |ijk|_2).$$

By proceeding in a similar way as in the case 2 and by using the left triprolongable factors, we verify that the set of values taken by ijk is

$$\{010, 011, 012, 120, 121, 122, 200, 201, 202\}.$$

Consequently, for all the factors u satisfying these conditions, $\psi^{rel}(u)$ takes all the values of the set

$$\{(1, 0, -1), (0, 0, 0), (1, -1, 0), (0, 1, -1), (-1, 1, 0), (-1, 0, 1), (0, -1, 1)\}.$$

After all, we have:

$$\Psi_n^{rel}(\mathbf{t}_3) = \{(1, 0, -1), (0, 0, 0), (1, -1, 0), (0, 1, -1), (-1, 1, 0), (-1, 0, 1), (0, -1, 1)\}.$$

■

Proposition 4.2. *For all $k \geq 1$, $\rho^{ab}(3k+1) = 6$.*

Proof: Let u be a factor of $\mathbf{t}_3$ of length $3k + 1$, $k \geq 1$. Then, u synchronizes in the form $i\mu_3(v)$, $\mu_3(v)j$ or $ij\mu_3(v)kl$, i, j, k, $l \in \mathcal{A}_3$, ij, $kl \in \{01, 12, 20\}$ and $v \in F(\mathbf{t}_3)$. The prefix $\mathbf{t}_{3[3k+1]}$ is in the form $\mu_3(v_1)i$, $i \in \mathcal{A}_3$. We have:

Case 1: $i = 0$. Then, $\mathbf{t}_{3[3k+1]} = \mu_3(v_1)0$. Let us determine $\Psi^{rel}_{3k+1}(\mathbf{t}_3)$.

- Let v_2 be a factor of $\mathbf{t}_3$ such that $u = i\mu_3(v_2)$. By using the left triprolongable factors of length $3k$, we verify that the values taken by i are 0, 1 and 2. Consequently, $\psi^{rel}(u)$ takes all the values of $\{(0, 0, 0), (-1, 1, 0), (-1, 0, 1)\}$. In the same way, we verify that if $u = \mu_3(v_2)j$, $\psi^{rel}(u)$ browses all the elements of the set $\{(0, 0, 0), (-1, 1, 0), (-1, 0, 1)\}$.

- Let u be a factor of $\mathbf{t}_3$ with the form $u = ij\mu_3(v_2)kl$. We write $\mathbf{t}_{3[3k+1]}$ in the form $\mu_3(v'_1)\mu_3(m)0$, $m \in \mathcal{A}_3$. It is known that each factor of the form $ij\mu_3(v_2)kl$ is the left extension of a factor of the form $j\mu_3(v_2)kl$ whose the set of values taken by jkl is

$$\{001, 012, 020, 101, 112, 120, 201, 212, 220\}.$$

Thus, those taken by $ijkl$ is

$$\{2001, 2012, 2020, 0101, 0112, 0120, 1201, 1212, 1220\}.$$

So, $\psi^{rel}(u)$ browses all the elements of the set

$$\{(0, -1, 1), (-1, 1, 0), (0, 0, 0), (-2, 1, 1), (-1, 0, 1), (0, 1, -1)\}.$$

Finally, we get

$$\Psi^{rel}_{3k+1}(\mathbf{t}_3) = \{(0, -1, 1), (-1, 1, 0), (0, 0, 0), (-2, 1, 1), (-1, 0, 1), (0, 1, -1)\}.$$

Case 2: $i = 1$. Then, $\mathbf{t}_{3[3k+1]} = \mu_3(v_1)1$. Consider the different forms of u.

- Let u be a factor of $\mathbf{t}_3$ of the form $u = i\mu_3(v_2)$. As in the case 1, we verify that i takes the values 0, 1 and 2. Therefore, $\psi^{rel}(u)$ browses all the elements of $\{(0, 0, 0), (1, -1, 0), (0, -1, 1)\}$.

- Let u be a factor of $\mathbf{t}_3$ of the form $u = ij\mu_3(v_2)kl$. Then, we write $\mathbf{t}_{3[3k+1]}$ in the form $\mathbf{t}_{3[3k+1]} = \mu_3(v'_1)\mu_3(m)1$, $m \in \mathcal{A}_3$. By proceeding in a similar way as in the case 1, we verify that the set of values taken by $ijkl$ is

$$\{2001, 2012, 2020, 0101, 0112, 0120, 1201, 1212, 1220\}.$$

Thus, $\psi^{rel}(u)$ browses all of the elements of the set

$$\{(1, 0, -1), (1, -1, 0), (0, 0, 0), (-1, 0, 1), (0, -1, 1), (1, -2, 1)\}.$$

After all, we have:

$$\Psi^{rel}_{3k+1}(t_3) = \{(1, 0, -1), (1, -1, 0), (0, 0, 0), (-1, 0, 1), (0, -1, 1), (1, -2, 1)\}.$$

Case 3: $i = 2$. Then, $t_{3[3k+1]} = \mu_3(v_1)2$. By proceeding in a similar way as in the previous case we get:

$$\Psi^{rel}_{3k+1}(t_3) = \{(1, 1, -2), (0, 1, -1), (1, 0, -1), (-1, 1, 0), (0, 0, 0), (1, -1, 0)\}. \blacksquare$$

Proposition 4.3. *For all* $k \geq 1$, $\rho^{ab}(3k + 2) = 6$.

Proof: Let u be a factor of length $3k + 2$ of t_3, $k \geq 1$. Then, u can be written in the form $i\mu_3(v_2)j$, $ij\mu_3(v_2)$ or $\mu_3(v_2)kl$, $i, j, k, l \in \mathcal{A}_3$, $v_2 \in F(t_3)$. Otherwise, the prefix $t_{3[3k+2]}$ is in the form $\mu_3(v_1)ij$.

Case 1: $ij = 01$. Then, $t_{3[3k+1]} = \mu_3(v_1)01$. Let us determine the set $\Psi^{rel}_{3k+2}(t_3)$.

- Let u be a factor of t_3 of the form $i\mu_3(v_2)j$. Then, v_1 and v_2 have the same length. So, $\psi^{rel}(u) = \psi(ij) - \psi(01)$. With right triprolongable factors of length $k - 1$, we verify that the set of values taken by ij is $\{00, 01, 02, 10, 11, 12, 20, 21, 22\}$. So, $\psi^{rel}(u)$ takes all the values of the set

$$\{(1, -1, 0), (0, 0, 0), (0, -1, 1), (-1, 1, 0), (-1, 0, 1), (-1, -1, 2)\}.$$

- Let u be a factor of t_3 of the form $ij\mu_3(v_2)$. Then, v_1 and v_2 have the same length. So, $\psi^{rel}(u) = \psi(ij) - \psi(01)$. The factor ij is the suffix of the image of a letter. It takes the values 01, 12 and 20. Thus, $\psi^{rel}(u)$ takes all the values of the set

$$\{(0, 0, 0), (0, -1, 1), (-1, 0, 1)\}.$$

In a same way, we verify that if u has the form $\mu_3(v_2)kl$, $\psi^{rel}(u)$ takes all the values of the set $\{(0, 0, 0), (0, -1, 1), (-1, 0, 1)\}$.
Finally, we get:

$$\Psi^{rel}_{3k+2}(t_3) = \{(1, -1, 0), (0, 0, 0), (0, -1, 1), (-1, 1, 0), (-1, 0, 1), (-1, -1, 2)\}$$

Case 2: $ij = 12$. Then, $t_{3[3k+2]} = \mu_3(v_1)12$. Let us determine the set $\Psi^{rel}_{3k+2}(t_3)$.

- Let u be a factor of t_3 of the form $u = i\mu_3(v_2)j$. As in the previous case the set of values taken by ij is $\{00, 01, 02, 10, 11, 12, 20, 21, 22\}$. Therefore, $\psi^{rel}(u)$ takes all the values of the set

$$\{(2, -1, -1), (1, 0, -1), (0, 1, -1), (0, 0, 0), (0, -1, 1), (1, -1, 0)\}.$$

- Let u be a factor of t_3 of the form $u = ij\mu_3(v_2)$. Then, we show as in the case 1 that $\psi^{rel}(u)$ takes all the values of the set $\{(1, 0, -1), (1, -1, 0), (0, 0, 0)\}$.
After all, we have

$$\Psi^{rel}_{3k+2}(t_3) = \{(2, -1, -1), (1, 0, -1), (0, 1, -1), (0, 0, 0), (0, -1, 1), (1, -1, 0)\}.$$

Case 3: $ij = 20$. Then, $\mathbf{t}_{3[3k+2]} = \mu_3(v_1)20$. As in the previous cases we verify that:

$$\Psi_{3k+2}^{rel}(\mathbf{t}_3) = \{(1, 0, -1), (0, 1, -1), (0, 0, 0), (-1, 2, -1), (-1, 1, 0), (-1, 0, 1)\} \blacksquare$$

Theorem 4.1. *The Abelian complexity function of* $\mathbf{t}_3$ *is given by:*

$$\rho_{\mathbf{t}_3}^{ab}(n) = \begin{cases} 1 & si \ n = 0 \\ 3 & si \ n = 1 \\ 7 & si \ n = 3k, \ k \geq 1 \\ 6 & sinon \end{cases}$$

Proof: The result follows from Propositions 4.1, 4.2 and 4.3. $\blacksquare$

The ternary Thue-Morse word $\mathbf{t}_3$ is 3-automatic. Its Abelian complexity is the eventually periodic word $(\rho^{ab}(n))_{n \geq 0} = 136(766)^{\omega}$. Thus, we note that the word $\mathbf{t}_3$ responds to the conjecture of Parreau, Rigo, Rowland and Vandomme: *Any k-automatic word admits a l-Abelian complexity function which is k-automatic.* The reader can find this conjecture and more information on the concepts in [15].

References

1. Allouche, J.P., Arnold, A., Berstel, J., Brlek, S., Jockush, W., Plouffe, S., Sagan, B.E.: A relative of the thue-morse sequence. Disc. Math. **139**, 455–461 (1995)
2. Allouche, J.P., Peyrière, J., Wen, Z.-X., Wen, Z.-Y.: Hankel determinants of the thue-morse sequense. Ann. Inst. Fourier **48**(1), 1–27 (1998)
3. Allouche, J.P., Shallit, J.: The Ubiquitous Prouhet-Thue-Morse Sequence. In: Ding, C., Helleseth, T., Niederreiter, H. (eds.) Sequences and their Applications. Discrete Mathematics and Theoretical Computer Science, pp. 1–16. Springer, London (1999)
4. Balková, L., Břinda, K., Turek, O.: Abelian complexity of infinite words associated with non-simple parry number. Theor. Comput. Sci. **412**, 6252–6260 (2011)
5. Cassaigne, J.: Facteurs spéciaux. Bull. Belg. Math. **4**, 67–88 (1997)
6. Cassaigne, J., Kaboré, I.: Abelian complexity and frequencies of letters in infinite words. Int. J. Found. Comput. Sci. **27**(05), 631–649 (2016)
7. Cassaigne, J., Richomme, G., Saari, K., Zamboni, L.Q.: Avoiding abelian powers in binary wordswith bounded abelian complexity. Int. J. Found. Comput. Sci. **22**(2011), 905–920 (2011)
8. Chen, J., Lü, X., Wu, W.: On the k-abelian complexity of the cantor sequence. arXiv: 1703.04063 (2017)
9. Coven, E.M., Hedlund, G.A.: Sequences with minimal block growth. Math. Syst. Theo. **7**, 138–153 (1973)
10. Curie, J., Rampersad, N.: Recurrent words with constant abelian complexity. Adv. Appl. Math. **47**, 116–124 (2011)
11. Gottschalk, W.H.: Substitution on minimal sets. Trans. Amer. Math. Soc. **109**, 467–491 (1963)
12. Lü, X., Chen, J., Wen, Z., Wu, W.: On the abelian complexity of the rudin-shapiro sequence. J. Math. Anal. Appl. **451**, 822–838 (2017)

13. Madill, B., Rampersad, N.: The abelian complexity of the paperfolding word. Disc. Math. **313**, 831–838 (2013)
14. Massé, A.B., Brlek, S., Garon, A., Labbé, S.: Combinatorial properties of f-palindromes in the thue-morse sequence. Pure Math. Appl. **19**, 39–52 (2008)
15. Parreau, A., Rigo, M., Rowland, E., Vandomme, E.: A new approach to the $2-$regularity of the $l-$abelian complexity of $2-$automatic sequences. Electron. J. Combin. **22**, 1–27 (2015)
16. Richomme, G., Saari, K., Zamboni, L.Q.: Balance and abelian complexity of the tribonacci word. Adv. Appl. Math. **45**, 212–231 (2010)
17. Richomme, G., Saari, K., Zamboni, L.Q.: Abelian complexity in minimal subshifts. J. London Math. Soc. **83**, 79–95 (2011)
18. Štarosta, S.: Generalised thue-morse word and palindromic richness. Kybernetika **48**(3), 361–370 (2012)
19. Thue, A.: Über unendliche zeichenreihen. Norske Vid. Selsk. Skr. Mat. Nat. Kl. **7**, 1–22 (1906)
20. Thue, A.: Über die gegenseilige lage gleicher teile gewisser zeichenreihen. Norske Vid. Selsk. Skr. Mat. Nat. Kl. **1**, 139–158 (1912)
21. Turek, O.: Balance and abelian complexity of a certain class of infinite ternary words. RAIRO Theor. Inf. Appl. **44**, 313–337 (2010)
22. Turek, O.: Abelian complexity and abelian co-decomposition. Theor. Comput. Sci. **469**, 77–91 (2013)
23. Turek, O.: Abelian complexity of the tribonacci word. J. Integer Seq. **18**, 212–231 (2015)

A Set of Sequences of Complexity $2n + 1$

J. Cassaigne[1,2,3], S. Labbé[1,2,3]($\boxtimes$), and J. Leroy[1,2,3]

[1] Institut de mathématiques de Marseille, CNRS UMR 7373, Campus de Luminy, Case 907, 13288 Marseille Cedex 09, France
julien.cassaigne@math.cnrs.fr, sebastien.labbe@labri.fr
[2] CNRS, LaBRI, UMR 5800, 33400 Talence, France
j.leroy@ulg.ac.be
[3] Institut de mathématique, Université de Liège, Allée de la découverte 12 (B37), 4000 Liège, Belgium

Abstract. We prove the existence of a ternary sequence of factor complexity $2n + 1$ for any given vector of rationally independent letter frequencies. Such sequences are constructed from an infinite product of two substitutions according to a particular Multidimensional Continued Fraction algorithm. We show that this algorithm is conjugate to a well-known one, the Selmer algorithm. Experimentations (Baldwin, 1992) suggest that their second Lyapunov exponent is negative which presages finite balance properties.

Keywords: Substitutions · Factor complexity · Selmer · Continued fraction · Bispecial

1 Introduction

Words of complexity $2n+1$ were considered in [2] with the condition that there is exactly one left and one right special factor of each length. These words are called Arnoux-Rauzy sequences and are a generalization of Sturmian sequences on a ternary alphabet. It is known that the frequencies of any Arnoux-Rauzy word are well defined and belong to the Rauzy Gasket [3], a fractal set of Lebesgue measure zero. Thus the above condition on the number of special factors is very restrictive for the possible letter frequencies.

Sequences of complexity $p(n) \leq 2n+1$ include Arnoux-Rauzy words, codings of interval exchange transformations and more [12]. For any given letter frequencies one can construct sequences of factor complexity $2n + 1$ by the coding of a 3-interval exchange transformation. It is known that these sequences are unbalanced [14]. Thus the question of finding balanced ternary sequences of factor complexity $2n+1$ for all letter frequencies remains. This article intends to give a positive answer to this question for almost all vectors of letter frequencies (with respect to Lebesgue measure).

J. Leroy—FNRS post-doctoral fellow.

S. Brlek et al. (Eds.): WORDS 2017, LNCS 10432, pp. 144–156, 2017.
DOI: 10.1007/978-3-319-66396-8_14

In recent years, multidimensional continued fraction algorithms were used to obtain ternary balanced sequences with low factor complexity for any given letter frequency vector. Indeed the Brun algorithm leads to balanced sequences [10] and it was shown that the Arnoux-Rauzy-Poincaré algorithm leads to sequences of factor complexity $p(n) \leq \frac{5}{2}n + 1$ [5].

In 2015, the first author introduced a new Multidimensional Continued Fraction algorithm [9] based on the study of Rauzy graphs. In this work, we formalize the algorithm, its matrices, substitutions and associated cocycles and S-adic words. We show that S-adic words obtained from these substitutions have complexity $2n + 1$. We also show that the algorithm is conjugate to the Selmer algorithm, a well-known Multidimensional Continued Fraction algorithm. We believe that almost all sequences generated by the algorithm are balanced.

2   A Bidimensional Continued Fraction Algorithm

On $\Lambda = \mathbb{R}^3_{\geq 0}$, the bidimensional continued fraction algorithm introduced by the first author [9] is

$$F_C(x_1, x_2, x_3) = \begin{cases} (x_1 - x_3, x_3, x_2), & \text{if } x_1 \geq x_3; \\ (x_2, x_1, x_3 - x_1), & \text{if } x_1 < x_3. \end{cases}$$

More information on Multidimensional Continued Fraction Algorithms can be found in [8,13].

2.1   The Matrices

Alternatively, the map F_C can be defined by associating nonnegative matrices to each part of a partition of Λ into $\Lambda_1 \cup \Lambda_2$ where

$$\Lambda_1 = \{(x_1, x_2, x_3) \in \Lambda \mid x_1 \geq x_3\},$$
$$\Lambda_2 = \{(x_1, x_2, x_3) \in \Lambda \mid x_1 < x_3\}.$$

The matrices are given by the rule $M(\mathbf{x}) = C_i$ if and only if $\mathbf{x} \in \Lambda_i$ where

$$C_1 = \begin{pmatrix} 1 & 1 & 0 \\ 0 & 0 & 1 \\ 0 & 1 & 0 \end{pmatrix} \quad \text{and} \quad C_2 = \begin{pmatrix} 0 & 1 & 0 \\ 1 & 0 & 0 \\ 0 & 1 & 1 \end{pmatrix}.$$

The map F_C on Λ and the projective map f_C on $\Delta = \{\mathbf{x} \in \Lambda \mid \|\mathbf{x}\|_1 = 1\}$ are then defined as:

$$F_C(\mathbf{x}) = M(\mathbf{x})^{-1}\mathbf{x} \quad \text{and} \quad f_C(\mathbf{x}) = \frac{F_C(\mathbf{x})}{\|F_C(\mathbf{x})\|_1}.$$

Many of its properties can be found in [11] and the density function of the invariant measure of f_C was computed in [1].

2.2   The Cocycle

The algorithm F_C defines a *cocycle* $M_n : \Lambda \to SL(3, \mathbb{Z})$ by

$$M_0(\mathbf{x}) = I \quad \text{and} \quad M_n(\mathbf{x}) = M(\mathbf{x})M(F_C\mathbf{x})M(F_C^2\mathbf{x}) \cdots M(F_C^{n-1}\mathbf{x})$$

satisfying the cocycle property $M_{n+m}(\mathbf{x}) = M_n(\mathbf{x}) \cdot M_m(F_C^n\mathbf{x})$.

For example starting with $\mathbf{x} = (1, e, \pi)^T$, the first iterates (approximate to the nearest hundredth) under F_C are

$$\begin{pmatrix} 1.00 \\ 2.72 \\ 3.14 \end{pmatrix} \xrightarrow{F_C} \begin{pmatrix} 2.72 \\ 1.00 \\ 2.14 \end{pmatrix} \xrightarrow{F_C} \begin{pmatrix} 0.58 \\ 2.14 \\ 1.00 \end{pmatrix} \xrightarrow{F_C} \begin{pmatrix} 2.14 \\ 0.58 \\ 0.42 \end{pmatrix} \xrightarrow{F_C} \begin{pmatrix} 1.72 \\ 0.42 \\ 0.58 \end{pmatrix} \xrightarrow{F_C} \begin{pmatrix} 1.14 \\ 0.58 \\ 0.42 \end{pmatrix}$$

The associated cocycle at $\mathbf{x} = (1, e, \pi)^T$ when $n = 5$ is

$$\begin{aligned} M_5(\mathbf{x}) &= M(\mathbf{x})M(F_C\mathbf{x})M(F_C^2\mathbf{x})M(F_C^3\mathbf{x})M(F_C^4\mathbf{x}) \\ &= C_2 C_1 C_2 C_1 C_1 \\ &= \begin{pmatrix} 0\ 1\ 0 \\ 1\ 0\ 0 \\ 0\ 1\ 1 \end{pmatrix} \begin{pmatrix} 1\ 1\ 0 \\ 0\ 0\ 1 \\ 0\ 1\ 0 \end{pmatrix} \begin{pmatrix} 0\ 1\ 0 \\ 1\ 0\ 0 \\ 0\ 1\ 1 \end{pmatrix} \begin{pmatrix} 1\ 1\ 0 \\ 0\ 0\ 1 \\ 0\ 1\ 0 \end{pmatrix} \begin{pmatrix} 1\ 1\ 0 \\ 0\ 0\ 1 \\ 0\ 1\ 0 \end{pmatrix} = \begin{pmatrix} 0\ 1\ 1 \\ 1\ 2\ 1 \\ 1\ 2\ 2 \end{pmatrix}. \end{aligned}$$

2.3   The Substitutions

Let $\mathcal{A} = \{1, 2, 3\}$. The substitutions on $\mathcal{A}^*$ are given by the rule $\sigma(\mathbf{x}) = c_i$ if and only if $\mathbf{x} \in \Lambda_i$ for $i = 1, 2$ where

$$c_1 = \begin{cases} 1 \mapsto 1 \\ 2 \mapsto 13 \\ 3 \mapsto 2 \end{cases} \quad \text{and} \quad c_2 = \begin{cases} 1 \mapsto 2 \\ 2 \mapsto 13 \\ 3 \mapsto 3 \end{cases}$$

One may check that C_i is the incidence matrix of c_i for $i = 1, 2$. For any word $w \in \mathcal{A}^*$, we denote $\vec{w} = (|w|_1, |w|_2, |w|_3) \in \mathbb{N}^3$ where $|w|_i$ means the number of occurrences of the letter i in w. Therefore, for all $\mathbf{x} \in \Lambda$, $\sigma(\mathbf{x}) : \mathcal{A}^* \to \mathcal{A}^*$ is a monoid morphism such that its incidence matrix is $M(\mathbf{x})$, i.e., $\overrightarrow{\sigma(\mathbf{x})(w)} = M(\mathbf{x}) \cdot \vec{w}$.

2.4   S-adic Words

Let S be a set of morphisms. A word $\mathbf{w}$ is said to be *S-adic* if there is a sequence $\mathbf{s} = (\tau_n : A_{n+1}^* \to A_n^*)_{n \in \mathbb{N}} \in S^{\mathbb{N}}$ and a sequence $\mathbf{a} = (a_n) \in \prod_{n \in \mathbb{N}} A_n$ such that $\mathbf{w} = \lim_{n \to +\infty} \tau_0 \tau_1 \cdots \tau_{n-1}(a_n)$. The pair $(\mathbf{s}, \mathbf{a})$ is called an *S-adic representation* of $\mathbf{w}$ and the sequence $\mathbf{s}$ a *directive sequence* of $\mathbf{w}$. The S-adic representation is said to be *primitive* whenever the directive sequence $\mathbf{s}$ is primitive, i.e., for all $r \geq 0$, there exists $r' > r$ such that all letters of A_r occur in all images $\tau_r \tau_{r+1} \cdots \tau_{r'-1}(a)$, $a \in A_{r'}$. Observe that if $\mathbf{w}$ has a primitive S-adic representation, then $\mathbf{w}$ is uniformly recurrent. For all n, we set $\mathbf{w}^{(n)} = \lim_{m \to +\infty} \tau_n \tau_{n+1} \cdots \tau_{m-1}(a_m)$.

2.5 S-adic Words Associated with the Algorithm F_C

The algorithm F_C defines the function $\sigma_n : \Lambda \to \text{End}(\mathcal{A}^*)$, $\sigma_n(\mathbf{x}) = \sigma(F_C^n \mathbf{x})$
When the sequence $(\sigma_n(\mathbf{x}))_{n \in \mathbb{N}}$ contains infinitely many occurrences of c_1 and
c_2, this defines a C-adic word, $C = \{c_1, c_2\}$,

$$W(\mathbf{x}) = \lim_{n \to \infty} \sigma_0(\mathbf{x})\sigma_1(\mathbf{x}) \cdots \sigma_n(\mathbf{x})(1).$$

Indeed, let $w_n = \sigma_0(\mathbf{x}) \cdots \sigma_n(\mathbf{x})(1)$. As c_1 and c_2 occur infinitely often, there
exist infinitely many indices m such that $\sigma_{m+1}(\mathbf{x}) = c_1$ and $\sigma_{m+2}(\mathbf{x}) = c_2$. For
all $n \geq m + 2$, let $z = \sigma_{m+3}(\mathbf{x}) \cdots \sigma_n(\mathbf{x})(1)$. Since $\{1, 2\}\mathcal{A}^*$ is stable under both
c_1 and c_2, we have $z \in \{1, 2\}\mathcal{A}^*$, so that $c_1 c_2(z) \in \{13, 12\}\mathcal{A}^*$. Then 1 is a proper
prefix of $c_1 c_2(z) = \sigma_{m+1}(\mathbf{x}) \cdots \sigma_n(\mathbf{x})(1)$, and therefore w_m is a proper prefix of
w_n. It follows that the limit of (w_n) exists.
    For example, using vector $\mathbf{x} = (1, e, \pi)^T$, we have

$$\sigma(\mathbf{x})\sigma(F_C \mathbf{x})\sigma(F_C^2 \mathbf{x})\sigma(F_C^3 \mathbf{x})\sigma(F_C^4 \mathbf{x}) = c_2 c_1 c_2 c_1 c_1 = \begin{cases} 1 \mapsto 23 \\ 2 \mapsto 23213 \\ 3 \mapsto 2313 \end{cases},$$

whose incidence matrix is $M_5(\mathbf{x})$. The associated infinite C-adic word is

$$W(\mathbf{x}) = 2323213232323132323213232321323231323232 \cdots .$$

Lemma 1. *Let* $\mathbf{x} \in \Delta$. *The following conditions are equivalent.*

(i)   the entries of $\mathbf{x}$ *are rationally independent,*
(ii) the directive sequence of $W(\mathbf{x})$ *is primitive,*
(iii) the directive sequence of $W(\mathbf{x})$ *does not belong to* $C^*\{c_1^2, c_2^2\}^\omega$.

Furthermore, the vector of letter frequencies of 1, 2 *and* 3 *in* $W(\mathbf{x})$ *is* $\mathbf{x}$.

Proof. Let us first prove that (ii) and (iii) are equivalent. Assume that $\mathbf{s} = (\tau_n) \in$
$C^*\{c_1^2, c_2^2\}^\omega$. Then there exists $r \in \mathbb{N}$ such that for all $i \in \mathbb{N}$, $\tau_{r+2i} = \tau_{r+2i+1}$,
and $\tau_{r+2i}\tau_{r+2i+1}$ is either c_1^2 or c_2^2. Observe that $c_1^2(1) = 1$, $c_2^2(3) = 3$, and
$c_1^2(3) = c_2^2(1) = 13$. Let $r' > r$. If $r' - r$ is even, then $\tau_r \tau_{r+1} \ldots \tau_{r'-1}(1)$ does not
contain the letter 2. If $r' - r$ is odd, then $\tau_r \tau_{r+1} \ldots \tau_{r'-1}(2)$ does not contain the
letter 2. Therefore the directive sequence $\mathbf{s}$ is not primitive.
    Conversely, if $\mathbf{s} \notin C^*\{c_1^2, c_2^2\}^\omega$, then $\mathbf{s}$ contains infinitely many occurrences of
words in $\{c_1 c_2^{2i+1} c_1^j c_2^k c_1^l c_2^m, c_2 c_1^{2i+1} c_2^j c_1^k c_2^l c_1^m : i \in \mathbb{N}, j, k, l, m \in \mathbb{N}\setminus\{0\}\}$. It can be
checked that all the matrices of these substitutions have positive entries, so that
$\mathbf{s}$ is primitive.
    Let us now assume that (iii) does not hold. Then, as above, there exists
$r \in \mathbb{N}$ such that for all $i \in \mathbb{N}$, $\tau_{r+2i} = \tau_{r+2i+1}$. Note that, if $\mathbf{y} = (y_1, y_2, y_3)$, then
$C_1^{-2}\mathbf{y} = (y_1 - y_2 - y_3, y_2, y_3)$ and $C_2^{-2}\mathbf{y} = (y_1, y_2, y_3 - y_1 - y_2)$. In both cases,
the middle entry is unchanged, and the sum of the two other entries decreases
by at least y_2. Let $F_C^r(\mathbf{x}) = (y_1, y_2, y_3)$ and $F_C^{r+2i}(\mathbf{x}) = (z_1, z_2, z_3)$. Then $z_2 = y_2$

and $z_1 + z_3 \leq y_1 + y_3 - iy_2$. This is possible for all i only if $y_2 = 0$, and then $\ell' F_C^r(\mathbf{x}) = 0$, where ℓ' is the row vector $\ell' = (0,1,0)$. Then $\ell\mathbf{x} = 0$ where $\ell = \ell' M(F_C^{r-1}(\mathbf{x}))^{-1} \dots M(\mathbf{x})^{-1}$ is a nonzero integer row vector, showing that the entries of $\mathbf{x}$ are rationally dependent.

Finally, let us assume that (iii) holds and (i) does not hold. Observe first that, if $F_C^r(\mathbf{x})$ has a zero entry for some r, then either $F_C^r(\mathbf{x})$ or $F_C^{r+1}(\mathbf{x})$ has a zero middle entry, and from this point on the directive sequence can be factored over $\{c_1^2, c_2^2\}$, contradicting (iii). From now on we assume that all entries of $F_C^n(\mathbf{x})$ are positive for all n

Let ℓ_0 be a nonzero integer row vector such that $\ell_0\mathbf{x} = 0$. The directive sequence can be factored over $\{c_1 c_2^k c_1, c_2 c_1^k c_2 : k \in \mathbb{N}\}$. Let us consider the sequence (n_m) such that $n_0 = 0$ and $\tau_{n_m} \dots \tau_{n_{m+1}-1}$ is in this set for all $m \in \mathbb{N}$. Let $\ell_m = \ell_0 M(\mathbf{x}) \dots M(F_C^{n_m-1}(\mathbf{x}))$. Then ℓ_m is a nonzero integer row vector such that $\ell_m F_C^{n_m}(\mathbf{x}) = 0$, and ℓ_{m+1} is either $\ell_m C_1 C_2^k C_1$ or $\ell_m C_2 C_1^k C_2$ for some k.

Assume that $\ell_m = (a, b, c)$. Then ℓ_{m+1} is one of

$$\ell_m C_1 C_2^{2k} C_1 = (a', a'+b, a'+c) \text{ with } a' = a + kb,$$
$$\ell_m C_1 C_2^{2k+1} C_1 = (a'-b, a', a'-c) \text{ with } a' = a + (k+1)b + c,$$
$$\ell_m C_2 C_1^{2k} C_2 = (c'+a, c'+b, c') \text{ with } c' = c + kb,$$
$$\ell_m C_2 C_1^{2k+1} C_2 = (c'-a, c', c'-b) \text{ with } c' = c + (k+1)b + a.$$

Define D_m as the difference between the maximum and the minimum entry of ℓ_m. Note that, as $F_C^{n_m}(\mathbf{x})$ has positive entries, the maximum entry of ℓ_m is positive and the minimum entry is negative. Then $D_m = \max(|b-a|, |c-b|, |c-a|)$. In the first two cases $D_{m+1} = \max(|b|, |c|, |c-b|)$. If a is (inclusively) between b and c, which must then have opposite signs, then $D_{m+1} = D_m = |c-b|$. Otherwise $D_{m+1} < D_m$. Similarly, in the other two cases $D_{m+1} = \max(|a|, |b|, |b-a|)$, and $D_{m+1} = D_m$ if c is inclusively between a and b, while $D_{m+1} < D_m$ otherwise.

The sequence of positive integers (D_m) is non-increasing. To reach a contradiction, we need to show that it decreases infinitely often.

If for large enough m all transitions between ℓ_m and ℓ_{m+1} are of the first type, then (iii) is not satisfied. Similarly, if for large enough m all transitions are of the third type, then (iii) is not satisfied. So we must either have infinitely often transitions of the second or fourth type, or infinitely often a transition of the first type followed by a transition of the third type.

Assume first that the transition between ℓ_m and ℓ_{m+1} is of the second type. Then $\ell_{m+1} = (a'-b, a', a'-c)$ and ℓ_{m+2} is one of

$$\ell_{m+1} C_1 C_2^{2k'} C_1 = (a'', a''+a', a''+a'-c) \text{ with } a'' = a' - b + k'a',$$
$$\ell_{m+1} C_1 C_2^{2k'+1} C_1 = (a''-a', a'', a''-a'+c) \text{ with } a'' = a' - b + (k'+1)a' + a' - c,$$
$$\ell_{m+1} C_2 C_1^{2k'} C_2 = (c''+a'-b, c''+a', c'') \text{ with } c'' = a' - c + k'a',$$
$$\ell_{m+1} C_2 C_1^{2k'+1} C_2 = (c''-a'+c, c'', c''-a') \text{ with } c'' = a' - c + (k'+1)a' + a' - b.$$

If $D_{m+1} = D_m$, then a is between b and c which must have opposite signs. Then a' is strictly between $a' - b$ and $a' - c$, which implies in all four cases that $D_{m+2} < D_{m+1}$. So we always have $D_{m+2} < D_m$.

The case where the transition between ℓ_m and ℓ_{m+1} is of the fourth type is similar. Assume now that this transition is of the first type, and the transition between ℓ_{m+1} and ℓ_{m+2} is of the third type. Then $\ell_{m+1} = (a', a' + b, a' + c)$ and $\ell_{m+2} = (c'' + a', c'' + a' + b, c'')$ with $c'' = a' + c + k'(a' + b)$. If $D_{m+1} = D_m$, then a is between b and c which must have opposite signs, so that a' is strictly between $a' + b$ and $a' + c$, which implies that $D_{m+2} < D_{m+1}$. So again we always have $D_{m+2} < D_m$, and this concludes the proof.

3   Factor Complexity of Primitive $\mathcal{C}$-adic Words

Let $\mathbf{w}$ be a (infinite) word over some alphabet A. We let $\mathrm{Fac}(\mathbf{w})$ denote the set of factors of $\mathbf{w}$, i.e., $\mathrm{Fac}(\mathbf{w}) = \{u \in A^* \mid \exists i \in \mathbb{N} : \mathbf{w}_i \cdots \mathbf{w}_{i+|u|-1} = u\}$. The *extension set* of $u \in \mathrm{Fac}(\mathbf{w})$ is the set $E(u, \mathbf{w}) = \{(a, b) \in A \times A \mid aub \in \mathrm{Fac}(\mathbf{w})\}$. We represent it by an array of the form

$$
E(u, \mathbf{w}) \quad = \quad
\begin{array}{c|ccc}
 & & \cdots\ j\ \cdots & \\ \hline
 & \cdots & & \\
i & & \times & \\
 & \cdots & &
\end{array}
\quad,
$$

where a symbol $\times$ in position (i, j) means that (i, j) belongs to $E(u, \mathbf{w})$. When the context is clear we omit the information on $\mathbf{w}$ and simply write $E(u)$. We also represent it as an undirected bipartite graph, called the *extension graph*, whose set of vertices is the disjoint union of $\pi_1(E(u, \mathbf{w}))$ and $\pi_2(E(u, \mathbf{w}))$ (π_1 and π_2 respectively being the projection on the first and on the second component) and its edges are the pairs $(a, b) \in E(u, \mathbf{w})$. A factor u of $\mathbf{w}$ is said to be *bispecial* whenever $\#\pi_1(E(u, \mathbf{w})) > 1$ and $\#\pi_2(E(u, \mathbf{w})) > 1$. A bispecial factor $u \in \mathrm{Fac}(\mathbf{w})$ is said to be *ordinary* if there exists $(a, b) \in E(u, \mathbf{w})$ such that $E(u, \mathbf{w}) \subset (\{a\} \times A) \cup (A \times \{b\})$.

To simplify proofs, we consider $\mathcal{C}' = \{c_{11}, c_{22}, c_{122}, c_{211}, c_{121}, c_{212}\}$, where

$$
c_{11} = c_1^2 : \begin{cases} 1 \mapsto 1 \\ 2 \mapsto 12 \\ 3 \mapsto 13 \end{cases}
\quad
c_{122} = c_1 c_2^2 : \begin{cases} 1 \mapsto 12 \\ 2 \mapsto 132 \\ 3 \mapsto 2 \end{cases}
\quad
c_{121} = c_1 c_2 c_1 : \begin{cases} 1 \mapsto 13 \\ 2 \mapsto 132 \\ 3 \mapsto 12 \end{cases}
$$

$$
c_{22} = c_2^2 : \begin{cases} 1 \mapsto 13 \\ 2 \mapsto 23 \\ 3 \mapsto 3 \end{cases}
\quad
c_{211} = c_2 c_1^2 : \begin{cases} 1 \mapsto 2 \\ 2 \mapsto 213 \\ 3 \mapsto 23 \end{cases}
\quad
c_{212} = c_2 c_1 c_2 : \begin{cases} 1 \mapsto 23 \\ 2 \mapsto 213 \\ 3 \mapsto 13 \end{cases}
$$

Any (primitive) $\mathcal{C}$-adic word is a (primitive) $\mathcal{C}'$-adic word and conversely. We let ε denote the empty word. We have the following result, where uniqueness follows from the fact that $\tau(A)$ is a code.

Lemma 2 (Synchronization). *Let* $\mathbf{w}$ *be a* $\mathcal{C}'$*-adic word with directive sequence* $(\tau_n)_{n\in\mathbb{N}} \in \mathcal{C}'^{\mathbb{N}}$. *If* $u \in \mathrm{Fac}(\mathbf{w})$ *is a non-empty bispecial factor, then*

1. *If* $\tau_0 = c_{11}$, *there is a unique word* $v \in \mathrm{Fac}(\mathbf{w}^{(1)})$ *such that* $u = \tau_0(v)1$.
2. *If* $\tau_0 = c_{22}$, *there is a unique word* $v \in \mathrm{Fac}(\mathbf{w}^{(1)})$ *such that* $u = 3\tau_0(v)$.
3. *If* $\tau_0 = c_{122}$, *there is a unique word* $v \in \mathrm{Fac}(\mathbf{w}^{(1)})$ *such that* $u \in 2\tau_0(v)\{1,\varepsilon\}$.
4. *If* $\tau_0 = c_{211}$, *there is a unique word* $v \in \mathrm{Fac}(\mathbf{w}^{(1)})$ *such that* $u \in \{3,\varepsilon\}\tau_0(v)2$.
5. *If* $\tau_0 = c_{121}$, *there is a unique word* $v \in \mathrm{Fac}(\mathbf{w}^{(1)})$ *such that* $u \in \{2,\varepsilon\}\tau_0(v)\{1,13\}$.
6. *If* $\tau_0 = c_{212}$, *there is a unique word* $v \in \mathrm{Fac}(\mathbf{w}^{(1)})$ *such that* $u \in \{3,13\}\tau_0(v)\{2,\varepsilon\}$.

Furthermore, v *is a bispecial factor of* $\mathbf{w}^{(1)}$ *and is shorter than* u.

Let $\mathbf{w}$, u and v be as in Lemma 2. The word v is called the *bispecial antecedent of* u under τ_0. Similarly, u is called a *bispecial extended image of* v under τ_0. Since the bispecial antecedent of a non-empty bispecial word is always shorter, for any bispecial factor u of $\mathbf{w}$, there is a unique sequence $(u_i)_{0\le i\le n}$ such that

- $u_0 = u$, $u_n = \varepsilon$ and $u_i \ne \varepsilon$ for all $i < n$;
- for all $i < n$, $u_{i+1} \in \mathrm{Fac}(\mathbf{w}^{(i+1)})$ is the bispecial antecedent of u_i.

All bispecial factors of the sequence $(u_i)_{0\le i<n}$ are called the *bispecial descendants* of ε in $\mathbf{w}^{(n)}$.

As any bispecial factor of a primitive $\mathcal{C}$-adic word is a descendant of the empty word, to understand the extension sets of any bispecial word in $\mathbf{w}$, we need to know the possible extension sets of ε in $\mathbf{w}^{(n)}$ and to understand how the extension set of a bispecial factor governs the extension sets of its bispecial extended images.

Lemma 3. *If* $\mathbf{w}$ *is a primitive* $\mathcal{C}$*-adic word with directive sequence* $(\tau_n)_{n\in\mathbb{N}} \in \mathcal{C}'^{\mathbb{N}}$, *then the extension set of* ε *is one of the following, depending on* τ_0.

$\tau_0 = c_{11}$	1	2	3
1	×	×	×
2	×		
3	×		

$\tau_0 = c_{122}$	1	2	3
1		×	×
2	×	×	
3			×

$\tau_0 = c_{121}$	1	2	3
1		×	×
2	×		
3		×	×

$\tau_0 = c_{22}$	1	2	3
1			×
2			×
3	×	×	×

$\tau_0 = c_{211}$	1	2	3
1			×
2	×	×	×
3		×	

$\tau_0 = c_{212}$	1	2	3
1			×
2	×		×
3	×	×	

Proof. The directive sequence being primitive, all letters of $\mathcal{A}$ occur in $\mathbf{w}^{(1)}$. The result then follows from the fact that all morphisms τ in $\mathcal{C}'$ are either left proper $(\tau(\mathcal{A}) \subset a\mathcal{A}^*$ for some letter $a)$ or right proper $(\tau(\mathcal{A}) \subset \mathcal{A}^*a$ for some letter $a)$. $\qquad \blacksquare$

The next lemma describes how the extension set of a bispecial word determines the extension set of any of its bispecial extended images.

Lemma 4. *Let* $\mathbf{w}$ *be a* $\mathcal{C}'$-*adic word with directive sequence* $(\tau_n)_{n \in \mathbb{N}} \in \mathcal{C}'^{\mathbb{N}}$. *If* $u \in \mathrm{Fac}(\mathbf{w})$ *is the bispecial extended image of* $v \in \mathrm{Fac}(\mathbf{w}^{(1)})$ *and if* $x, y \in \mathcal{A}^*$ *are such that* $u = x\tau_0(v)y$, *then*

1. *if* $\tau_0(\mathcal{A}) \subset i\mathcal{A}^*$ *for some letter* $i \in \mathcal{A}$, *we have*

$$E(u, \mathbf{w}) = \{(a, b) \mid \exists (a', b') \in E(v, \mathbf{w}^{(1)}) : \tau_0(a') \in \mathcal{A}^*ax \wedge \tau_0(b')i \in yb\mathcal{A}^*\};$$

2. *if* $\tau_0(\mathcal{A}) \subset \mathcal{A}^*i$ *for some letter* $i \in \mathcal{A}$, *we have*

$$E(u, \mathbf{w}) = \{(a, b) \mid \exists (a', b') \in E(v, \mathbf{w}^{(1)}) : i\tau_0(a') \in \mathcal{A}^*ax \wedge \tau_0(b') \in yb\mathcal{A}^*\}.$$

Proof. Let us prove the first equality, the second one being symmetric.

For the inclusion $\supseteq$, consider $(a', b') \in E(v)$ such that $\tau_0(a') \in \mathcal{A}^*ax$ and $\tau_0(b')i \in yb\mathcal{A}^*$. Let $c \in \mathcal{A}$ be such that $a'vb'c$ is a factor of $\mathbf{w}^{(1)}$. Then $\tau_0(a'vb'c) \in \tau_0(a'vb')i\mathcal{A}^* \subseteq \mathcal{A}^*ax\tau_0(v)yb\mathcal{A}^*$ is a factor of w and we have $(a, b) \in E(u)$.

For the inclusion $\subseteq$, consider $(a, b) \in E(u)$. Using Lemma 2, the word ax (resp., yb) is the suffix (resp., prefix) of a word $\tau_0(x')$, $x' \in \mathcal{A}^+$ (resp., $\tau_0(y')$, $y' \in \mathcal{A}^+$) such that $x'vy' \in \mathrm{Fac}(\mathbf{w}^{(1)})$. Furthermore, still using Lemma 2, x is a strict suffix of $\tau_0(a')$, where $x' \in \mathcal{A}^*a'$ and y is a prefix of $\tau_0(b')$, where $y' \in b'\mathcal{A}^*$. If y is a strict prefix of $\tau_0(b')$, then (a', b') is an extension of v such that $\tau_0(a') \in \mathcal{A}^*ax$ and $\tau_0(b') \in yb\mathcal{A}^*$. Otherwise, if $\tau_0(b') = y$, we have $b = i$ since $\tau_0(\mathcal{A}) \subset i\mathcal{A}^*$ and (a', b') is an extension of v such that $\tau_0(a') \in \mathcal{A}^*ax$ and $\tau_0(b')i = yi$, which concludes the proof. $\qquad \blacksquare$

Lemma 4 can be more easily understood using the tabular representation of the extension sets. Indeed, for the first case $(\tau_0(\mathcal{A}) \subset i\mathcal{A}^*)$, the extensions of $u = x\tau(v)y$ can be obtained as follows: (1) replace any left extensions a by $\tau(a)$ and any right extension b by $\tau(b)i$; (2) remove the suffix x from the left extensions whenever it is possible (otherwise, delete the row) and remove the prefix y from the right extensions whenever it is possible (otherwise, delete the column); (3) keep only the last letter of the left extensions and the first letter of the right extensions; (4) permute and merge the rows and columns with the same label. The second case $(\tau_0(\mathcal{A}) \subset \mathcal{A}^*i)$ is similar.

Let us make this more clear on an example and consider the extension set $E(v) = \{(1, 3), (2, 1), (2, 2), (2, 3), (3, 2)\}$. This extension set corresponds to the extension set of the empty word whenever the last applied substitution is c_{211}

(see Lemma 3). Using Lemma 4, the extension sets of $2c_{122}(v)$ and $2c_{121}(v)1$ are obtained as follows (arrow labels indicate above step number):

$$E(v)$$

1	2	3
$\times$		
$\times$	$\times$	$\times$
	$\times$	

$\xrightarrow{(1)}$

$$E(c_{122}(v))$$

	12	132	2
212			$\times$
2132	$\times$	$\times$	$\times$
22			$\times$

$\xrightarrow{(2) \text{ and } (3)}$

$$E(2c_{122}(v))$$

	1	1	2
1			$\times$
3	$\times$	$\times$	$\times$
2		$\times$	

$\xrightarrow{(4)}$

$$E(2c_{122}(v))$$

	1	2
1		$\times$
2	$\times$	
3	$\times$	$\times$

$$E(v)$$

1	2	3
$\times$		
$\times$	$\times$	$\times$
	$\times$	

$\xrightarrow{(1)}$

$$E(c_{121}(v))$$

	131	1321	121
13			$\times$
132	$\times$	$\times$	$\times$
12			$\times$

$\xrightarrow{(2) \text{ and } (3)}$

$$E(2c_{121}(v)1)$$

	3	3	2
$.$			
3	$\times$	$\times$	$\times$
1		$\times$	

$\xrightarrow{(4)}$

$$E(2c_{121}(v)(1)$$

	2	3
1		$\times$
3	$\times$	$\times$

The proof of Proposition 6 will essentially consists in describing how ordinary bispecial words occur. The next lemma allows to understand when bispecial words have ordinary bispecial extended images.

Lemma 5. *Let $\mathbf{w}$ be a C'-adic word with directive sequence $(\tau_n)_{n\in\mathbb{N}} \in C'^{\mathbb{N}}$. Let $u \in \mathrm{Fac}(\mathbf{w})$ be a non-empty bispecial factor and v be its bispecial antecedent. We have the following.*

1. *If $\tau_0 \in \{c_{11}, c_{22}\}$, then $E(u) = E(v)$;*
2. *if $v = \varepsilon$ and $\tau_0 \in \{c_{121}, c_{212}\}$, then u is ordinary;*
3. *if $\tau_0 \in \{c_{122}, c_{121}, c_{212}\}$, if $E(v) \subseteq (A \times \{1,2\}) \cup \{(a,3)\}$ for some letter $a \in A$ with $E(v) \cap \{(a,1), (a,2)\} \neq \emptyset$ and if $E(v) \backslash \{(a,3)\}$ is the extension set of an ordinary bispecial word, then u is ordinary;*
4. *if $\tau_0 \in \{c_{211}, c_{121}, c_{212}\}$, if $E(v) \subseteq (\{2,3\} \times A) \cup \{(1,a)\}$ for some letter $a \in A$ with $E(v) \cap \{(2,a), (3,a)\} \neq \emptyset$ and if $E(v) \backslash \{(1,a)\}$ is the extension set of an ordinary bispecial word, then u is ordinary;*
5. *if v is ordinary, then u is ordinary*

Proof. Items 1 and 5 directly follow from Lemma 4. Item 2 can be checked by hand using Lemmas 3 and 4. Let us prove Item 3, Item 4 being symmetric.

Let us say that two extension sets E and E' are equivalent whenever there exist two permutations p_1 and p_2 of A such that $E = \{(p_1(a), p_2(b)) \mid (a,b) \in E'\}$. If $\tau_0 = c_{122}$, then $u \in \{2\sigma(v), 2\sigma(v)1\}$ by Lemma 2. We make use of Lemma 4. If $u = 2\sigma(v)$, then the extension set of u is equivalent to the one obtained from $E(v)$ by merging the columns with labels 1 and 2. If $u = 2\sigma(v)1$, then the extension set of u is equivalent to the one obtained from $E(v)$ by deleting the column with label 3. In both cases, u is ordinary.

The same reasoning applies when $\tau_0 \in \{c_{121}, c_{212}\}$: depending on the word x such that $u \in A^*\sigma(v)x$, either we delete the column with label 3, or we merge the columns with labels 1 and 2.

Recall that an infinite word is a *tree word* if the extension graph of any of its bispecial factors is a tree. Obviously, if u is an ordinary bispecial word, its extension graph is a tree. If $\mathbf{w} \in \mathcal{A}^{\mathbb{N}}$ is a tree word in which all letters of $\mathcal{A}$ occur, then $\mathbf{w}$ has factor complexity $p(n) = (\mathrm{Card}(\mathcal{A}) - 1)n + 1$ for all n [6].

Proposition 6. *Any primitive C-adic word is a uniformly recurrent tree word. In particular, any primitive C-adic word has factor complexity $p(n) = 2n + 1$.*

Proof. Any primitive $\mathcal{C}$-adic word has a primitive $\mathcal{C}$-adic representation, hence is uniformly recurrent.

To show that the extension graphs of all bispecial factors are trees, we make use of Lemma 5. If u is a bispecial factor of $\mathbf{w}$, it is a descendant of $\varepsilon \in \mathrm{Fac}(\mathbf{w}^{(n)})$ for some n. If $\tau_n \in \{c_{11}, c_{22}\}$, then from Lemmas 3 and 5, all descendants of ε are ordinary. The extension graph of u is thus a tree.

For $\tau_n \in \{c_{122}, c_{211}, c_{121}, c_{212}\}$, we represent the extension sets of the descendants of ε in the graphs represented in Figs. 1 and 2. Observe that the situation is symmetric for c_{122} and c_{211} and for c_{121} and c_{212} so we only represent the graphs for c_{122} and c_{121}. Furthermore, in these graphs, we do not represent the extension sets of ordinary bispecial factors as the property of being ordinary is preserved by taking bispecial extended images (Lemma 5). Given an extension set of some bispecial word v, if u is a bispecial extended image of v such that $u = x\tau(v)y$, we label the edge from $E(v)$ to $E(u)$ by $x \cdot \tau \cdot y$. Finally, for all v, we have $E(c_{11}(v)1) = E(v)$ and $E(3c_{22}(v)) = E(v)$, but for the sake of clarity, we do not draw the loops labeled by $c_{11} \cdot 1$ and by $3 \cdot c_{22}$. We conclude the proof by observing that the extension graphs of all descendants are trees.

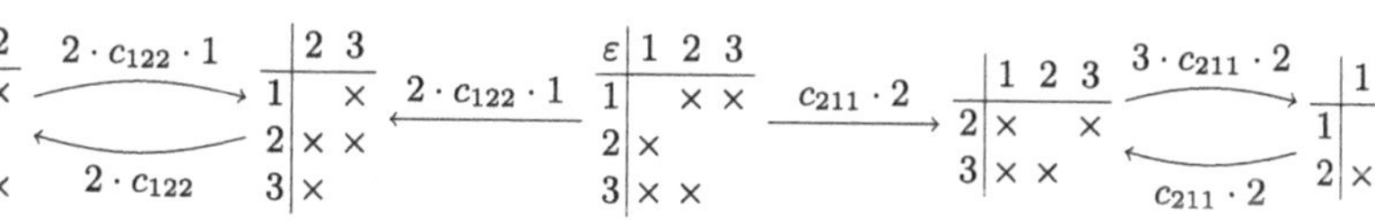

Fig. 1. Non-ordinary bispecial descendants of $\varepsilon \in \mathrm{Fac}(\mathbf{w}^{(n)})$ whenever $\tau_n = c_{122}$.

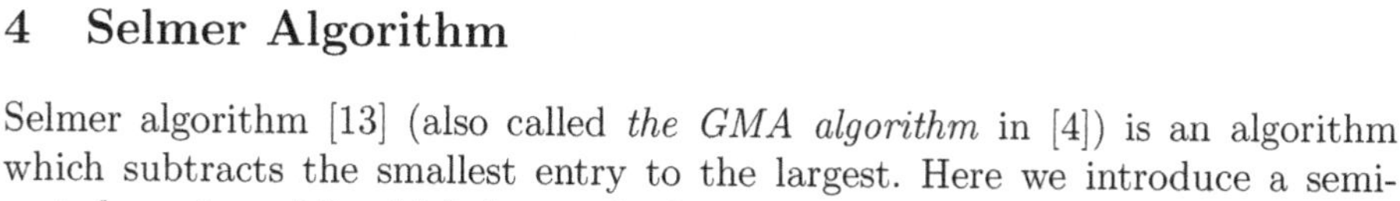

Fig. 2. Non-ordinary bispecial descendants of $\varepsilon \in \mathrm{Fac}(\mathbf{w}^{(n)})$ whenever $\tau_n = c_{121}$.

4   Selmer Algorithm

Selmer algorithm [13] (also called *the GMA algorithm* in [4]) is an algorithm which subtracts the smallest entry to the largest. Here we introduce a semi-sorted version of it which keeps the largest entry at index 1. On $\Gamma = \{\mathbf{x} = (x_1, x_2, x_3) \in \mathbb{R}_{\geq 0}^3 \mid \max(x_2, x_3) \leq x_1 \leq x_2 + x_3\}$, it is defined as

$$F_S(x_1, x_2, x_3) = \begin{cases} (x_2, x_1 - x_3, x_3) & \text{if } x_2 \geq x_3, \\ (x_3, x_2, x_1 - x_2) & \text{if } x_2 < x_3. \end{cases}$$

The partition of Γ into $\Gamma_1 \cup \Gamma_2$ is

$$\Gamma_1 = \{(x_1, x_2, x_3) \in \Gamma \mid x_2 \geq x_3\},$$
$$\Gamma_2 = \{(x_1, x_2, x_3) \in \Gamma \mid x_2 < x_3\}.$$

For semi-sorted Selmer algorithm, the matrices and associated substitutions are

$$S_1 = \begin{pmatrix} 0\,1\,1 \\ 1\,0\,0 \\ 0\,0\,1 \end{pmatrix}, \quad S_2 = \begin{pmatrix} 0\,1\,1 \\ 0\,1\,0 \\ 1\,0\,0 \end{pmatrix} \quad \text{and} \quad s_1 = \begin{cases} 1 \mapsto 2 \\ 2 \mapsto 1 \\ 3 \mapsto 31 \end{cases}, \quad s_2 = \begin{cases} 1 \mapsto 3 \\ 2 \mapsto 12 \\ 3 \mapsto 1 \end{cases}.$$

The matrices are given by the rule $M(\mathbf{x}) = S_i$ if and only if $\mathbf{x} \in \Gamma_i$. The map F_S on Γ is then defined as: $F_S(\mathbf{x}) = M(\mathbf{x})^{-1}\mathbf{x}$. The substitutions on $\mathcal{A}^*$ are given by the rule $\sigma(\mathbf{x}) = s_i$ if and only if $\mathbf{x} \in \Gamma_i$ for $i = 1, 2$.

5   Conjugacy of F_C and F_S

The numerical computation of Lyapunov exponents made in [11] indicate that exponents for the unsorted Selmer algorithm and F_C have statistically equal values. The next proposition gives an explanation for this observation.

Proposition 7. *Algorithms $F_C : \Lambda \to \Lambda$ and $F_S : \Gamma \to \Gamma$ are topologically conjugate.*

Proof. Let $z : \Lambda \to \Gamma$ be the homeomorphism defined by $\mathbf{x} \mapsto Z\mathbf{x}$ with

$$Z = \begin{pmatrix} 1\,1\,1 \\ 1\,1\,0 \\ 0\,1\,1 \end{pmatrix}.$$

We verify that C_i is conjugate to S_i through matrix Z for $i = 1, 2$:

$$S_1 Z = \begin{pmatrix} 1\,2\,1 \\ 1\,1\,1 \\ 0\,1\,1 \end{pmatrix} = ZC_1 \quad \text{and} \quad S_2 Z = \begin{pmatrix} 1\,2\,1 \\ 1\,1\,0 \\ 1\,1\,1 \end{pmatrix} = ZC_2.$$

Thus we have $z \circ F_C = F_S \circ z$.

An infinite word $u \in A^{\mathbb{N}}$ is said to be *finitely balanced* if there exists a constant $C > 0$ such that for any pair v, w of factors of the same length of u, and for any letter $i \in A$, $||v|_i - |w|_i| \leq C$.

Based on [7, Theorem 6.4], and considering that computer experiments suggest that the second Lyapunov exponent of Selmer algorithm is negative ($\theta_1 \approx \log(1.200) \approx 0.182$ and $\theta_2 \approx \log(0.9318) \approx -0.0706$ in [4, p. 1522], $\theta_1 \approx 0.18269$ and $\theta_2 \approx -0.07072$ in [11]), we believe that the following conjecture holds.

Conjecture 8. For almost every $\mathbf{x} \in \Delta$, the word $W(\mathbf{x})$ is finitely balanced.

5.1 Substitutive Conjugacy

Let z_l and z_r be the following two substitutions:

$$z_l : \begin{cases} 1 \mapsto 12 \\ 2 \mapsto 123 \\ 3 \mapsto 13 \end{cases} \quad \text{and} \quad z_r : \begin{cases} 1 \mapsto 21 \\ 2 \mapsto 231 \\ 3 \mapsto 31 \end{cases}.$$

The substitution z_l is left proper while z_r is right proper. Moreover they are conjugate through the equation

$$z_l(w) \cdot 1 = 1 \cdot z_r(w) \qquad \text{for every } w \in \mathcal{A}^*.$$

Notice that Z is the incidence matrix of both z_l and z_r.

The substitutions c_i are not conjugate to s_i but are related through substitutions z_l and z_r for $i = 1, 2$:

$$s_1 \circ z_l = z_r \circ c_1 = (1 \mapsto 21, 2 \mapsto 2131, 3 \mapsto 231),$$
$$s_2 \circ z_r = z_l \circ c_2 = (1 \mapsto 123, 2 \mapsto 1213, 3 \mapsto 13).$$

We deduce that

Proposition 9. *S-adic sequences when $S = \{s_1, s_2\}$ restricted to the application of the semi-sorted Selmer algorithm F_S on totally irrational vectors $\mathbf{x} \in \Gamma$ have factor complexity $2n + 1$.*

The problem of finding an analogue of F_C in dimension $d \geq 4$ (i.e. projective dimension $d - 1$), generating S-adic sequences with complexity $(d-1)n + 1$ for almost every vector of letter frequencies is still open.

Acknowledgments. We are thankful to Valérie Berthé for her enthusiasm toward this project and for the referees for their thorough reading and pertinent suggestions.

References

1. Arnoux, P., Labbé, S.: On some symmetric multidimensional continued fraction algorithms. Ergodic Theor. Dyn. Syst., 1–26 (2017). doi:10.1017/etds.2016.112
2. Arnoux, P., Rauzy, G.: Représentation géométrique de suites de complexité $2n+1$. Bull. Soc. Math. France **119**(2), 199–215 (1991)
3. Arnoux, P., Starosta, Š.: The Rauzy gasket. In: Barral, J., Seuret, S. (eds.) Further Developments in Fractals and Related Fields, pp. 1–23. Birkhäuser/Springer, New York (2013). doi:10.1007/978-0-8176-8400-6_1, Trends in Mathematics
4. Baldwin, P.R.: A convergence exponent for multidimensional continued-fraction algorithms. J. Stat. Phys. **66**(5–6), 1507–1526 (1992)
5. Berthé, V., Labbé, S.: Factor complexity of S-adic words generated by the Arnoux-Rauzy-Poincaré algorithm. Adv. Appl. Math. **63**, 90–130 (2015). doi:10.1016/j.aam.2014.11.001
6. Berthé, V., De Felice, C., Dolce, F., Leroy, J., Perrin, D., Reutenauer, R.G.: Acyclic, connected and tree sets. Monatsh. Math. **176**(4), 521–550 (2015). doi:10.1007/s00605-014-0721-4

7. Berthé, V., Delecroix, V.: Beyond substitutive dynamical systems: S-adic expansions. In: Numeration and Substitution 2012, pp. 81–123 (2014). RIMS Kôkyûroku Bessatsu, B46, Res. Inst. Math. Sci. (RIMS), Kyoto

8. Brentjes, A.J.: Multidimensional Continued Fraction Algorithms. Mathematisch Centrum, Amsterdam (1981)

9. Cassaigne, J.: Un algorithme de fractions continues de complexité linéaire, DynA3S meeting, LIAFA, Paris, 12th October 2015. http://www.irif.fr/dyna3s/Oct2015

10. Delecroix, V., Hejda, T., Steiner, W.: Balancedness of Arnoux-Rauzy and Brun words. In: Karhumäki, J., Lepistö, A., Zamboni, L. (eds.) WORDS 2013. LNCS, vol. 8079, pp. 119–131. Springer, Heidelberg (2013). doi:10.1007/978-3-642-40579-2_14

11. Labbé, S.: 3-dimensional Continued Fraction Algorithms Cheat Sheets, November 2015. http://arxiv.org/abs/arxiv:1511.08399

12. Leroy, J.: An S-adic characterization of minimal subshifts with first difference of complexity $1 \leq p(n+1) - p(n) \leq 2$. Discrete Math. Theor. Comput. Sci. **16**(1), 233–286 (2014)

13. Schweiger, F.: Multidimensional Continued Fractions. Oxford University Press, New York (2000)

14. Zorich, A.: Deviation for interval exchange transformations. Ergodic Theor. Dyn. Syst. **17**(6), 1477–1499 (1997)

The Word Entropy and How to Compute It

Sébastien Ferenczi[1]([✉]), Christian Mauduit[1], and Carlos Gustavo Moreira[2]

[1] Aix Marseille Université, CNRS, Centrale Marseille, Institut de Mathématiques de Marseille, I2M - UMR 7373, 163, avenue de Luminy, 13288 Marseille Cedex 9, France
ssferenczi@gmail.com, mauduit@iml.univ-mrs.fr
[2] Instituto de Matemática Pura e Aplicada, Estrada Dona Castorina 110, Rio de Janeiro, RJ 22460-320, Brazil
gugu@impa.br

Abstract. The complexity function of an infinite word counts the number of its factors. For any positive function f, its *exponential rate of growth* $E_0(f)$ is $\lim_{n\to\infty} \inf \frac{1}{n} \log f(n)$. We define a new quantity, the *word entropy* $E_W(f)$, as the maximal exponential growth rate of a complexity function smaller than f. This is in general smaller than $E_0(f)$, and more difficult to compute; we give an algorithm to estimate it. The quantity $E_W(f)$ is used to compute the Hausdorff dimension of the set of real numbers whose expansions in a given base have complexity bounded by f.

Keywords: Word complexity · Positive entropy

1 Definitions

Let A be the finite alphabet $\{0, 1, \ldots, q-1\}$, If $w \in A^{\mathbb{N}}$, and $L(w)$ the set of finite factors of w; for any non-negative integer n, we write $L_n(w) = L(w) \cap A^n$. The classical complexity function is described for example in [2].

Definition 1. *The complexity function of $w \in A^{\mathbb{N}}$ is defined for any non-negative integer n by $p_w(n) = |L_n(w)|$.*

Our work concerns the study of infinite words w the complexity function of which is bounded by a given function f from $\mathbb{N}$ to $\mathbb{R}^+$. More precisely, if f is such a function, we put

$$W(f) = \{w \in A^{\mathbb{N}}, p_w(n) \leq f(n), \forall n \in \mathbb{N}\}.$$

Definition 2. *If f is a function from $\mathbb{N}$ to $\mathbb{R}^+$, we call exponential rate of growth of f the quantity*

$$E_0(f) = \lim_{n\to\infty} \inf \frac{1}{n} \log f(n)$$

and word entropy of f the quantity

$$E_W(f) = \sup_{w \in W(f)} E_0(p_w).$$

© Springer International Publishing AG 2017
S. Brlek et al. (Eds.): WORDS 2017, LNCS 10432, pp. 157–163, 2017.
DOI: 10.1007/978-3-319-66396-8_15

Of course, if $E_0 = 0$ then E_W is zero also. Thus the study of E_W is interesting only when f has exponential growth: we are in the little-explored field of *word combinatorics in positive entropy*, or exponential complexity. For an equivalent theory in zero entropy, see [3,4].

2 First Properties of E_0 and E_W

The basic study of these quantities is carried out in [5], where the following results are proved.

If f is itself a complexity function (i.e. $f = p_w$ for some $w \in A^{\mathbb{N}}$), then $E_W(f) = E_0(f)$. But *in general E_W may be much smaller than E_0*.

We define mild regularity conditions for f: f is said to satisfy $(\mathcal{C})$ if the sequence $(f(n))_{n \geq 1}$ is strictly increasing, there exists $n_0 \in \mathbb{N}$ such that $\forall n \geq n_0 \Rightarrow f(2n) \leq f(n)^2$, $f(n+1) \leq f(1)f(n)$, and the sequence $\left(\frac{1}{n} \log f(n)\right)_{n \geq 1}$ converges.

But for each $1 < \theta \leq q$, and $n_0 \in \mathbb{N}$ such that $\theta^{n_0+1} > n_0 + q - 1$, we define the function f by $f(1) = q$, $f(n) = n + q - 1$ for $1 \leq n \leq n_0$ and $f(n) = \theta^n$ for $n > n_0$. We have $E_0(f) = \log \theta$ and it is proved that

$$E_W(f) \leq \frac{1}{n_0} \log(n_0 + q - 1),$$

which can be made arbitrarily small, independently of θ, while f satisfies $(\mathcal{C})$.

We define stronger regularity conditions for f.

Definition 3. *We say that a function f from $\mathbb{N}$ to $\mathbb{R}^+$ satisfies the conditions $(\mathcal{C}^*)$ if (i) for any $n \in \mathbb{N}$ we have $f(n+1) > f(n) \geq n+1$; (ii) for any $(n, n') \in \mathbb{N}^2$ we have $f(n + n') \leq f(n)f(n')$.*

But even with $(\mathcal{C}^*)$ we may have $E_W(f) < E_0(f)$. Indeed, let f be the function defined by $f(n) = \lceil 3^{n/2} \rceil$ for any $n \in \mathbb{N}$. Then it is easy to check that f satisfies conditions $(\mathcal{C}^*)$ and that $E_0(f) = \lim_{n \to \infty} \frac{1}{n} \log f(n) = \log(\sqrt{3})$. On the other hand, we have $f(1) = 2$, $f(2) = 3$; thus the language has no 00 or no 11, and this implies that $E_W(f) \leq \log(\frac{1+\sqrt{5}}{2}) < E_0(f)$.

At least, under these conditions, we have the important

Theorem 4. *If f is a function from $\mathbb{N}$ to $\mathbb{R}^+$ satisfying the conditions $(\mathcal{C}*)$, then $E_W(f) > \frac{1}{2} E_0(f)$.*

It is also shown in [5] that the constant $\frac{1}{2}$ is optimal.

Finally, it will be useful to know that

Theorem 5. *For any function f from $\mathbb{N}$ to $\mathbb{R}^+$, there exists $w \in W(f)$ such that for any $n \in \mathbb{N}$ we have $p_w(n) \geq \exp(E_W(f)n)$.*

3   Algorithm

In general $E_W(f)$ is much more difficult to compute than $E_0(f)$; now we will give an algorithm which allows us to estimate with arbitrary precision $E_W(f)$ from finitely many values of f, if we know already $E_0(f)$ and have some information on the speed with which this limit is approximated.

We assume that f satisfies conditions $\mathcal{C}^*$. We don't loose too much generality with this assumption, since if the function f which satisfies the weaker conditions $\mathcal{C}$, we can replace it by the function $\tilde{f}$ given recursively by

$$\tilde{f}(n) := \min\{f(n), \min_{1 \le k < n} \tilde{f}(k)\tilde{f}(n-k)\},$$

which satisfies conditions $\mathcal{C}^*$, such that $\tilde{f}(n) \le f(n), \forall n \in \mathbb{N}$ and $W(\tilde{f}) = W(f)$.

Theorem 6. *There is an algorithm which gives, starting from f and ε, a quantity h such that $(1-\varepsilon)h \le E_W(f) \le h$. h depends explicitly on ε, $E_0(f)$, N, $f(1)$, ..., $f(N)$, for an integer N which depends explicitly on ε, $E_0(f)$, and an integer n_0, larger than an explicit function of ε and $E_0(f)$, and such that*

$$\frac{\log f(n)}{n} < (1 + \frac{E_0(f)\varepsilon}{210(4 + 2E_0(f))})E_0(f), \quad for \quad n_0 \le n < 2n_0.$$

We shall now give the algorithm. f is given and henceforth we omit to mention it in $E_0(f)$ and $E_W(f)$. Also given is $\varepsilon \in (0,1)$.

Description of the algorithm

– Let

$$\delta := \frac{E_0 \varepsilon}{105(4 + 2E_0)} < \frac{\varepsilon}{210}.$$

– Let

$$K := \lceil \delta^{-1} \rceil + 1.$$

– Choose a positive integer

$$n_0 \ge K \vee \frac{4K^2}{420^3 E_0}$$

such that

$$\frac{\log f(n)}{n} < (1 + \frac{\delta}{2})E_0, \forall n \ge n_0;$$

in view of conditions $\mathcal{C}^*$, this last condition is equivalent to $\frac{\log f(n)}{n} < (1 + \frac{\delta}{2})E_0, n_0 \le n < 2n_0$.

– Choose intervals so large that all the lengths of words we manipulate stay in one of them. Namely, for each $t \geq 0$, let

$$n_{t+1} := \exp(K((1+\delta)^2 E_0 n_t + E_0)).$$

We take

$$N := n_K.$$

– Choose a set $Y \subset A^N$: for each possible Y, we define $L_n(Y) = \cup_{\gamma \in Y} L(\gamma)$, $q_n(Y) := |L_n(Y)|$, for $1 \leq n \leq N$. We look at those Y for which $q_n(Y) \leq f(n), \forall n \leq N$, and choose one among them such that

$$\min_{1 \leq n \leq N} \frac{\log q_n(Y)}{n}$$

is maximum.
– By Lemma 7 below, on one of the large intervals we have defined, namely $[n_r, n_{r+1}]$, $\frac{\log q_n(Y)}{n}$ will be almost constant. Let

$$h := \frac{\log q_{n_r}(Y)}{n_r}.$$

Here is the lemma we needed; henceforth, Y is fixed and we omit to mention it in the $q_n(Y)$:

Lemma 7. *There exists $r < K$, such that*

$$\frac{\log q_{n_r}}{n_r} < (1+\delta)\frac{\log q_{n_{r+1}}}{n_{r+1}}.$$

Proof. Otherwise $\frac{\log q_{n_0}}{n_0} \geq (1+\delta)^K \frac{\log q_{n_K}}{n_K}$: as $K > \frac{1}{\delta}$, $(1+\delta)^K$ would be close to e for δ small enough, and is larger than $\frac{9}{4}$ as $\delta < \frac{1}{2}$; thus, as $\frac{\log q_{n_K}}{n_K} \geq E_W$ by the proof of Proposition 8 below, we have $\frac{\log q_{n_0}}{n_0} \geq \frac{9}{4}E_W$, but $q_{n_0} \leq f(n_0)$ hence $\frac{\log q_{n_0}}{n_0} < (1+\frac{\delta}{2})E_0$, and this contradicts $E_0 \leq 2E_W$, which is true by Theorem 4.

We prove now that indeed h is a good approximation of the word entropy.

Proposition 8.

$$h \geq E_W.$$

Proof. We prove that

$$\min_{1 \leq n \leq N} \frac{\log q_n}{n} \geq E_W.$$

We know by Theorem 5 that there is $\hat{w} \in W(f)$ with $p_n(\hat{w}) \geq \exp(E_W n)$, for all $n \geq 1$. For such a word $\hat{w}$, let $X := L_N(\hat{w}) \subset A^N$. We have, for each n with $1 \leq n \leq N$, $L_n(X) = L_n(\hat{w})$ and $f(n) \geq \#L_n(\hat{w}) = p_n(\hat{w}) \geq \exp(E_W n)$. Thus X is one of the possible Y, and the result follows from the maximality of $\min_{1 \leq n \leq N} \frac{\log q_n}{n}$.

What remains to prove is the following proposition (which, understandably, does not use the maximality of $\min_{1 \leq n \leq N} \frac{\log q_n}{n}$).

Proposition 9.

$$(1 - \varepsilon)h \leq E_W.$$

Proof. Our strategy is to build a word w such that, for all $n \geq 1$,

$$\exp((1 - \varepsilon)hn) \leq p_n(w) \leq f(n),$$

which gives the conclusion by definition of E_W. To build the word w, we shall define an integer m, and build successive subsets of $L_m(Y)$; for such a subset Z, we order it (lexicographically for example) and define $w(Z)$ to be the *Champernowne word* on Z: namely, if $Z = \{\beta_1, \beta_2, ..., \beta_t\}$, we build the infinite word

$$w(Z) := \beta_1\beta_2 \ldots \beta_t\beta_1\beta_1\beta_1\beta_2\beta_1\beta_3 \ldots \beta_{t-1}\beta_t\beta_1\beta_1\beta_1\beta_1 \ldots \beta_t\beta_t\beta_t \ldots$$

made by concatenation of all words in Z followed by the concatenations of all pairs of words of Z followed by the concatenations of all triples of words of Z, etc.

The word $w(Z)$ will satisfy $\exp((1 - \varepsilon)hn) \leq p_n(w(Z))$ for all n as soon as

$$|Z| \geq \exp((1 - \varepsilon)hm),$$

since, for every positive integer k, we will have at least $|Z|^k$ factors of length km in $w(Z)$.

The successive (decreasing) subsets Z of $L_m(Y)$ we build will all have cardinality at least $\exp((1 - \varepsilon)hm)$, and the words $w(Z)$ will satisfy $p_n(w(Z)) \leq f(n)$ for n in an interval which will increase at each new set Z we build, and ultimately contains all the integers.

We give only the main ideas of the remaining proof. In the first stage we define two lengths of words, $\hat{n}$ and $m > \frac{\hat{n}}{2\varepsilon}$, which will be both in the interval $[n_r, n_{r+1}]$, and a set Z_1 of words of length m of the form $\gamma\theta$, for words γ of length $\hat{n}$, such that the word $\gamma\theta\gamma$ is in $L_{m+\hat{n}}(Y)$. This is done by looking precisely at twin occurrences of words.
Let $\tilde{\varepsilon} = \frac{\varepsilon}{15} = \frac{7(4+2E_0)\delta}{E_0} > 14\delta$; then we can get such a set Z_1 with $|Z_1| \geq \exp((1 - \tilde{\varepsilon})h(m + \hat{n}))$.

In the second stage, we define a new set $Z_2 \subset Z_1$ in which all the words have the same prefix γ_1 of length $6\tilde{\varepsilon}hm$, and all the words have the same suffix γ_2 of length $6\tilde{\varepsilon}hm$, with $|Z_2| \geq |Z_1| \exp(-12\tilde{\varepsilon}hm - 2\delta h\hat{n})$, and $2\delta h\hat{n} \leq (1 - \tilde{\varepsilon})\hat{n}$, thus

$$|Z_2| \geq \exp((1 - 13\tilde{\varepsilon})hm).$$

As a consequence of the definition of Z_2, all words of Z_2 have the same prefix of length $\hat{n}$, which is a prefix γ_0 of γ_1; as Z_2 is included in Z_1, any word of Z_2 is of the form $\gamma_0\theta$, amd the word $\gamma_0\theta\gamma_0$ is in $L_{m+\hat{n}}(Y)$.

At this stage we can prove

Claim. $p_{w(Z_2)}(n) \leq f(n)$ for all $1 \leq n \leq \hat{n} + 1$.

Let us shrink again our set of words.

Lemma 10. *For a given subset Z of Z_2, there exists $Z' \subset Z$, $|Z'| \geq (1 - \exp(-(j-1)\frac{E_0}{2}))^j |Z|$, such that the total number of factors of length $\hat{n} + j$ of all words $\gamma_0 \theta \gamma_0$ such that $\gamma_0 \theta$ is in Z' is at most $f(\hat{n} + j) - j$.*

We start from Z_2 and apply successively Lemma 10 from $j = 2$ to $j = 6\tilde{\varepsilon}m$, getting $6\tilde{\varepsilon}m - 1$ successive sets Z'; at the end, we get a set Z_3 such that the total number of factors of length $\hat{n} + j$ of words $\gamma_0 \theta \gamma_0$ for $\gamma_0 \theta$ in Z_3 is at most $f(\hat{n} + j) - j$ for $j = 2, \ldots, 6\tilde{\varepsilon}m$, and $\frac{|Z_3|}{|Z_2|}$ is at least

$$\Pi_{2 \leq j \leq 6\tilde{\varepsilon}m - \hat{n}}(1 - \exp(-(j-1)\frac{E_0}{2}))^j \geq \Pi_{j \geq 2}(1 - \exp(-(j-1)\frac{E_0}{2}))^j,$$

which implies after computations that

$$|Z_3| \geq \exp((1 - 14\tilde{\varepsilon})hm).$$

We can now bound the number of short factors by using the factors we have just deleted and properties of γ_0, γ_1 and γ_2.

Claim. $p_{w(Z_3)}(n) \leq f(n)$ for all $1 \leq n \leq 6\tilde{\varepsilon}m$.

We shrink our set again.

Let $m \geq n > 6\tilde{\varepsilon}m$; in average a factor of length n of a word in Z_3 occurs in at most $\frac{m|Z_3|}{f(n)}$ elements of Z_3. We consider the $\frac{f(n)}{mn^2}$ factors of length n which occur the least often. In total, these factors occur in at most $\frac{m|Z_3|}{f(n)} \frac{f(n)}{mn^2} = \frac{|Z_3|}{n^2}$ elements of Z_3. We remove these words from Z_3, for all $m \geq n > 6\tilde{\varepsilon}m$, obtaining a set Z_4 with $|Z_4| \geq \exp((1 - 15\tilde{\varepsilon})hm)$.

We can now control medium length factors, using again the missing factors we have just created, and γ_1 and γ_2, but not γ_0.

Claim. $p_{w(Z_4)}(n) \leq f(n)$ for all $1 \leq n \leq m$.

Finally we put $Z_5 = Z_4$ if $|Z_4| \leq \exp((1 - 4\tilde{\varepsilon})hm)$, otherwise we take for Z_5 any subset of Z_4 with $\lceil \exp((1 - 4\tilde{\varepsilon})hm) \rceil$ elements. In both cases we have

$$|Z_5| \geq \exp((1 - \varepsilon)hm).$$

For the long factors, we use mainly the fact that there are many missing factors of length m, but we need also some help from γ_1 and γ_2.

Claim. $p_{w(Z_5)}(n) \leq f(n)$ for all n.

In view of the considerations at the beginning of the proof of Proposition 9, Claim 3 completes the proof of that proposition, and thus of Theorem 6.

4   Application

We define

$$C(f) = \{x = \sum_{n \geq 0} \frac{w_n}{q^{n+1}} \in [0,1], w(x) = w_0 w_1 \cdots w_n \cdots \in W(f)\}.$$

We are interested in the Hausdorff dimensions of this set, see [1] for definitions; indeed, the main motivation for studying the word entropy is Theorem 4.8 of [5]:

Theorem 11.
The Hausdorff dimension of $C(f)$ is equal to $E_W(f)/\log q$.

References

1. Falconer, K.: Fractal Geometry: Mathematical Foundations and Applications. Wiley, Chichester (1990)
2. Ferenczi, S.: Complexity of sequences and dynamical systems. Discrete Math. **206**(1–3), 145–154 (1999). http://dx.doi.org/10.1016/S0012-365X(98)00400-2, (Tiruchirappalli 1996)
3. Mauduit, C., Moreira, C.G.: Complexity of infinite sequences with zero entropy. Acta Arith. **142**(4), 331–346 (2010). http://dx.doi.org/10.4064/aa142-4-3
4. Mauduit, C., Moreira, C.G.: Generalized Hausdorff dimensions of sets of real numbers with zero entropy expansion. Ergodic Theor. Dynam. Syst. **32**(3), 1073–1089 (2012). http://dx.doi.org/10.1017/S0143385711000137
5. Mauduit, C., Moreira, C.G.: Complexity and fractal dimensions for infinite sequences with positive entropy (2017)

First Steps in the Algorithmic Reconstruction of Digital Convex Sets

Paolo Dulio[1], Andrea Frosini[2], Simone Rinaldi[3], Lama Tarsissi[4(⊠)], and Laurent Vuillon[4]

[1] Dipartimento di Matematica "F. Brioschi", Politecnico di Milano,
Piazza Leonardo da Vinci 32, 20133 Milano, Italy
paolo.dulio@polimi.it

[2] Dipartimento di Matematica e Informatica "U. Dini", Università di Firenze,
Viale Morgagni 65, 50134 Firenze, Italy
andrea.frosini@unifi.it

[3] Dipartimento di Ingegneria dell'Informazione e Scienze Matematiche,
Università di Siena, Via Roma, 56, 53100 Siena, Italy
rinaldi@unisi.it

[4] Laboratoire de Mathématiques, Université de Savoie Mont Blanc,
CNRS UMR 5127, 73376 Le Bourget du Lac, France
lama.tarsissi@uiv-smb.fr, laurent.vuillon@univ-smb.fr

Abstract. Digital convex (DC) sets plays a prominent role in the framework of digital geometry providing a natural generalization to the concept of Euclidean convexity when we are dealing with polyominoes, i.e., finite and connected sets of points. A result by Brlek, Lachaud, Provençal and Reutenauer (see [4]) on this topic sets a bridge between digital convexity and combinatorics on words: the boundary word of a *DC* polyomino can be divided in four monotone paths, each of them having a Lyndon factorization that contains only Christoffel words.

The intent of this paper is to provide some local properties that a boundary words has to fulfill in order to allow a single point modifications that preserves the convexity of the polyomino.

Keywords: Digital convexity · Discrete geometry · Discrete tomography · Reconstruction problem

1 Introduction

Digital convex sets play a prominent role in the framework of digital geometry providing a natural generalization to the concept of Euclidean convexity. It is not so easy to define digital convex sets because for example in the papers of Sklansky [9] and Minsky and Papert [9] digital convex sets may contain many connected components. In order to have exactly one connected component, Chaudhuri and Rosenfeld [5] impose implicitly that a digital convex set must be a polyomino. Recall that a *polyomino* P is a simply connected union of unit squares, that is a union of unit squares without holes. In fact the two authors propose the notion

© Springer International Publishing AG 2017
S. Brlek et al. (Eds.): WORDS 2017, LNCS 10432, pp. 164–176, 2017.
DOI: 10.1007/978-3-319-66396-8_16

of DL-convexity (where DL means digital line) and by definition a region is DL-convex if, for any two squares belonging to it, there exists a digital straight line between them all of whose squares belong to the region. Thus for Chaudhuri and Rosenfeld the region must be a polyomino. Debled-Rennesson, Rémy and Rouyer-Degli have worked on the arithmetical properties of discrete segments in order to detect the convexity of polyominoes (see [8]). Another nice result on this topic using a bridge between digitally convex notion and combinatorics on words is stated by Brlek, Lachaud, Provençal and Reutenauer (see [4]). Indeed, a poly-omino P is described by its boundary word b. The boundary word b can be divided in 4 monotone paths and we compute the Lyndon factorization of each path. If each of these factorizations contains only Christoffel words then we have a digitally convex polyomino. We will recall these definitions and use technics to address the following problem: how to give a sequence of single square modifications of a digitally convex polyomino in order to remain at each step in the set of digitally convex polyominoes? This approach is usual in discrete tomography where we would like to reconstruct a polyomino from the horizontal and vertical projections. For example, if we consider HV-convex polyominoes which are polyominoes formed by horizontal and vertical bars of squares, Barcucci et al. [1] shown that the enumeration of all solutions with given projections could be done using switching components. Switching components are local modifications on the boundary of the polyomino which preserve the horizontal and vertical projections. In this paper, we point out the positions where by a single modification of a square the digital convexity of the whole polyomino is maintained. These local decompositions allow us to detect the possible positions of a switching component. And this is crucial in order to make only modifications in the set of digitally convex polyominoes.

2   Preliminaries

Let A be a finite alphabet and w be the word obtained by the concatenation of finite letters of A. We write $w = l_1 l_2 \ldots l_n$ where n represents the length of w and denoted by $|w|$. For all $l \in A$, the number of occurrences of this letter in a word w is denoted: $|w|_l$. The set of finite words is denoted A^*, the empty word is denoted by ϵ and by convention $A^+ = A^* \backslash \{\epsilon\}$. The word $\tilde{w} = l_n \ldots l_2 l_1$ is the reversal of $w = l_1 l_2 \ldots l_n$, where w is called a palindrome if $\tilde{w} = w$. Let p represent the period of the word w such that $w_{i+p} = w_i$ for all $1 \leq i \leq |w| - p$. The following notation $w^k = w^{k-1}.w$ represents the kth power of $w \in A^*$, where $w^0 = \epsilon$. A word w is said *primitive* if it is not the power of a nonempty word.

Two words w and w' in A^* are conjugate if there exists $u, v \in A^*$, such that: $w = uv$ and $w' = vu$. The set of all circular permutations of a word w of its letters is equivalent to the conjugacy class of the word that is defined as the set of all the conjugates of w.

We call $w' = w[i, j]$ a factor of w if it is a subword of w of length $j - i + 1$, starting from the ith position of w to its jth position, where $1 \leq i < j$. For brevity sake, we write $w[i]$ in place of $w[i, i]$. Respectively, if $w = uv$ where u and v are nonempty words in A^*, u is called prefix of w and v is its suffix.

2.1   Digital Convexity and Convexity on Polyominoes

Kim and Rosenfeld introduced different characterizations of discrete convex sets, where a set in Euclidean geometry is convex if and only if for any pair of points p_1, p_2 in a region R, the line segment joining them is completely included in R. In discrete geometry on square grids, this notion refers to the convexity of unit squares.

Let $x = (x_1, x_2)$ be an an element of $\mathbb{Z}^2$, we define two norms: $||x||_1 = \sum_{i=1}^2 |x_i|$ and $||x||_\infty = \max\{|x_1|, |x_2|\}$ to study the connectedness in $\mathbb{Z}^2$.

Definition 1. *A sequence of points* $p_0, p_1, \ldots, p_n \in \mathbb{Z}^2$ *is* $k-$*connected ;* $k \in \mathbb{N}^*$ *if* $||p_i - p_{i-1}||_\infty \le 1$ *and* $||p_i - p_{i-1}||_1 \le k$.

Definition 2. *A path P is* $k-$*connected if* $\forall x, y \in P, \exists p_0, p_1, \ldots, p_n$ *points that are* $k-$*connected such that* $p_0 = x$, $p_n = y$ *and* $p_i \in P\ \forall i$.

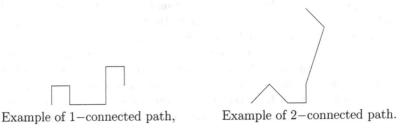

Example of 1$-$connected path,   Example of 2$-$connected path.

In this paper, we deal with polyominoes that are defined as following:

Definition 3. *We call* $P \in \mathbb{Z}^2$ *a polyomino if P is a* 1$-$*connected and finite with* $\mathbb{Z}^2/P$ *also* 1$-$*connected.*

Given a finite subset S of $\mathbb{Z}^2$, its *convex hull* is defined as the intersection of all Euclidean convex sets containing S. We say that a polyomino P is *digitally convex* if the convex hull denoted $conv(P)$ and $\mathbb{Z}^2$ are in P. In other words: $conv(P) \cap \mathbb{Z}^2 \in P$.

In Sect. 3, we define the boundary of a polyomino and we give a second equivalent definition for the convexity.

2.2   Christoffel and Lyndon Words

Now we provide the basic definitions and some known results about Christoffel and Lyndon words:

Christoffel Words. In 1771, Jean Bernoulli [2] was the first to give the definition of Christoffel words in the discrete plane and in 1990 Jean Berstel gave them this name with respect to Elwin. B. Christoffel (1829–1900).

Let a, b be two co-prime numbers, i.e. $\gcd(a, b) = 1$. The *Lower* Christoffel path of slope a/b is the 1$-$connected path joining the origin $O(0, 0)$ to the point (b, a) and respecting the following characteristics:

1. it is the nearest path below the Euclidean line segment joining these two points;
2. there are no points of $\mathbb{Z} \times \mathbb{Z}$ between the path and line segment.

Analogously, the *Upper* Christoffel path is the path that lies above the line segment. By convention, the Christoffel path is exactly the *Lower* Christoffel path.

The Christoffel word of slope a/b, denoted $C(\frac{a}{b})$, is a word defined on a binary alphabet $A = \{0, 1\}$ that codes the path. We obtain the Christoffel word by assigning a 0 for each horizontal step and a 1 for each vertical one for the Christoffel path of slope a/b. We get that the fraction $\frac{a}{b}$ is exactly: $\frac{|w|_1}{|w|_0}$. The following result is known.

Property 1. Let $w = C\left(\frac{a}{b}\right)$ be the Christoffel word of slope a/b, we write $w = 0w'1$, where w' is a palindrome.

We name w' the central part of w. Note that the lower and upper Christoffel words have the same central part.
We define the morphism $\rho : A^* \longrightarrow \mathbb{Q} \cup \{\infty\}$ by:

$$\rho(\epsilon) = 1 \text{ and } \rho(w) = \frac{|w|_1}{|w|_0} \ \forall \ w \neq \epsilon \in A^*;$$

where $\frac{1}{0} = \infty$. This morphism determines the slope for each given word in A^*.

Example 1. Consider the line segment joining the origin $O(0,0)$ to the point $(8,5)$. We have $a = 5, b = 8$ and $n = a + b = 13$. The Christoffel word of slope $5/8$ is: $C(\frac{5}{8}) = 00100101001010$.

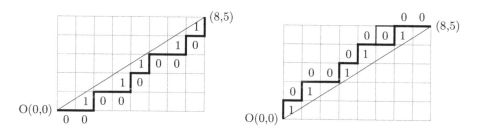

Fig. 1. The Lower and Upper Christoffel word of slope $5/8$ are 0010010100101 and 1010010100100, respectively.

The word $w = C(\frac{5}{8}) = 0\boxed{01001010010}1$, where the central part 01001010010 is a palindrome.

Another definition for the Christoffel word of slope a/b was introduced by Christoffel [7].

Definition 4. *Suppose a and b are relatively prime and* $(b, a) \neq (0, 1)$. *The label of a point* (i, j) *on the Christoffel path of slope* $\frac{a}{b}$ *is the number* $\frac{ia-jb}{b}$. *That is, the label of* (i, j) *is the vertical distance from the point* (i, j) *to the line segment from* $(0, 0)$ *to* (b, a).

Lemma 1. *Suppose w is a Christoffel word of slope a/b with a and b relatively prime. If* $\frac{s}{b}$ *and* $\frac{t}{b}$ *are two consecutive labels, as defined in Definition 4, on the Christoffel path from* $(0, 0)$ *to* (b, a), *then* $t \equiv s + a \mod (a + b)$. *Moreover, t takes as values each integer* $0, 1, 2, \ldots, a + b - 1$ *exactly once.*

This part shows that every Christoffel word can be expressed as the product of two Christoffel words in a unique way. This factorization is called the *standard factorization* and was introduced by Jean-Pierre Borel and Laubie [3] (Fig. 2).

Definition 5. *The* maximal point *of a given Christoffel word w, is the point P of w closest to the line segment. Analogously, the* minimal point *Q of w is the furthest from the line segment.*

By Lemma 1, we obtain that the maximal and minimal points of a Christoffel word are obtained when t takes the values 1 and -1, respectively.

Definition 6. *The* standard factorization *of the Christoffel word* $w = C(\frac{a}{b})$ *of slope a/b is the factorization* $w = (w_1, w_2)$, *where* w_1 *encodes the portion of the Christoffel path from* (0, 0) *to P and* w_2 *encodes the portion from P to* (b, a).

Example 2. The standard factorization of the Christoffel word of slope 5/8 is: $C(\frac{5}{8}) = (00100101, 00101)$ and is shown in the Fig. 2.

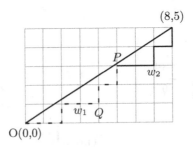

Fig. 2. The standard factorization of $C(\frac{5}{8}) = (00100101, 00101)$.

Theorem 1 *(Borel, Laubie [3]).* *A nontrivial Christoffel word w (i.e. different from* $(0)^n$ *or* $(1)^n$*) has a unique factorization* $w = (w_1, w_2)$ *with* w_1 *and* w_2 *Christoffel words.*

Lyndon Words. In 1954, Roger Lyndon introduced the *Lyndon words* by defining an order relation over all the words in A^*. The *Standard lexicographic sequence* order denoted $<_l$ is defined as the alphabetic order defined in a dictionary. Hence for the two words $w = 00101$ and $w' = 01001$, we have $w <_l w'$. The order relation between w and w' can be defined as follows:

$$w <_l w' \; if \; w' = w.u \; where \; u \in A^*, \; or$$
$$w = v0z \; and \; w' = v1z' \; where \; v, z \; and \; z' \in A^*.$$

Definition 7. *Let* $w = uv$ *with* $u, v \in A^+$, w *is a Lyndon word if it is the smallest between all its conjugates with respect to the lexicographic order.*

Note that Lyndon words are always primitive.

Example 3. The word $w = 00101$ of length 5 is a Lyndon word since it is the minimal element of the set of all conjugates of w. Hence, we can write: $00101 = \min\{00101, 01010, 01001, 10100, 10010\}$. While 0010100101 is not a Lyndon word since it is not primitive.

An important factorization is deduced from the definition of Lyndon words and given by the following theorem by Chen, Fox and Lyndon in [6]

Theorem 2. *Every non-empty word* w *admits a unique factorization as a lexicographically decreasing sequence of Lyndon words. We write* $w = w_1^{n_1} w_2^{n_2} \ldots w_k^{n_k}$, *such that* $w_k <_l \cdots <_l w_2 <_l w_1$, $n_i \geq 1$ *and* w_i *are Lyndon words for all* $1 \leq i \leq k$.

3   Theoretical Results

The definition of (digital) convexity of a connected set S implies that the set can be described by a word on a four letters alphabet Σ that codes its border, say $Bd(S)$. Each of the four letters in $\Sigma = \{1, \overline{1}, 0, \overline{0}\}$ provides a step along the border in one of the four different directions North, South, East and West, respectively.

To reach a standard coding of the border of $Bd(S)$, it can be noticed that a convex set touches the border of its minimal bounding rectangle in four bars, called *(N)orth, (S)outh, (E)ast* and *(W)est foot*. Moving counterclockwise on the border of the set, let us denote the ending corner of each foot by N, W, S and E according to the correspondent foot, as shown in Fig. 3. The word $Bd(S)$ starts from W and runs clockwise along the border of S in a closed path: $Bd(S)$ can be factorized into four non-void sub-paths WN, NE, ES and SW each using only two of the four steps in Σ to connect the related points; such a factorization is called *standard*. We say that a word in $\{0, 1\}$ is *WN-convex* if it is the NW path of a convex set. The NE, ES, and SW convexity can be defined similarly. Obviously, a path is convex if its standard factorization is made by four paths that are NW, NE, ES, and SW convex.

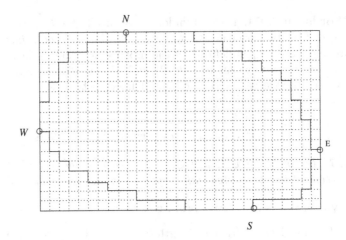

Fig. 3. A convex polyomino and its standard factorization. The word $w \in \{0,1\}^*$ coding the WN path is $w = 1110110110100100001$.

3.1 Perturbations on the WN Paths

From now on, we will consider the WN path only, assuming that all the properties hold for the other three paths up to rotations. In [4], the authors characterized the words that are border of a convex connected set by means of Lyndon and Christoffel words:

Property 2. A word w is WN-convex iff its unique Lyndon factorization $w_1^{n_1} w_2^{n_2} \ldots w_k^{n_k}$ is such that all w_i are primitive Christoffel words.

Such a result highlights the fact that a WN convex path is composed by line segments, i.e. Christoffel words, having a decreasing slope, so that they respect the lexicographical order and produce a Lyndon factorization. Furthermore, in the same paper, the authors pointed out that such a decomposition can be obtained in linear time.

In order to define a procedure that reconstructs convex sets from projections, we are interested in finding a set of loci of a WN convex path where it is possible to add one single point without loosing the convexity.

Let us consider a primitive Christoffel word w and define $min(w)$ to be the length of the prefix that reaches its minimal point as in Definition 5.

As an example, let $w = 00100101$ be a primitive Christoffel word; its minimal point is at position $(4, 1)$ reached by the prefix 00100 of w, so $min(w) = |00100| = 5$. Since we assume w to be primitive, then $min(w)$ is unique. Figure 4 shows a WN path and the minimal points of the Christoffel words.

By definition, we remark that if $k = min(w)$, then $w[k] = 0$, and $w[k+1] = 1$. The following property states that if we flip the elements of w at positions k and $k+1$ the obtained word w' is not a Christoffel word; on the other hand, it can be split into two words $w'[1, k]$ and $w'[k + 1, n]$, with $n = |w'|$, that are Christoffel words. Furthermore, position k is the only one allowing such a decomposition:

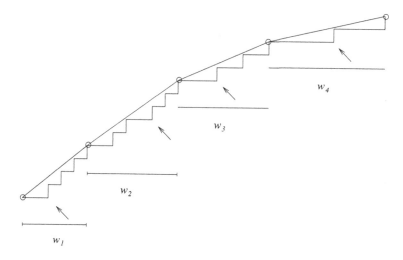

Fig. 4. A WN path and its decomposition into four Christoffel words w_1, w_2, w_3, and w_4 related to four line segments. The four minimal points of each segment are highlighted.

Proposition 1. *Let w be a primitive Christoffel word of length n and $k = min(w)$.*

(i) The words $u = w[1, k-1]\,1$ and $v = 0\,w[k+2, n]$, are two Christoffel words;
(ii) for each k' different from k, the words $u' = w[1, k'-1]1$ and $v' = 0w[k'+2, n]$ are not both Christoffel words.

The proof is a direct consequence of Property 1 and Lemma 1. We point out the following immediate and useful consequence:

Corollary 1. *Let w, u and v be as in Property 1. It holds*

(i) $\rho(u) > \rho(v)$;
(ii) $\rho(w[1, k']1) > \rho(u)$, for each $k' \neq k(= min(w))$, and $w[k'] = 0$. A symmetric result holds for v.

3.2   Definition of the *split* Operator

Proposition 1 allows us to define a *split operator*, that acts on the Christoffel word w and provides as output the concatenation of the two words u and v, by simply flipping the sequence 01 at position k and $k+1$ of w into the sequence 10, i.e., $split(w) = u\,v$. From now on, we consider the extension of the operator to *sequences* of Christoffel words, and we index it with the (index of the) sub-word where the split takes place, i.e., if $w = w_1\,w_2 \ldots w_n$ is a sequence of primitive Christoffel words, then $split_k(w) = w_1\,w_2 \ldots split(w_k) \ldots w_n$. Consecutive applications of the split operator to the word w will be indexed by the sequence of the indexes of the involved sub-words.

Given two words p_1 and p_2 of the same length l, we say that p_1 is *greater* than or equal to p_2 ($p_1 \geq p_2$), if for each $k \leq l$ it holds $|p_1[1 \cdots k]|_1 \geq |p_2[1 \cdots k]|_1$. The "$\geq$" relation is a natural partial ordering on words.

As an immediate consequence we have:

Property 3. Let $w = w_1 \, w_2 \ldots w_n$ be a sequence of Christoffel words. It holds:

(i) $split_k(w) \geq w$, with $k \leq n$;
(ii) the split operator commutes with respect to successive applications, i.e., $split_{k,h}(w) = split_{h,k}(w)$.

Attention must be paid when we are dealing with sequences of Christoffel words that are paths of a convex polyomino: the split operator provides an

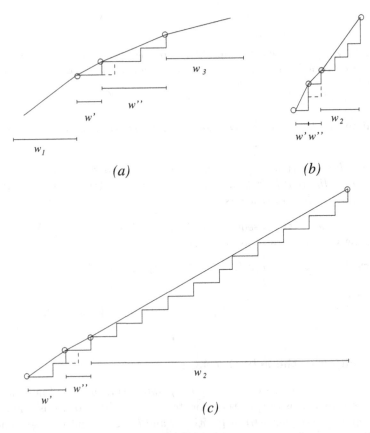

Fig. 5. Three WN paths of a convex polyomino: (a) the Lyndon factorization and the global convexity are preserved; (b), the global convexity is preserved, but not the Lyndon factorization that has to be modified including w'' in w_2, i.e., $w_2' = w'' w_2 = (01)(0101011)$; (c) the global WN convexity is lost. Observe that the word $w''w_2 = (001)(00100100100100101001001001001001) $ is not a Christoffel word. So, we need a second split in w_2 and the addition of a further point to obtain a WN path back.

efficient way to add one point on a line segment of the border of the polyomino without loosing the convexity on that segment, but it does not guarantee either to preserve the Lyndon factorization of the related word or its convexity.

We can classify the perturbations performed by the split operator on the factor w_i (i.e. $split(w_i) = u_i \, v_i$) of the Lyndon decomposition of a convex path into three different types, according to the values of the slopes of the consecutive Lyndon factors after the perturbation:

(a) the Lyndon factorization and the global convexity are preserved (see Fig. 5 (a)), i.e., the two new factors u_i and v_i globally preserve the slope decreasing of the line segments of the path;
(b) the Lyndon factorization is not preserved but the obtained path is still convex (see Fig. 5 (b)), i.e., $w_{i-1}u_iv_iw_{i+1}$, with w_{i+1} eventually void, is not a Lyndon factorization, so it does not preserve the slope decreasing of the line segments of the path, while the new Lyndon factorization does;
(c) neither the Lyndon factorization nor the convexity are preserved (see Fig. 5 (c)), i.e., $w_{i-1}u_iv_iw_{i+1}$, with w_{i+1} eventually void, is not a Lyndon factorization. Furthermore, the new Lyndon factorization is not composed by Christoffel words only.

3.3   Commutativity of the *split* Operator

In what follow, we are interested in showing under which assumptions the commutative behavior of the split operator is preserved in a WN path (as already underlined, by symmetry the found results hold for the remaining three kinds of paths). In particular, we are going to show that, if the split operator produces perturbations of type (a) on two consecutive Christoffel words in a same octant of a WN path, then the result of the two successive applications is independent from their appliance order.

Theorem 3. *Let w_1 and w_2 be two consecutive Christoffel words in the same octant of a WN path of a polyomino, and let $split(w_1) = u_1 \, v_1$ and $split(w_2) = u_2 \, v_2$. If $\rho(v_1) > \rho(w_2)$ and $\rho(w_1) > \rho(u_2)$ (i.e. the split operator provides two perturbations of type (a) on w_1 and w_2), then it holds $\rho(v_1) > \rho(u_2)$.*

Proof. Let $\rho(u_2) = \frac{b}{a}$ and $\rho(v_1) = \frac{b'}{a'}$, and assume without loss of generality that w_1 and w_2 lie in the upper octant, i.e., $a > b$, and $a' > b'$. Since the split operator acts on $min(w_1)$ and $min(w_2)$, then it holds

$$\frac{b-1}{a-1} < \rho(w_2) < \frac{b}{a}, \text{ and } \frac{b'-1}{a'-1} < \rho(w_1) < \frac{b'}{a'}. \qquad (1)$$

Let us proceed by contradiction, assuming that $\rho(v_1) < \rho(u_2)$, i.e., $ab' < ba'$. We first prove that the inequality

$$\frac{b'-1}{a'-1} < \frac{b-1}{a-1} \qquad (2)$$

is always satisfied: several cases have to be considered according to the mutual dimensions of the four parameters a, a', b, and b':

Case (1) $a = a'$. From $a = a'$ it follows $b' < b$ and consequently $b' - 1 < b - 1$, so Inequality 2 holds. The case $b = b'$ is symmetrical.

Case (2) $a < a'$. Two subcases arise: if $b' < b$, then we have $b' - 1 < b - 1$. Since we also have $a - 1 < a' - 1$, then Inequality 2 again is a direct consequence. On the other hand, let $b' > b$. In this case we show that a contradiction arises, i.e., $\rho(w_1) < \rho(u_2)$, against the hypothesis.

The Christoffel tree is isomorphic to the Stern-Brocot tree that contains all the irreducible fractions. The fractions are distributed all over the tree using the Farey addition, which is:

$$\frac{a}{b} \oplus \frac{c}{d} = \frac{a+c}{b+d}.$$

Let the two Christoffel words w_1 and w_2, of slopes respectively $\rho(w_1)$ and $\rho(w_2)$, be split into Christoffel sub-words as $w_1 = u_1 \, v_1$ and $w_2 = u_2 \, v_2$, as shown in Fig. 6.

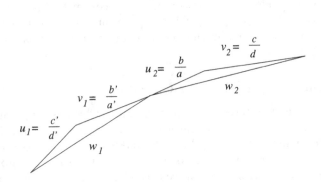

Fig. 6. The split of the Christoffel words w_1 and w_2 into (u_1, v_1) and (u_2, v_2).

We let, $\rho(u_1) = \frac{c'}{d'}$, $\rho(v_1) = \frac{b'}{a'}$, $\rho(u_2) = \frac{b}{a}$, and $\rho(v_2) = \frac{c}{d}$.

Using the construction of the Stern-Brocot tree, we know that there exist $k, t \in \mathbb{N}$ such that $c' = b' - k$ and $d' = a' - t$.

Since, by assumption, $b' > b$ and $a' > a$, there also exist $k', t' \in \mathbb{N}$ such that $b = b' - k'$ and $a = a' - t'$.

The following inequalities hold:

$$\frac{b'}{a'} < \frac{b}{a}$$
$$ab' < a'b$$
$$a'b' - a'b < a'b' - ab'$$
$$\frac{b' - b}{a' - a} < \frac{b'}{a'}$$
$$\frac{k'}{t'} < \frac{b'}{a'}$$
$$2a'k' - 2b't' < 0.$$

Reminding that we assumed to be confined in the first octant where the slopes of the Christoffel words are less than 1, then $b < a$ implies $b' - k' < a' - t'$. Therefore: $-(a' - t') < -(b' - k')$ and, since by the Stern-Brocot tree we have $\frac{c'}{d'} \oplus \frac{k}{t} = \frac{b'}{a'}$, then $\frac{k}{t} < 1$ and consequently $k < t$, then we get

$$-k(a' - t') < -t(b' - k') \text{ i.e. } -k(a' - t') + t(b' - k') < 0.$$

The inequalities gathered up to now lead to the following ones: $\rho(w_1) = \frac{b'+c'}{a'+d'} = \frac{2b'-k}{2a'-t}$ and $\rho(u_2) = \frac{b}{a} = \frac{b'-k'}{a'-t'}$ that are enough to prove that $\rho(w_1) < \rho(u_2)$ always holds against the hypothesis. In fact

$$2b'a' - 2b't' - ka' + kt' < 2a'b' - 2a'k' - b't + k't$$
$$2a'k' - 2b't' - k(a' - t') + t(b' - k') < 0$$

which is always true since $2a'k' - 2b't' < 0$ and $-k(a' - t') + t(b' - k') < 0$.

Case (3) $a > a'$. As above, two subcases arise:

- if $b < b'$, then since $\frac{1}{a} < \frac{1}{a'}$ we get $\frac{b}{a} < \frac{b'}{a'}$, i.e., a contradiction to the initial hypothesis.
- The other case concerns $b > b'$. Let us consider $a = a' + h$ and $b = b' + k$ (see Fig. 7), and consequently Inequality 2 can be written as

$$\frac{b' + k - 1}{a' + h - 1} - \frac{b' - 1}{a' - 1} > 0. \tag{3}$$

Since by hypothesis, we have $\frac{b'}{a'} = \frac{b-k}{a-h} < \frac{b}{a}$, hence $ab - ak - ab + hb < 0$ and $\frac{b}{a} < \frac{k}{h}$ as shown in Fig. 7. Therefore the following relations hold:

$$\frac{k}{h} > \frac{b}{a} > \frac{b'}{a'} > \frac{b' - 1}{a' - 1}.$$

By the last one, it holds $k(a'-1) > h(b'-1)$ and consequently $ka' - hb' - k + h > 0$ that is equivalent to Inequality 3.

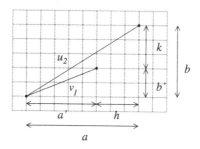

Fig. 7. An example of the case $a > a'$ and $b > b'$ in the proof of Theorem 3.

In the 3 cases, and assuming that $\rho(v_1) > \rho(u_2)$, we obtain Inequality 2. The following inequalities are deduced:

$$\frac{b'-1}{a'-1} < \frac{b-1}{a-1} < \rho(w_2) < \frac{b'}{a'} < \frac{b}{a} < \rho(w_1)$$

and, comparing with the first inequality of the second chain in Inequality 1, we get a contradiction.   □

Corollary 2. *After performing a sequence of perturbations of type* (a) *in a WN path, then the obtained path is still a WN path.*

This last result can be rephrased by saying that the split operator commutes in case of perturbations of type (a) inside the same octant. A further analysis has to be carried on in presence of perturbations of type (b), both in the same octant and in the whole quadrant.

Acknowledgment. This study has been partially supported by INDAM - GNCS Project 2017.

References

1. Barcucci, E., Del Lungo, A., Nivat, M., Pinzani, R.: Reconstructing convex polyominoes from horizontal and vertical projections. Theoret. Comput. Sci. **155**(2), 321–347 (1996). http://dx.doi.org/10.1016/0304-3975(94)00293-2
2. Bernoulli, J.: Sur une nouvelle espèce de calcul, pp. 255–284. Recueil pour les astronomes
3. Borel, J.P., Laubie, F.: Quelques mots sur la droite projective réelle. J. Théor. Nombres Bordeaux **5**(1), 23–51 (1993). http://jtnb.cedram.org/item?id=JTNB_1993_5_1_23_0
4. Brlek, S., Lachaud, J.O., ProvenÃSS, X., Reutenauer, C.: Lyndon+christoffel = digitally convex. Pattern Recogn. **42**(10), 2239–2246 (2009). Selected papers from the 14th IAPR International Conference on Discrete Geometry for Computer Imagery 2008, http://www.sciencedirect.com/science/article/pii/S0031320308004706
5. Chaudhuri, B.B., Rosenfeld, A.: On the computation of the digital convex hull and circular hull of a digital region. Pattern Recogn. **31**(12), 2007–2016 (1998)
6. Chen, K.T., Fox, R.H., Lyndon, R.C.: Free differential calculus, IV: the quotient groups of the lower central series. Ann. Math. **68**(2), 81–95 (1958). http://dx.doi.org/10.2307/1970044
7. Christoffel, E.: Observatio arithmetica. Annali di Matematica **6**, 145–152 (1875)
8. Debled-Rennesson, I., Jean-Luc, R., Rouyer-Degli, J.: Detection of the discrete convexity of polyominoes. In: Borgefors, G., Nyström, I., Baja, G.S. (eds.) DGCI 2000. LNCS, vol. 1953, pp. 491–504. Springer, Heidelberg (2000). doi:10.1007/3-540-44438-6_40
9. Sklansky, J.: Recognition of convex blobs. Pattern Recogn. **2**(1), 3–10 (1970)

Variants Around the Bresenham Method

J.-P. Borel[(✉)]

XLim, UMR 6172 - Université de Limoges - CNRS, 123 Avenue Albert Thomas,
87060 Limoges Cedex, France
borel@unilim.fr

Abstract. We present various ways of representing a segment in black-
ing or not some pixels on a computer screen. These methods include
well-known classical cases, such as the one proposed by Bresenham, or
the concept of Cutting Sequence. Our method relies on the concept of
active multi-pixel. The sequence of pixels in black or in various levels of
gray is coded on some alphabet, which depends on the multi-pixel, and
the structure of these encodings is discussed.

Keywords: Discrete geometry · Freeman codes · Bresenham method

AMS classification. 68R15.

1 Introduction

1.1 Drawing Lines

A screen can be schematized as a grid of small unit square pixels $\mathfrak{P}$, naturally
white but which can be colored in various ways. The most natural way to represent a segment is to blacken the squares encountered by this segment. It is not
the one that is chosen in practice, probably for aesthetic reasons. The method
adopted dates back to the 1960s and was proposed by Bresenham: a pixel $\mathfrak{P}$ is
blacked when the line (with a slope lower than 1) encounters the small horizontal
segment joining the midpoints of the horizontal sides of the square (See Fig. 1
ex. 2), and similarly with the small vertical segment for slopes greater than 1.

In practice, it will also be considered that the pixel can be grayed with various
levels of gray in finite number.

In the following, we will consider dynamically the segments, starting at the
origin $(0,0)$ and ending in some integer point (q,p), or half-lines originating
from $(0,0)$ and of any slope, whether rational or not. This slope will always be
positive, the other cases will naturally deduce from it. We use *lines* for these two
cases.

The unit pixels composing the screen grid are considered to be centered at
the integer points. The hypothesis of putting the integer points at the vertices
of the squares can also be made, and the results adapted.

Partially supported by Region Nouvelle Aquitaine.

S. Brlek et al. (Eds.): WORDS 2017, LNCS 10432, pp. 177–189, 2017.
DOI: 10.1007/978-3-319-66396-8_17

1.2 Our Concept: The Active Multi-pixel

Definition 1. *An active multi-pixel is an increasing sequence of closed sets:*

$$(0,0) \in \mathfrak{A}_n \subset \ldots \subset \mathfrak{A}_2 \subset \mathfrak{A}_1.$$

In practice, n corresponds to the maximum level, that is, black.

When $n = 1$, which corresponds to a single level of gray that will be black, we will call it an *active pixel*.

When $n = 1$ and $\mathfrak{A} \subset \mathfrak{P}$, we will call it a *strict active pixel*.

1.3 Representation of a Segment Using an Active Multi-pixel

The line is drawn by blackening the starting pixel at the origin and then with the gray level i the center pixel (n, m) as soon as the segment passes through the translated set $\mathfrak{A}_i + \overrightarrow{(n,m)}$, the i being the maximum index with this property.

In the case of an active pixel, all the pixels corresponding to the translates of the active pixels crossed by the segment are blackened.

1.4 Coding the Lines

The line can then be described by the finite or infinite sequence of the symbols corresponding to the translations going from one colored pixel to the next one, adding if necessary the gray level when $n \geq 2$. Thus the alphabet used is therefore a subset of $\mathfrak{T} \times \{1, 2, \ldots, n\}$ where $\mathfrak{T}$ is the set of translations of positive integer vectors. We obtain what is conventionally called the Freeman Code, [9], denoted here by FC, each letter in this finite or infinite sequence coding the translations, and possibly being indexed by the level of gray.

Subsequently, movements - which are translations of an integer vector - can be viewed in geographical terms from the four elementary movements: N for *to the North*, i.e., upwards, S, E, W which combine. These codes are known in two very classical cases, linked to the question of the possible neighbors of a pixel:

- the case with *four neighbors*, where the possible movements are N, S, E, W. This corresponds to the simplest case and appears at first, [9]. In the case which is ours (positive slope and movement to the right and upwards), only N and E are used;
- the case with *eight neighbors*, where we also allow the diagonal movements NE, NW, SW and SE, as found in [16]. Here too we need only N, E and NE.

In the following we will encounter more complex cases:

- the case with *sixteen neighbors*, where we add to the previous ones the eight possible movements of the rider in the chess game, NNE, NNO, etc. As before we need only movements N, E, NE, ENE, NNE. In practice, we will symbolize these five movements by the letters a, b, c, d, e respectively.

Similar approaches are found in many authors, for example in [8] or [17], as well as higher dimensional analogues, [5]. In all cases we describe the line using a word that codes it, and we get *discrete lines* in the sense of Réveilles [15].

2   Some Examples

2.1   General Examples

We give some examples of drawing for the segment joining the integer points $(0,0)$ and $(5,2)$, the chosen active multi-pixel is given on the right.

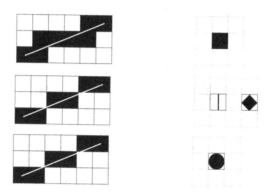

Fig. 1. Strict active pixels

The corresponding FC are respectively *abaaaba*, and *acaca* in the last two cases.

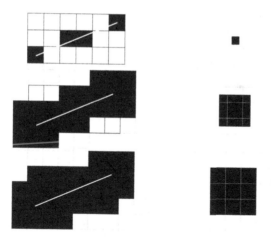

Fig. 2. Active pixels

The FC is *dad* in the first case, and we need more letters for the others. We denote by f, g the translations SSE and SE respectively. Then $bbfbbfbbgbbfbbgbbfbbfbb$ is a possible code in the second case (others are possible). The third one needs a new letter for the translation SSSE. The second and third active pixels can be used for drawing bold lines (Fig. 2).

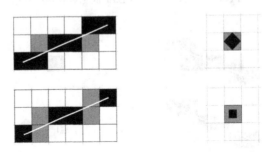

Fig. 3. Strict active multi-pixels with two levels of gray

Here letters must be indexed by the level of gray FC (1 = gray et 2 = black), and we get the FC $a_2b_1a_2a_2a_1b_2a_2$ et $a_1b_1a_2a_2a_1b_1a_2$ (Fig. 3).

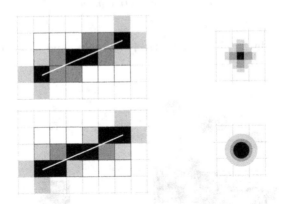

Fig. 4. Active multi-pixels with three levels of gray

2.2   Cutting Sequence

It is a matter of considering the active square pixel $\mathfrak{A} = \mathfrak{P}$, as given in the first example of Fig. 1. In practice, therefore, the pixels encountered by the line are blackened. The associated FC has been widely studied in a slightly different context, it is called classically *Cutting Sequence*, denoted here by CS, even if this term is normally used for the case where the origin is put on a vertex of the unit square. The reader will find in [1] or [2] broad presentations of this concept.

It is easy to build the CS, using the two following results.

Proposition 1. *Let u be the CS of the segment ending at the integer point (q, p). Then the CS of the segment ending at (nq, np) is equal to u^n.*

Proposition 2. *The CS of the segment ending at (q, p) is the word of length $|u| = p + q$ which can be obtained starting from the word $a^q b^p$ and iterating the following process:*

- *take the word $a^q b^p$;*
- *replace it by the new word obtained by the 3-2-decimation of its third power;*
- *stop when the new word is the same as the former one;*
- *if this last word is not a palindrome, then compare it with its reverse or mirror word (i.e., the word obtained by writing the letters in the reverse order) and replace any block ab corresponding to a block ba in the reverse word by c.*

The 3-2-decimation of the word u consists in keeping only the letters a whose rank between the a's of the word (u) is equal to 2 modulo 3, and making the same for the letters b.

This method is similar to those given in [10] or [3].

2.3   Bresenham Method

It consists in using as an active pixel the *diamond*, which connects the middles of the four edges of the unit square, as in Example 2 of Fig. 1. Indeed, crossing the vertical segment or crossing the diamond are equivalent properties for a straight line with a slope lower than 1, and we have the corresponding result when the slope is higher.

The computation of the FC can be done from the previous case, but with a small modification: in fact, it is the image by a simple transducer of u^2, see [4], where u is the classical CS, i.e., putting the origin on a vertex. This classic CS is obtained as before, but by taking 2-2 or 3-3-decimation.

2.4   A Particular Case: The Strict Active Pixel

It is clear that when the active pixel $\mathfrak{A}$ is contained in the unit square $\mathfrak{P}$, then the blackened or grayed pixels are among the blackened pixels of the CS. Thus we can only indicate for each of these pixels whether it is blackened or not ($n = 1$) or what is its level of gray, which amounts to coding this level on the alphabet $\{0, 1, \ldots, n\}$. For the segment that links $(0, 0)$ to (q, p) it is therefore of length $p + q$.

3   Properties of Active Pixels

For simplicity, active pixels with no intermediate gray levels will only be considered in this section.

3.1   Equivalent Pixels

Definition 1. *The positive convex hull of an active pixel $\mathfrak{A}$ is the intersection of the closed half-planes containing $\mathfrak{A}$, whose boundary line has a positive or zero slope.*

Two active pixels with the same positive convex hull are said to be equivalent.

It means that we add to the classical convex hull the SW et NE corners, as it can be seen on Fig. 5 below.

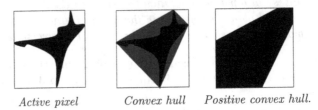

Active pixel	*Convex hull*	*Positive convex hull.*

Fig. 5. An example

A line or a segment of positive slope passes through an active pixel if and only if it passes through its positive convex hull. The FC of a segment or of an half-lines depends therefore only on the positive convex hull $\widetilde{\mathfrak{A}}$ of $\mathfrak{A}$. The two will then be confused in the following.

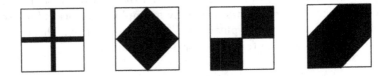

Fig. 6. Four equivalent active pixels

The second pixel in Fig. 6 corresponds to the classic case of the diamond (see for example [12]) and the fourth is its positive convex hull. It corresponds to the classical case of the four neighbors [16], as well as to the Bresenham method, as seen on the first form, [7].

3.2   Diameters

Definition 2. *The diameter seen from the angle θ of an active pixel is the length of its orthogonal projection on the line of polar angle $\theta + \frac{\pi}{2}$. It is denoted by δ_θ.*

We give three examples.

- When $\mathfrak{A}$ is a disk its diameters δ_θ does not depend on θ.
- When $\mathfrak{A} = \mathfrak{P}$ is the unit square $\delta_\theta = \sin\theta + \cos\theta$, so when $0 \le \theta \le \frac{\pi}{2}$ the diameters vary between 1 and $\sqrt{2}$, see Fig. 7.
- When we take the diamond, $\delta_\theta = \max(\sin\theta, \cos\theta)$ and the diameters vary between $1/\sqrt{2}$ and 1, see also Fig. 7.

Diameters δ_θ of equivalent active pixels are obviously the same for $0 \le \theta \le \frac{\pi}{2}$.

4   The Size of the Languages

Here, too, will be only considered to simplify strict active pixels. The results may be generalized, the constraint then holds on the exterior part $\mathfrak{A}_1$, and must be written in a slightly different way.

4.1   FC of a Line

Proposition 3. *The FC of any segment linking two integer points or any half-line starting from an integer point is written on a finite alphabet, unless it is empty.*

A segment meets the two active pixels associated with its extremities, so its FC is non-empty and of course finite. The same is true for an half-line with rational slope. Finally, for an half-line with an irrational slope, two cases may occur essentially:

- $\mathfrak{A}$ is the point $(0,0)$, and the FC is empty or does not exist;
- $\mathfrak{A}$ contains a small open disk around the origin, and the line encounters an infinite number of translated active pixels, with a bounded return time. It implies that the FC uses a finite number of letters.

4.2   FC in the Neighborhood of a Line

Proposition 4. *For any positive number ρ, there exists some neighborhood $\rho \pm \varepsilon$ and some finite alphabet $\mathcal{A}$ such that all the FC of any segment or half-line whose slope belongs to this interval can be written using $\mathcal{A}$.*

4.3   The Set of All the FC

Proposition 5. *The set of all FC of segments and half-lines can be written on a unique finite alphabet $\mathcal{A}$ if and only if the diameters δ_θ of the active pixel $\mathfrak{A}$ satisfy:*

- $\delta_{-\frac{\pi}{4}} = \sqrt{2}$;
- $\delta_{\frac{\pi}{4}} \ge \frac{\sqrt{2}}{2}$.

Then we can choose $\mathcal{A} = \{a, b, c, d, e\}$, *corresponding to the five elementary moves E, N, NE, ENE, NNE.*

The first condition means exactly that the two points $(-1/2, -1/2)$ and $(1/2, 1/2)$ belongs to the convex hull of the active strict pixel, that is to say that $\mathfrak{A}$ reaches the four edges of the unit square $\mathfrak{P}$.

The proof of this Proposition uses the notion of *visible* active pixel: a translated active pixel $\mathfrak{A} + \overrightarrow{(q, p)}$ is visible from $\mathfrak{A}$ when there exists some segment linking these two active pixels and not crossing any other translated active pixel.

We easily get that the FC only uses the alphabet of visible active pixels, and a density argument shows that all its letters must be used for some segments or half-lines. Then some elementary geometrical considerations gives Proposition 5.

The active strict pixel $\mathfrak{A}$ given in Fig. 5 is an example of a five letters alphabet. In this case each occurence of the letter d or e correspond to some discontinuity in the drawing.

4.4   Some Examples

- For any CS we need letters a and b except for horizontal or vertical segments or half-lines. Letter c only appears when the line passes though a semi-integrer point, i.e., a point on the grid. It corresponds to a segment or half-line with an odd rational slope, i.e., with odd numerator and denominator.
- For Bresenham Method, the letters a and c both appears for a slope between 0 and 1, and the letter b appears only for rational slope with an odd numerator and an even denominator. If we choose to remove the point $(1/2, 1)$ from the diamond, letter b no longer appears.
  For a slope greater to 1 we get the same result permuting letters a and b.
- If we consider the disk of diameter 1 as the strict active pixel, the three letters a, b, c are needed for all FC, except for rational slope with sufficiently small numerator and denominator in irreducible form.

5   How to Compute the Freeman Code

5.1   Automatic Computation of the FC

We have seen before in Proposition 2 that the CS can be easily computed, so we are interested in the possibility to use this CS to compute the FC associated with arbitrary active multi-pixel. However, it is not possible except in some special cases.

Theorem 1. *There exist a given transducer $\mathcal{T}$, depending only on the active multi-pixel, such that the FC of any line is the image by $\mathcal{T}$ of its CS u (half-line) or some power u^n (segment) if and only if the following properties are true:*

- *the global part $\mathfrak{A}_1$ of the active multi-pixel satisfies the properties of Proposition 5;*

– *for all $1 \leq i \leq n$ the sets $\mathfrak{A}_i$ are polygons whose vertices have rational coordinates with odd denominators.*

This result is proven in [4], in a sligthly different context and for a strict active pixel only. The general proof is similar.

5.2   The General Case: The Matrix of Factors

We consider only strict active pixels for simplicity reasons. The dynamical system corresponding to any line is an intervalle exchange transformation, the number of intervals involved being the number of different letters. These transformations have been introduced by Keane and Rauzy in the 70's, see [11] or [14], and intensively studied. Using this general principle, we can get the FC corresponding to any strict active pixel.

The process is shown in the case corresponding to the segment joining $(7,3)$, and when the active pixel is the disk of diameter 1, generalization is easy.

Step 1: build a square matrix of letters, putting seven times letter a then three times letter b vertically in the first column, and shifting upward the b's in the successive columns:

$$
\begin{array}{cccccccccc}
a & a & a & b & a & a & b & a & a & b \\
a & a & b & a & a & a & b & a & a & b \\
a & a & b & a & a & b & a & a & a & b \\
a & a & b & a & a & b & a & a & b & a \\
a & b & a & a & a & b & a & a & b & a \\
a & b & a & a & b & a & a & a & b & a \\
a & b & a & a & b & a & a & b & a & a \\
b & a & a & a & b & a & a & b & a & a \\
b & a & a & b & a & a & a & b & a & a \\
b & a & a & b & a & a & b & a & a & a \\
\end{array}
$$

The coefficients $c_{i,j}$ of this matrix are equal to b when $i + 3j \equiv 1, 2, 3 \pmod{10}$ and a elsewhere. Rows are exactly the factors of length 10 of the parallel half-lines, written in lexicographic order. Such a matrix is connected with the Burrows-Wheeler transformation, and has been studied by many authors, see [6,13] for example. We get easily the following properties.

Proposition 6.   – *Two consecutive rows differ exactly on a block ab (in the upper row) which gives a block ba (in the lower row). Then the corresponding intermediate row is given by replacing this block by a single letter c.*
  – *The classical CS corresponds to the common part of the consecutive rows ending by different letters. In our case, it corresponds to the third and the forth rows, which gives aabaabaa for the classical CS.*
  – *The CS corresponds to the row in the middle when $p + q$ is odd, or the intermediate row between the two middle rows when $p + q$ is even.*

Step 2: As said in Proposition 6 we get the CS by taking the two middle rows ($p + q = 10$ is even), so here we take the fifth and the sixth rows:

$$a\,b\,a\,a\,a\,b\,a\,a\,b\,a$$
$$a\,b\,a\,a\,b\,a\,a\,a\,b\,a$$

and we replace the block by c, so the CS is:

$$a\,b\,a\,a\,c\,a\,a\,b\,a$$

Step 3: For a given strict active pixel, look at its diameter $\delta_{\arctan\frac{3}{7}}$, in our example 1 as we have chosen the unit disk. The key point of this method is to show that we have to cancel (in our case) exactly one letter on each side, so we underline the two extremal letters a and b in the first column, then the corresponding ones in the others columns:

$$\underline{a}\,a\,a\,\underline{b}\,a\,a\,b\,a\,a\,b$$
$$a\,a\,b\,\underline{a}\,a\,a\,\underline{b}\,a\,a\,b$$
$$a\,a\,b\,a\,a\,b\,\underline{a}\,a\,a\,\underline{b}$$
$$a\,a\,\underline{b}\,a\,a\,b\,a\,a\,b\,\underline{a}$$
$$a\,b\,\underline{a}\,a\,a\,\underline{b}\,a\,a\,b\,a$$
$$a\,b\,a\,a\,b\,\underline{a}\,a\,a\,\underline{b}\,a$$
$$a\,\underline{b}\,a\,a\,b\,a\,a\,b\,\underline{a}\,a$$
$$b\,\underline{a}\,a\,a\,\underline{b}\,a\,a\,b\,a\,a$$
$$b\,a\,a\,b\,\underline{a}\,a\,a\,\underline{b}\,a\,a$$
$$\underline{b}\,a\,a\,b\,a\,a\,b\,\underline{a}\,a\,a$$

Then look at the two middle rows, as $p + q = 10$ is even:

$$a\,b\,\underline{a}\,a\,a\,\underline{b}\,a\,a\,b\,a$$
$$a\,b\,a\,a\,b\,\underline{a}\,a\,a\,\underline{b}\,a$$

and replace all the two letters-blocks whose second letter is underlined in any of these two lines by the letter c. Then we get the FC:

$$a\,c\,a\,c\,a\,c\,a$$

6 The Role of Diameters

6.1 The Average Visual Thickness

If we consider a continuous black line (segment) of length ℓ and width δ, then its black surface is equal to the product $\ell \times \delta$. The surface of a discrete line - union of black pixels - is equal to the number N of black pixels, each unit pixel being of surface 1. If we consider that the visual effect of a gray pixel at level i is the ratio i/n, we get the global visual effect, or visual surface, of the line:

$$\frac{1}{n} \sum_{i=1}^{n} N_i$$

where N_i is the number of gray pixels at the level i.

It is natural to say that the length of the discrete line is the distance between the centers of the two pixels blackened at the extremes. Choose the following definition, by analogy to the continuous case:

Definition 3. *Using the previous notations, the mean visual thickness of a discrete line is the ratio between its visual area and its length:*

$$\frac{1}{n\ell} \sum_{i=1}^{n} N_i.$$

Then we get:

Proposition 7. *The average visual thickness of a discrete line of polar angle θ and obtained using an active multi-pixel $\mathfrak{A}_i$ is:*

$$\sum_{i=1}^{N} \delta_{\theta,i} + O(\frac{1}{\ell'})$$

where ℓ' is the length of the associated irreducible segment.

Roughly speaking we have to count the integer points inside a rectangle, which can be made by looking at the orthogonal projection of these points on the direction of polar angle $\theta + \frac{\pi}{2}$. These projections are regularly spaced by $\frac{1}{\ell}$.

6.2   Size of the Diameters

More generally, if we look only at strict active pixels satisfying to Proposition 5:

– when the three letters alphabet $\{a, b, c\}$ can be used, the diameters δ_θ take any value in the domain whose limits are the curves $f_1(\theta) = \sin \theta + \cos \theta$ and $f_2(\theta) = \max(\sin \theta; \cos \theta)$: the number of blackened pixels is clearly between $p + q$ et $\max\{p; q\}$;
– when we use the five letters alphabet, the upper limit is the same, i.e., f_1, and the lower limit is $f_3(\theta) = \max(|\cos \theta - \sin \theta|; \frac{1}{2}(\cos \theta + \sin \theta))$, corresponding to the diamond whose diagonals have lengths $\sqrt{2}$ and $\frac{\sqrt{2}}{2}$.

These three curves can be seen on Fig. 7.

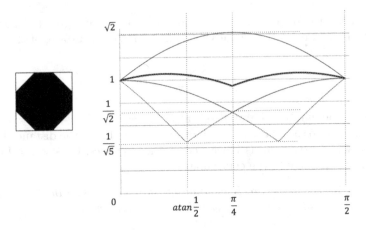

Fig. 7. Octogonal strict active pixel and variations of the diameters (bold curve)

6.3 Irregularity of the Thickness

We expect that the thickness of the representation of a line should not depend on its inclination. If we limit ourselves to the case of a strict active pixel, this means that its diameter must not depend on the polar angle θ. This is clear for the disk of diameter 1. On the other hand, it is then not possible to build the FC from the CS by a simple process, as we have seen in Theorem 1.

In both classical cases CS or FC corresponding to Bresenham Method, the ratio between the maximal and the minimal values of the diameters is equal to $\sqrt{2}$.

It is therefore interesting to find a strict active pixel for which the variation of δ_θ is small, and whose calculation of the FC from the CS is easy, using Theorem 1. It is therefore a polygon with few sides and fairly regular to be round enough. A compromise is to take the octagon build on the one third middle segment of each of the four edges of $\mathfrak{P}$. In that case we have (bold curve in Fig. 7):

$$\delta_\theta = \frac{1}{3}(\sin\theta + \cos\theta) + \frac{2}{3}\max(\sin\theta; \cos\theta).$$

The ratio between the maximal and the minimal value of the diameters is equal $\frac{\sqrt{5}}{2}$, which corresponds to a global variation of 11,8%, which can be compared to 41,2% in the two classical cases CS and Bresenham Method. Remark that for any octagon we get at least 8,2%. Moreover, we can build the FC from the CS using a transducer with 11 states, following the method given in [4].

References

1. Allouche, J.P., Shallit, J.: Automatic Sequences: Theory and Applications. Cambridge University Press, Cambridge (2003)
2. Berstel, J., Seébold, P.: Sturmian words. In: Lothaire, M. (ed.) Algebraic Combinatorics on Words. Cambridge University Press, Cambridge (2002)

3. Borel, J.P.: How to build Billiard words using decimations. RAIRO-Inf. Theor. Appl. **44**(1), 59–77 (2010)
4. Borel, J.P.: Various Methods for drawing Digitized Lines (2016)
5. Borel, J.P., Reutenauer, C.: Palindromic factors of Billiard words. Theoret. Comput. Sci. **340**(2), 334–348 (2005)
6. Borel, J.P., Reutenauer, C.: On Christoffel classes. RAIRO-Inf. Theor. Appl. **40**, 15–27 (2006)
7. Bresenham, J.E.: Algorithm for computer control of a digital plotter. IBM Syst. J. **4**(1), 25–30 (1965)
8. Crisp, D., Moran, W., Pollington, A., Shive, P.: Substitution invariant cutting sequences. J. Théorie des Nombres Bordeaux **5**, 123–137 (1993)
9. Freeman, H.: On the encoding of arbitrary geometric configuration. IRE Trans. Electron. Comput. **10**, 260–268 (1961)
10. Justin, J., Pirillo, G.: Decimations and Sturmian words. Theor. Inform. Appl. **31**, 271–290 (1997)
11. Keane, M.: Intervalle exchange transformations. Math. Z. **141**, 25–31 (1975)
12. Koplowitz, J.: On the performance of chain codes for quantization of line drawings. IEEE Trans. Pattern Anal. Mach. Intell. **PAMI-3**, 357–393 (1981)
13. Mantaci, S., Restivo, A., Sciortino, M.: Burrows-Wheeler transform and Sturmian words. Inform. Proc. Lett. **86**, 241–246 (2003)
14. Rauzy, G.: Echanges d'intervalles et transformations induites. Acta Arith. **34**, 315–328 (1979)
15. Reveilles, J.P.: Géométrie discrète, calcul en nombres entiers et algorithmique. Ph.D. thesis, University Louis Pasteur - Strasbourg, France (1991)
16. Rosenfeld, A.: Digital straight line segments. IEEE Trans. Comput. **32**(12), 1264–1269 (1974)
17. Series, C.: The geometry of Markoff numbers. Math. Intell. **7**, 20–29 (1985)

Combinatorics of Cyclic Shifts in Plactic, Hypoplactic, Sylvester, and Related Monoids

Alan J. Cain[(⊠)] and António Malheiro

Departamento de Matemática and Centro de Matemática e Aplicações, Faculdade de Ciências e Tecnologia, Universidade Nova de Lisboa, 2829–516 Caparica, Portugal
{a.cain,ajm}@fct.unl.pt

Abstract. The cyclic shift graph of a monoid is the graph whose vertices are elements of the monoid and whose edges link elements that differ by a cyclic shift. For certain monoids connected with combinatorics, such as the plactic monoid (the monoid of Young tableaux) and the sylvester monoid (the monoid of binary search trees), connected components consist of elements that have the same evaluation (that is, contain the same number of each generating symbol). This paper discusses new results on the diameters of connected components of the cyclic shift graphs of the finite-rank analogues of these monoids, showing that the maximum diameter of a connected component is dependent only on the rank. The proof techniques are explained in the case of the sylvester monoid.

Keywords: Cyclic shift · Plactic monoid · Sylvester monoid · Binary search tree · Cocharge

1 Introduction

In a monoid M, two elements s and t are related by a cyclic shift, denoted $s \sim t$, if and only if there exist $x, y \in M$ such that $s = xy$ and $t = yx$. In the plactic monoid (the monoid of Young tableaux, denoted plac; see [9, Chap. 5]), elements that have the same evaluation (that is, which contain the same number of each symbol) can be obtained from each other by iterated application of cyclic shifts [8, Sect. 4]. Furthermore, in the plactic monoid of rank n (denoted plac_n), it is known that $2n - 2$ applications of cyclic shifts are sufficient [3, Theorem 17].

To restate these results in a new form, define the *cyclic shift graph* $K(M)$ of a monoid M to be the undirected graph with vertex set M and, for all $s, t \in M$, an edge between s and t if and only if $s \sim t$. Connected components of $K(M)$ are $\sim^*$-classes (where $\sim^*$ is the reflexive and transitive closure of $\sim$), since they

---

A.J. Cain—The first author was supported by an Investigador FCT fellowship (IF/01622/2013/CP1161/CT0001).

Both authors—This work was partially supported by the Fundação para a Ciência e a Tecnologia (Portuguese Foundation for Science and Technology) through the project UID/MAT/00297/2013 (Centro de Matemática e Aplicações), and the project PTDC/MHC-FIL/2583/2014.

S. Brlek et al. (Eds.): WORDS 2017, LNCS 10432, pp. 190–202, 2017.
DOI: 10.1007/978-3-319-66396-8_18

consist of elements that are related by iterated cyclic shifts. Thus the results discussed above say that each connected component of $K(\mathsf{plac})$ consists of precisely the elements with a given evaluation, and that the diameter of a connected component of $K(\mathsf{plac}_n)$ is at most $2n - 2$. Note that connected components are of unbounded size, despite there being a bound on diameters that is dependent only on the rank.

The plactic monoid is celebrated for its ubiquity, appearing in many diverse contexts (see the discussion and references in [9, Chap. 5]). It is, however, just one member of a family of 'plactic-like' monoids that are closely connected with combinatorics. These monoids include the hypoplactic monoid (the monoid of quasi-ribbon tableaux) [7,10], the sylvester monoid (binary search trees) [5], the taiga monoid (binary search trees with multiplicities) [11], the stalactic monoid (stalactic tableaux) [6,11], and the Baxter monoid (pairs of twin binary search trees) [4]. These monoids, including the plactic monoid, arise in a parallel way. For each monoid, there is a so-called insertion algorithm that allows one to compute a combinatorial object (of the corresponding type) from a word over the infinite ordered alphabet $\mathcal{A} = \{1 < 2 < 3 < \ldots\}$; the relation that relates pairs of words that give the same combinatorial object is a congruence (that is, it is compatible with multiplication in the free monoid $\mathcal{A}^*$). The monoid arises by factoring the free monoid $\mathcal{A}^*$ by this congruence; thus elements of the monoid (equivalence classes of words) are in one-to-one correspondence with the combinatorial objects. Table 1 lists these monoids and their corresponding objects.

Table 1. Monoids and corresponding combinatorial objects.

Monoid	Symbol	Combinatorial object	Citation
Plactic	plac	Young tableau	[9, Chap. 5]
Hypoplactic	hypo	Quasi-ribbon tableau	[10]
Stalactic	stal	Stalactic tableau	[6]
Sylvester	sylv	Binary search tree	[5]
Taiga	taig	Binary search tree with multiplicities	[11, Sect. 5]
Baxter	baxt	Pair of twin binary search trees	[4]

Analogous questions arise for the cyclic shift graph of each of these monoids. In a forthcoming paper [1], the present authors make a comprehensive study of connected components in the cyclic shift graphs of each of these monoids. For several of these monoids, it turns out that each connected component of its cyclic shift graph consists of precisely the elements with a given evaluation, and that the diameters of connected component in the rank-n case are bounded by a quantity dependent on the rank. (Again, it should be emphasized that there is no bound on the *size* of these connected components.) In each case, the authors either establish the exact value of the maximum diameter or give bounds;

Table 2. Properties of connected component of the cyclic shift graph for rank-n monoids: whether they are characterized by evaluation, and known values and bounds for their maximum diameters.

Monoid	Char. by evaluation	Maximum diameter			
		Known value	Conjecture	Known bounds	
				Lower	Upper
plac_n	Y	?	$n-1$	$n-1$	$2n-3$
hypo_n	Y	$n-1$	—	—	—
stal_n	N	$\begin{cases} n-1 & \text{if } n < 3 \\ n & \text{if } n \geq 3 \end{cases}$	—	—	—
sylv_n	Y	?	$n-1$	$n-1$	n
taig_n	Y	?	$n-1$	$n-1$	n
baxt_n	N	?	?	?	?

Table 2 summarizes the results from [1]. Also, although these monoids are multihomogeneous (words in $\mathcal{A}^*$ representing the same element contain the same number of each symbol), the authors also exhibit a rank 4 multihomogeneous monoid for which there is no bound on the diameter of connected components. Thus it seems to be the underlying combinatorial objects that ensure the bound on diameters. This also is of interest because cyclic shifts are a possible generalization of conjugacy from groups to monoids; thus the combinatorial objects are here linked closely to the algebraic structure of the monoid.

The present paper illustrates these results by focussing on the sylvester monoid (denoted sylv or sylv_n in the rank-n case). (The authors previously proved that each connected component of $K(\text{sylv})$ consists of precisely the elements with a given evaluation [2, Sect. 3].) Sect. 2 recalls the definition and necessary facts about the sylvester monoid. Section 3 gives a complete proof that there is a connected component in $K(\text{sylv}_n)$ with diameter at least $n - 1$; this establishes the lower bound on the maximum diameter shown in Table 2. The complete proof that every connected component of $K(\text{sylv}_n)$ has diameter at most n, establishing the upper bound on the maximum diameter shown in Table 2, is very long and complicated. Thus Sect. 4 gives the proof for connected components consisting of elements that contain each symbol from $\{1, \ldots, k\}$ (for some k) exactly once; this avoids many of the complexities of the general case.

2   Binary Search Trees and the Sylvester Monoid

This section gathers the relevant definitions and background on the sylvester monoid; see [5] for further reading.

Recall that $\mathcal{A}$ denotes the infinite ordered alphabet $\{1 < 2 < \ldots\}$. Fix a natural number n and let $\mathcal{A}_n = \{1 < 2 < \ldots < n\}$ be the set of the first n natural numbers, viewed as a finite ordered alphabet. A word $u \in \mathcal{A}^*$ is *standard* if it contains each symbol in $\{1, \ldots, |u|\}$ exactly once.

A (*right strict*) *binary search tree* (BST) is a rooted binary tree labelled by symbols from $\mathcal{A}$, where the label of each node is greater than or equal to the label of every node in its left subtree, and strictly less than the label of every node in its right subtree. An example of a binary search tree is:

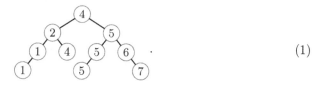

$$(1)$$

The following algorithm inserts a new symbol into a BST, adding it as a leaf node in the unique place that maintains the property of being a BST.

Algorithm 1. *Input: A binary search tree T and a symbol $a \in \mathcal{A}_n$.*

If T is empty, create a node and label it a. If T is non-empty, examine the label x of the root node; if $a \leq x$, recursively insert a into the left subtree of the root node; otherwise recursively insert a into the right subtree of the root note. Output the resulting tree.

For $u \in \mathcal{A}^*$, define $\mathrm{P_{sylv}}(u)$ to be the right strict binary search tree obtained by starting with the empty tree and inserting the symbols of the word u one-by-one using Algorithm 1, proceeding *right-to-left* through u. For example, $\mathrm{P_{sylv}}(5451761524)$ is (1). Define the relation $\equiv_{\mathrm{sylv}}$ by

$$u \equiv_{\mathrm{sylv}} v \iff \mathrm{P_{sylv}}(u) = \mathrm{P_{sylv}}(v),$$

for all $u, v \in \mathcal{A}^*$. The relation $\equiv_{\mathrm{sylv}}$ is a congruence, and the *sylvester monoid*, denoted sylv, is the factor monoid $\mathcal{A}^*/\equiv_{\mathrm{sylv}}$; the *sylvester monoid of rank n*, denoted sylv_n, is the factor monoid $\mathcal{A}_n^*/\equiv_{\mathrm{sylv}}$ (with the natural restriction of $\equiv_{\mathrm{sylv}}$). Each element $[u]_{\equiv_{\mathrm{sylv}}}$ (where $u \in \mathcal{A}^*$) can be identified with the binary search tree $\mathrm{P_{sylv}}(u)$. The monoid sylv is presented by $\langle \mathcal{A} \mid \mathcal{R}_{\mathrm{sylv}} \rangle$, where

$$\mathcal{R}_{\mathrm{sylv}} = \{(cavb, acvb) : a \leq b < c,\ v \in \mathcal{A}^*\};$$

the monoid sylv_n is presented by $\langle \mathcal{A}_n \mid \mathcal{R}_{\mathrm{sylv}} \rangle$, where the set of defining relations $\mathcal{R}_{\mathrm{sylv}}$ is naturally restricted to $\mathcal{A}_n^* \times \mathcal{A}_n^*$. Notice that sylv and sylv_n are multihomogeneous.

A *reading* of a binary search tree T is a word formed from the symbols that appear in the nodes of T, arranged so that the child nodes appear before parents. A word $w \in \mathcal{A}^*$ is a reading of T if and only if $\mathrm{P_{sylv}}(w) = T$. The words in $[u]_{\equiv_{\mathrm{sylv}}}$ are precisely the readings of $\mathrm{P_{sylv}}(u)$.

A binary search tree T with k nodes is *standard* if it has exactly one node labelled by each symbol in $\{1, \ldots, k\}$; clearly T is standard if and only if all of its readings are standard words.

The *left-to-right postfix traversal*, or simply the *postfix traversal*, of a rooted binary tree T is the sequence that 'visits' every node in the tree as follows: it recursively performs the postfix traversal of the left subtree of the root of T, then recursively performs the postfix traversal of the right subtree of the root of T, then visits the root of T. The *left-to-right infix traversal*, or simply the *infix traversal*, of a rooted binary tree T is the sequence that 'visits' every node in the tree as follows: it recursively performs the infix traversal of the left subtree of the root of T, then visits the root of T, then recursively performs the infix traversal of the right subtree of the root of T. Thus the postfix and infix traversals of any binary tree with the same shape as (1) visit nodes as shown on the left and right below:

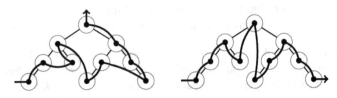

The following result is immediate from the definition of a binary search tree, but it is used frequently:

Proposition 2. *For any binary search tree T, if a node x is encountered before a node y in an infix traversal, then $x \leq y$.*

In this paper, a *subtree* of a binary search tree will always be a rooted subtree. Let T be a binary search tree and x a node of T. The *complete subtree of T at* x is the subtree consisting of x and every node below x in T. The *path of left child nodes in T from x* is the path that starts at x and enters left child nodes until a node with empty left subtree is encountered.

Let B be a subtree of T. Then B is said to be *on* the path of left child nodes from x if the root of B is one of the nodes on this path. The *left-minimal* subtree of B in T is the complete subtree at the left child of the left-most node in B; the *right-maximal* subtree of B in T is the complete subtree at the right child of the right-most node in B.

In diagrams of binary search trees, individual nodes are shown as round, while subtrees are shown as triangles. An edge emerging from the top of a triangle is the edge running from the root of that subtree to its parent node. A vertical edge joining a node to its parent indicates that the node may be either a left or right child. An edge emerging from the bottom-left of a triangle is the edge to that subtree's left-minimal subtree; an edge emerging from the bottom-right of a triangle is the edge to that subtree's right-maximal subtree.

3 Lower Bound on Diameters

Let $u \in A_n^*$ be a standard word. The *cocharge sequence* of u, denoted cochseq(u), is a sequence (of length u) calculated from u as follows:

1. Draw a circle, place a point $*$ somewhere on its circumference, and, starting from $*$, write u anticlockwise around the circle.
2. Label the symbol 1 with 0.
3. Iteratively, after labelling some i with k, proceed clockwise from i to $i + 1$. If the symbol $i + 1$ is reached *before* $*$, label $i + 1$ by $k + 1$. Otherwise, if the symbol $i + 1$ is reached *after* $*$, label $i + 1$ by k.
4. The sequence whose i-th term is the label of i is cochseq(u).

For example, for the word $u = 1246375$, the labelling process is shown on the right, and it follows that cochseq(u) = $(0, 0, 0, 1, 1, 2, 2)$. Notice that the first term of a cocharge sequence is always 0, and that each term in the sequence is either the same as its predecessor or greater by 1. Thus the i-th term in the sequence always lies in the set $\{0, 1, \ldots, i - 1\}$.

Labelling

The usual notion of 'cocharge' is obtained by summing the cocharge sequence (see [9, Sect. 5.6]).

Lemma 3. *1. Let $u \in A_n^*$ and $a \in A_n \setminus \{1\}$ be such that ua is a standard word. Then* cochseq(ua) *is obtained from* cochseq(au) *by adding 1 to the a-th component.*

2. Let $x, y \in A_n^*$ be such that $xy \in A_n^*$ is a standard word. Then corresponding components of cochseq(xy) *and* cochseq(yx) *differ by at most 1.*

Proof. Consider how a is labelled during the calculation of cochseq(ua) and cochseq(au):

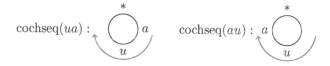

In the calculation of cochseq(ua), the symbol $a-1$ receives a label k, and then a is reached *after* $*$ is passed; hence a also receives the label k. In the calculation of cochseq(au), the symbols $1,\ldots,a-1$ receive the same labels as they do in the calculation of cochseq(ua), but after labelling $a-1$ by k the symbol a is reached *before* $*$ is passed; hence a receives the label $k+1$; after this point, labelling proceeds in the same way. This proves part (1). For part (2), notice that one of x and y does not contain the symbol 1; the result is now an immediate consequence of part (1).

Proposition 4. *Let $u,v \in \mathcal{A}_n^*$ be standard words such that $u \equiv_{\mathsf{sylv}} v$. Then* cochseq($u$) = cochseq($v$).

Proof. It suffices to prove the result when w and w' differ by a single application of a defining relation $(cavb, acvb) \in \mathcal{R}_{\mathsf{sylv}}$ where $a \leq b < c$. In this case, $w = pcavbq$ and $w' = pacvbq$, where $p,q,v \in \mathcal{A}_n^*$ and $a,b,c \in \mathcal{A}_n$ with $a \leq b < c$. Since w and w' are standard words, $a < b$.

Consider how labels are assigned to the symbols a, b, and c when calculating the cocharge sequence of w:

cochseq(w) :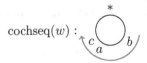

Among these three symbols, a will receive a label first, then b, then c. Thus, after a, the labelling process will pass $*$ at least once to visit b and only then visit c. Thus if we interchange a and c, we do not alter the resulting labelling. Hence cochseq(w) = cochseq(w').

For any standard binary tree T in sylv_n, define cochseq(T) to be cochseq(u) for any standard word $u \in \mathcal{A}_n^*$ such that $T = \mathsf{P}_{\mathsf{sylv}}(u)$. By Proposition 4, cochseq(T) is well-defined.

Proposition 5. *The connected component of $K(\mathsf{sylv}_n)$ consisting of standard elements has diameter at least $n-1$.*

Proof. Let $t = 12\cdots(n-1)n$ and $u = n(n-1)\cdots 21$, and let

$$T = \mathsf{P}_{\mathsf{sylv}}(t) = \qquad\qquad\qquad \text{and } U = \mathsf{P}_{\mathsf{sylv}}(u) =$$

Since T and U have the same evaluation, they are $\sim^*$-related by [2, Sect. 3], and so in the same connected component of $K(\mathsf{sylv}_n)$. Let $T = T_0, T_1, \ldots, T_{m-1}, T_m = U$ be a path in $K(\mathsf{sylv}_n)$ from T to U. Then for $i = 0, \ldots, m-1$, we have $T_i \sim T_{i+1}$. That is, there are words $u_i, v_i \in \mathcal{A}_n^*$ such that $T_i = \mathsf{P}_{\mathsf{sylv}}(u_i v_i)$ and $T_{i+1} = \mathsf{P}_{\mathsf{sylv}}(v_i u_i)$. By Lemma 3(2), $\mathrm{cochseq}(T_i)$ and $\mathrm{cochseq}(T_{i+1})$ differ by adding 1 or subtracting 1 from certain components. Hence corresponding components of $\mathrm{cochseq}(T)$ and $\mathrm{cochseq}(U)$ differ by at most m. Since $\mathrm{cochseq}(T) = (0, 0, \ldots, 0, 0)$ and $\mathrm{cochseq}(U) = (0, 1, \ldots, n-2, n-1)$, it follows that $m \geq n-1$. Hence T and U are a distance at least $n-1$ apart in $K(\mathsf{sylv}_n)$.

4   Upper Bounds on Diameters

Proposition 6. *Any two standard elements of* sylv_n *are a distance at most* n *apart in* $K(\mathsf{sylv}_n)$.

Proof. Since sylv_m embeds into sylv_n for all $m \leq n$, and since $K(\mathsf{sylv}_m)$ is the subgraph of $K(\mathsf{sylv}_n)$ induced by sylv_m, this result follows from Lemma 8 below.

Lemma 8 proves that in $K(\mathsf{sylv}_n)$ there is a path of length at most n between two standard elements with the same number of nodes. First, however, the strategy used to construct such a path is illustrated in the following example.

Example 7. Let

The aim is to build a sequence $T = T_0 \sim T_1 \sim T_2 \sim T_3 \sim T_4 \sim T_5 = U$. Consider the postfix traversal of U. The 5 steps in this traversal are shown below on the right, together with the relevant cases in the proof of Lemma 8. The parts of U that have been visited already at each step are outlined. The idea is that the h-th cyclic shift leads to a tree T_h where copies of the outlined parts of U appear on the path of left child nodes from the root of T_h. Note that cyclic shifts never break up the subwords (outlined) that represent the already-built subtrees. (The difficulty in the general proof is showing that a suitable cyclic shift exists at each step.)

$T = T_0 = \mathsf{P}_{\mathsf{sylv}}(13254) \sim \mathsf{P}_{\mathsf{sylv}}(54132)$

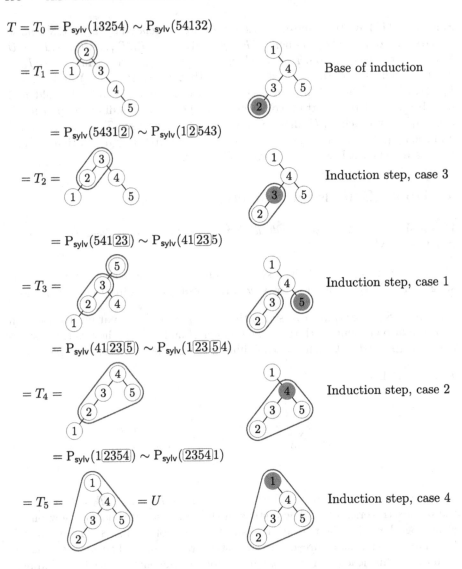

Lemma 8. *Let $T, U \in \mathsf{sylv}_n$ be standard and have n nodes. Then there is a sequence $T = T_0, T_1, \ldots, T_n = U$ with $T_h \sim T_{h+1}$ for $h = 0, \ldots, n-1$.*

Proof. This proof is only concerned with standard BSTs; thus for brevity nodes are identified with their labels. Notice that each of T and U has exactly one node labelled by each symbol in $\mathcal{A}_n$.

Consider the left-to-right postfix traversal of U; there are exactly n steps in this traversal. Let u_h be the node visited at the h-th step of this traversal.

For $h = 1, \ldots, n$, let $U_h = \{u_1, \ldots, u_h\}$ and let $U_h^\top$ be the set of nodes in U_h that do not lie below any other node in U_h. Since a later step in a postfix traversal is never below an earlier one, it follows that $u_h \in U_h^\top$ for all h. Let B_h be the subtree of U consisting of u_h and every node that is below u_h.

The aim is to construct inductively the required sequence. Let $h = 1, \ldots, n$ and suppose $U_h^\top = \{u_{i_1}, \ldots, u_{i_k}\}$ (where $i_1 < \ldots < i_k = h$). Then the tree T_h will satisfy the following conditions:

P1 The subtree B_{i_k} appears at the root of T_h.
P2 The subtrees $B_{i_k}, \ldots, B_{i_1}$ appear, in that order (but not necessarily consecutively), on the path of left child nodes from the root of T_h.

(Note that conditions P1 and P2 do not apply to T_0.)

Base of induction. Set $T_0 = T$. Take any reading of T_0 and factor it as wu_1w'. Let $T_1 = \mathsf{P}_{\mathsf{sylv}}(w'wu_1)$. Clearly T_1 has root node u_1. Since B_1 consists only of the node u_1 (since u_1 is the first node in U visited by the postfix traversal and is thus a leaf node), T_1 satisfies P1. Further, T_1 trivially satisfies P2. Finally, note that $T_0 \sim T_1$. (For an illustration, see the definition of T_1 in Example 7.)

Induction step. The remainder of the sequence of trees is built inductively. Suppose that the tree T_h satisfies P1 and P2; the aim is to find T_{h+1} satisfying P1 and P2 with $T_h \sim T_{h+1}$. There are four cases, depending on the relative positions of u_h and u_{h+1} in U:

1. u_h is a left child node and u_{h+1} is in the right subtree of the parent of u_h;
2. u_h is the right child of u_{h+1}, and u_{h+1} has non-empty left subtree;
3. u_h is the left child of u_{h+1} (which implies, by the definition of the postfix traversal, that u_{h+1} has empty right subtree);
4. u_h is the right child of u_{h+1}, and u_{h+1} has empty left subtree.

Case 1. Suppose that, in U, the node u_h is a left child node and u_{h+1} is in the right subtree of the parent of u_h. (For an illustration of this case, see the step from T_2 to T_3 in Example 7.) Then B_{h+1} consists only of the node u_{h+1}, since by the definition of a postfix traversal u_{h+1} is a leaf node. Furthermore, $U_{h+1}^\top = U_h^\top \cup \{u_{h+1}\}$.

By P1, B_h appears at the root of T_h. By Proposition 2 applied to U, the symbol u_{h+1} is greater than every node of B_h, so u_{h+1} must be in the right-maximal subtree of B_h in T_h.

As shown in Fig. 1, let λ be a reading of the left-minimal subtree of B_h. Let δ be a reading of the right-maximal subtree of B_h outside of the complete subtree at u_{h+1}. Let α and β be readings of the left and right subtrees of u_{h+1}, respectively. Note that the subtrees B_i for $u_i \in U_h^\top$ are contained in λ.

Thus $T_h = \mathsf{P}_{\mathsf{sylv}}(\alpha\beta u_{h+1}\delta\lambda B_h)$. Let $T_{h+1} = \mathsf{P}_{\mathsf{sylv}}(\delta\lambda B_h\alpha\beta u_{h+1})$; note that $T_h \sim T_{h+1}$.

In computing T_{h+1}, the symbol u_{h+1} is inserted first and becomes the root node. Since B_{h+1} consists only of the node u_{h+1}, the tree T_{h+1} satisfies P1. Since every symbol in B_h and λ is strictly less that every symbol in α, β, or δ, the trees B_h and λ are re-inserted on the path of left child nodes from the root of

$T_h =$  $= T_{h+1}$

Fig. 1. Induction step, case 1.

T_{h+1}. Thus all the subtrees B_i for $u_i \in U_{h+1}^\top$ are on the path of left child nodes from the root, and so T_{h+1} satisfies P2.

Case 2. Suppose that in U, the node u_h is the right child of u_{h+1}, and u_{h+1} has non-empty left subtree. (For an illustration of this case, see the step from T_3 to T_4 in Example 7.) Let u_g be the left child of u_{h+1}. Thus B_{h+1} consists of u_{h+1} with left subtree B_g and right subtree B_h. By the definition of the postfix traversal, u_g was visited before the h-th step, but no node above u_g has been visited. That is, $u_g \in U_h^\top$. Hence $U_{h+1}^\top = (U_h^\top \setminus \{u_g, u_h\}) \cup \{u_{h+1}\}$.

By P1, B_h appears at the root of T_h; by P2, B_g is next subtree B_{i_j} on the path of left child nodes from the root of T_h (and is thus in the left-minimal subtree of B_h). By Proposition 2 applied to U, the symbol u_{h+1} is the unique symbol that is greater than every node of B_g and less than every node of B_h. Then the node u_{h+1} may be in one of two places in T_h, leading to the two sub-cases below. In both cases, as shown in Fig. 2, let λ be a reading of the left subtree of B_g and let δ be a reading of the right-maximal subtree of B_h; note that the subtrees B_i for $u_i \in U_h^\top \setminus \{u_g, u_h\}$ are contained in λ.

1. Suppose u_{h+1} is the unique node on the path of left child nodes between B_g and B_h. In this case, as shown in Fig. 2(top), $T_h = \mathsf{P_{sylv}}(\lambda B_g u_{h+1} \delta B_h)$. Let $T_{h+1} = \mathsf{P_{sylv}}(\delta B_h \lambda B_g u_{h+1})$; note that $T_h \sim T_{h+1}$.
2. Suppose u_{h+1} is the unique node in the right-maximal subtree of B_g and there are no nodes between B_g and B_h on the path of left child nodes. In this case, as shown in Fig. 2(bottom), $T_h = \mathsf{P_{sylv}}(u_{h+1} \lambda B_g \delta B_h)$. Let $T_{h+1} = \mathsf{P_{sylv}}(\lambda B_g \delta B_h u_{h+1})$; note that $T_h \sim T_{h+1}$.

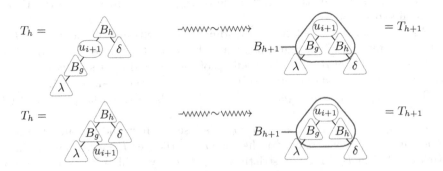

Fig. 2. Induction step, case 2, two sub-cases.

$T_h =$ 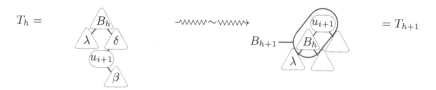 $= T_{h+1}$

Fig. 3. Induction step, case 3.

$T_h =$ 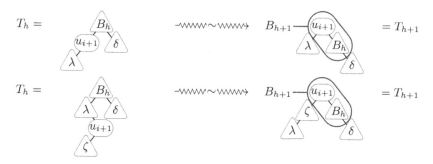 $= T_{h+1}$

$T_h =$ $= T_{h+1}$

Fig. 4. Induction step, case 4, two sub-cases.

In computing T_{h+1}, for both sub-cases, the symbol u_{h+1} is inserted first and becomes the root node. Every symbol in B_g and λ is less than u_{h+1}, so these trees are re-inserted into the left subtree of u_{h+1}. Every symbol in B_h and δ is greater than u_{h+1}, so these trees are re-inserted into the right subtree of u_{h+1}. Since B_{h+1} consists of u_{h+1} with B_g as its left subtree and B_h as its right subtree, the subtree B_{h+1} appears at the root and so T_{h+1} satisfies P1. All the other subtrees B_i for $u_i \in U_{h+1}^\top$ are contained in λ, so T_{h+1} satisfies P2.

Case 3. Suppose u_h is the left child of u_{h+1}. Then, by the definition of the postfix traversal, u_{h+1} has empty right subtree in U, and so B_{h+1} consists of u_{h+1} with left subtree B_h and right subtree empty. (For an illustration of this case, see the step from T_1 to T_2 in Example 7.) Proceeding in a similar way to the previous cases, one sees that, as in Fig. 3, $T_h = \mathsf{P}_{\mathsf{sylv}}(\beta u_{h+1}\delta\lambda B_h)$. Let $T_{h+1} = \mathsf{P}_{\mathsf{sylv}}(\delta\lambda B_h \beta u_{h+1})$; then $T_h \sim T_{h+1}$ and T_{h+1} satisfies P1 and P2.

Case 4. Suppose that, in U, the node u_h is the right child of u_{h+1}, and u_{h+1} has empty left subtree. (For an illustration of this case, see the step from T_4 to T_5 in Example 7.) Thus B_{h+1} consists of the node u_{h+1} with empty left subtree and right subtree U_h. Proceeding in a similar way to the previous cases, one sees that there are two sub-cases, as in Fig. 4:

1. $T_h = \mathsf{P}_{\mathsf{sylv}}(\lambda u_{h+1}\delta B_h)$. Let $T_{h+1} = \mathsf{P}_{\mathsf{sylv}}(\delta B_h \lambda u_{h+1})$.
2. $T_h = \mathsf{P}_{\mathsf{sylv}}(\zeta u_{h+1}\lambda\delta B_h)$. Let $T_{h+1} = \mathsf{P}_{\mathsf{sylv}}(\lambda\delta B_h \zeta u_{h+1})$.

In both sub-cases, $T_h \sim T_{h+1}$ and T_{h+1} satisfies P1 and P2.

Conclusion. Thus there is a sequence $T = T_0, T_1, \ldots, T_n = U$ with $T_h \sim T_{h+1}$ and T_{h+1} satisfying P1 and P2 for $h = 0, \ldots, h - 1$. In particular, T_n satisfies P1 and so the subtree $B_n = U$ appears in T_n, with its root at the root of T_n. Hence $T_n = U$.

References

1. Cain, A.J., Malheiro, A.: Combinatorics of cyclic shifts in plactic, hypoplactic, sylvester, and related monoids (in preparation)
2. Cain, A.J., Malheiro, A.: Deciding conjugacy in sylvester monoids and other homogeneous monoids. Int. J. Algebra Comput. **25**(5), 899–915 (2015). doi:10.1007/978-3-642-40579-2_11
3. Choffrut, C., Mercaş, R.: The lexicographic cross-section of the plactic monoid is regular. In: Karhumäki, J., Lepistö, A., Zamboni, L. (eds.) WORDS 2013. LNCS, vol. 8079, pp. 83–94. Springer, Heidelberg (2013). doi:10.1007/978-3-642-40579-2_11
4. Giraudo, S.: Algebraic and combinatorial structures on pairs of twin binary trees. J. Algebra **360**, 115–157 (2012). doi:10.1016/j.jalgebra.2012.03.020
5. Hivert, F., Novelli, J.C., Thibon, J.Y.: The algebra of binary search trees. Theoret. Comput. Sci. **339**(1), 129–165 (2005). doi:10.1016/j.tcs.2005.01.012
6. Hivert, F., Novelli, J.C., Thibon, J.Y.: Commutative combinatorial Hopf algebras. J. Algebraic Combin. **28**(1), 65–95 (2007). doi:10.1007/s10801-007-0077-0
7. Krob, D., Thibon, J.Y.: Noncommutative symmetric functions IV: quantum linear groups and hecke algebras at $q = 0$. J. Algebraic Combin. **6**(4), 339–376 (1997). doi:10.1023/A:1008673127310
8. Lascoux, A., Schützenberger, M.P.: Le monoïde plaxique. In: Noncommutative structures in algebra and geometric combinatorics, pp. 129–156. No. 109 in Quaderni de "La Ricerca Scientifica", CNR, Rome (1981). http://igm.univ-mlv.fr/berstel/Mps/Travaux/A/1981-1PlaxiqueNaples.pdf
9. Lothaire, M.: Algebraic Combinatorics on Words. No. 90 in Encyclopedia of Mathematics and its Applications. Cambridge University Press, Cambridge (2002)
10. Novelli, J.C.: On the hypoplactic monoid. Discrete Math. **217**(1–3), 315–336 (2000). doi:10.1016/S0012-365X(99)00270-8
11. Priez, J.B.: A lattice of combinatorial Hopf algebras: binary trees with multiplicities. In: Formal Power Series and Algebraic Combinatorics. The Association. Discrete Mathematics & Theoretical Computer Science, Nancy (2013). http://www.dmtcs.org/pdfpapers/dmAS0196.pdf

Palindromic Length in Free Monoids and Free Groups

Aleksi Saarela$^{(\boxtimes)}$

Department of Mathematics and Statistics, University of Turku,
20014 Turku, Finland
amsaar@utu.fi

Abstract. Palindromic length of a word is defined as the smallest number n such that the word can be written as a product of n palindromes. It has been conjectured that every aperiodic infinite word has factors of arbitrarily high palindromic length. A stronger variant of this conjecture claims that every aperiodic infinite word has also prefixes of arbitrarily high palindromic length. We prove that these two conjectures are equivalent. More specifically, we prove that if every prefix of a word is a product of n palindromes, then every factor of the word is a product of $2n$ palindromes. Our proof quite naturally leads us to compare the properties of palindromic length in free monoids and in free groups. For example, the palindromic lengths of a word and its conjugate can be arbitrarily far apart in a free monoid, but in a free group they are almost the same.

Keywords: Combinatorics on words · Palindrome · Free group

1 Introduction

Palindromes are a common topic in combinatorics on words. Some examples of subtopics are palindromic richness [11] and palindrome complexity [1]. In this article, we are interested in palindromic factorizations of words. Every word can be trivially written as a product of palindromes, because every letter is a palindrome. However, studying minimal palindromic factorizations is a highly nontrivial topic that has been studied in many articles, for example by Ravsky [15]. The length of a minimal palindromic factorization of a word, that is, the smallest number n such that the word can be written as a product of n palindromes, is called the palindromic length of the word.

Frid, Puzynina and Zamboni [10] made the following conjecture about the palindromic lengths of factors of infinite words.

Conjecture 1. Every aperiodic infinite word has factors of arbitrarily high palindromic length.

They also proved the conjecture for a large class of words, including all words that are k-power-free for some k. They actually proved that all words in this class have not only factors but also prefixes of arbitrarily high palindromic length. This leads to the following stronger version of the conjecture.

© Springer International Publishing AG 2017
S. Brlek et al. (Eds.): WORDS 2017, LNCS 10432, pp. 203–213, 2017.
DOI: 10.1007/978-3-319-66396-8_19

Conjecture 2. Every aperiodic infinite word has prefixes of arbitrarily high palindromic length.

Let us mention here some related results, many of which have been inspired by the conjectures. The complexity of determining the palindromic length of a word is known to be $O(n \log n)$ [8,14]. Words of palindromic length at most two are sometimes called *symmetric* or *palindrome pairs*, and they have appeared in many articles [5,6,12,13]. Variations of palindromic length called left greedy palindromic length and right greedy palindromic length were defined and studied by Bucci and Richomme [7].

In this article, we prove the equivalence of Conjectures 1 and 2. More specifically, we prove that the maximal palindromic length of factors of a word can be at most twice as large as the maximal palindromic length of prefixes of the word, and this result is at least very close to optimal. We also give other results on palindromic length. Conjecture 1 remains a very interesting open problem.

Palindromes and palindromic length can also be defined in a free group in a natural way. To avoid confusion, we talk about FG-palindromes and FG-palindromic length in the case of free groups. These concepts were studied by Bardakov, Shpilrain and Tolstykh [3]. They proved that in every nonabelian free group, there are elements with arbitrarily high FG-palindromic length. Palindromic length has been defined and studied in many other groups as well, see, for example, the paper by Bardakov and Gongopadhyay about finitely generated solvable groups [2] or the paper by Fink about wreath products [9].

Because a free monoid of words is a subset of a free group, both the ordinary palindromic length and FG-palindromic length are defined for words. However, there does not seem to be any research on the relation of these two concepts. We take the first steps in this direction, inspired by the fact that some of our results on palindromic length can be formulated by using free groups and, specifically, inverses of palindromes. We prove that the ratio of the palindromic length and the FG-palindromic length of a word can be arbitrarily large, and we study the relation of palindromic length, FG-palindromic length, conjugacy, and edit distance. Combinatorial and algorithmic analysis of FG-palindromic length seems like an interesting topic for future research.

2   Preliminaries

Throughout the article, let Σ be an alphabet. The set of all words over Σ is denoted by Σ^* and it is a free monoid. The empty word is denoted by ε and the length of a word $w \in \Sigma^*$ by $|w|$.

The set of all infinite words over Σ is denoted by Σ^ω. An infinite word w is *ultimately periodic* if there are words $u, v \in \Sigma^*$ such that $w = uv^\omega = uvvv\cdots$. If w is not ultimately periodic, it is *aperiodic*.

The set of factors of a finite or infinite word w is denoted by Fact(w) and the set of prefixes by Pref(w).

If $a_1, \ldots, a_n \in \Sigma$, then the *reverse* of the word $w = a_1 \cdots a_n$ is $w^R = a_n \cdots a_1$. If $w = w^R$, then w is a *palindrome*.

The *palindromic length* of a word w, denoted by $|w|_{\text{pal}}$, is the smallest number n such that w can be written as a product of n palindromes. Because every letter is a palindrome, $|w|_{\text{pal}} \leq |w|$. The palindromic length of ε is zero, the palindromic length of every nonempty palindrome is one, and the palindromic length of every other word is at least two.

Example 3. The reverse of the word *reverses* is *sesrever*, so it is not a palindrome. It is a product of the two palindromes *rever* and *ses*, so its palindromic length is two.

The *palindromic width* of a language L is

$$|L|_{\text{pal}} = \sup\{|u|_{\text{pal}} \mid u \in L\}$$

(this terminology actually comes from studying palindromicity in groups; the case of free groups is discussed below). Conjecture 1 can now be reformulated as claiming that $|\text{Fact}(w)|_{\text{pal}} = \infty$ for every aperiodic infinite word w, and Conjecture 2 can be reformulated as claiming that $|\text{Pref}(w)|_{\text{pal}} = \infty$ for every aperiodic infinite word w.

The free monoid Σ^* can be extended to a free group. For any subset S of the free group, let S^* be the monoid generated by S, let S^{-1} be the set of inverses of elements of S, and let $S^{\pm 1} = S \cup S^{-1}$. For example, $(\Sigma^{\pm 1})^*$ is the whole free group, and $(\Sigma^*)^{\pm 1}$ is the set of all words and their inverses. The term "word" always refers to an element of Σ^*.

Every element x of the free group $(\Sigma^{\pm 1})^*$ can be written uniquely in a reduced form $x = a_1 \cdots a_n$, where $n \geq 0$, $a_1, \ldots, a_n \in \Sigma^{\pm 1}$, and $a_{i-1}a_i \neq \varepsilon$ for all $i \in \{2, \ldots, n\}$. The *reverse* of x is then $x^R = a_n \cdots a_1$. This is an extension of the definition of the reverse of a word. If $x = x^R$, then x is an *FG-palindrome*. A word is an FG-palindrome if and only if it is a palindrome.

Reversal is an antimorphism, that is, $(xy)^R = y^R x^R$ for all $x, y \in (\Sigma^{\pm 1})^*$. It follows that if $x = a_0 \cdots a_n$, where $a_0, \ldots, a_n \in \Sigma^{\pm 1}$ (but not necessarily $a_{i-1}a_i \neq \varepsilon$ for all $i \in \{1, \ldots, n\}$), and if $a_i = a_{n-i}$ for all $i \in \{0, \ldots, n\}$, then x is an FG-palindrome. The converse is not true; for example, the empty word is an FG-palindrome, but it can be written as aa^{-1}, which does not "look like" a palindrome.

The *FG-palindromic length* of an element x is the smallest number n such that x can be written as a product of n FG-palindromes. This definition is not compatible with the definition of palindromic length, because there are words whose palindromic length is larger than their FG-palindromic length.

Example 4. The palindromic length of *abca* is four, but it is a product of three FG-palindromes:

$$abca = aba \cdot a^{-2} \cdot aca.$$

When studying palindromicity in free groups (and not in free monoids), FG-palindromes are usually called just palindromes, but in this article, the term "palindrome" always refers to a word. Similarly, FG-palindromic length is sometimes called just palindromic length, but because it is different from the usual palindromic length of words, it is important to use different terms in this article.

3   Palindromic Lengths of Factors and Prefixes

We start with an easy lemma, which was also proved in [12]. Lemmas of similar flavor can be found in [4].

Lemma 5. *Let $x, y \in \Sigma^*$. If two of the words x, y, xy are palindromes, then the third one is a product of two palindromes.*

Proof. If x and y are palindromes, then the claim is clear.

If x and xy are palindromes, then $xy = (xy)^R = y^R x^R = y^R x$, so y and y^R are conjugates, meaning that there are words p, q such that $y = pq$ and $y^R = qp$. Then $qp = y^R = (pq)^R = q^R p^R$, so $q = q^R$ and $p = p^R$, and thus $y = pq$ is a product of two palindromes.

If y and xy are palindromes, then the claim can be proved in a symmetric way. Alternatively, we can notice that y^R and $y^R x^R$ are palindromes, so x^R is a product of two palindromes by the previous case, and therefore also x is a product of two palindromes.   □

If x is a product of m palindromes and y is a product of n palindromes, then xy is a product of $m + n$ palindromes, so we have the inequality $|xy|_{\mathrm{pal}} \leq |x|_{\mathrm{pal}} + |y|_{\mathrm{pal}}$. The following generalization of Lemma 5 gives two other similar "triangle inequalities" for palindromic length.

Lemma 6. *Let $x, y \in \Sigma^*$. Then*

$$|y|_{\mathrm{pal}} \leq |x|_{\mathrm{pal}} + |xy|_{\mathrm{pal}} \qquad and \qquad |x|_{\mathrm{pal}} \leq |y|_{\mathrm{pal}} + |xy|_{\mathrm{pal}}.$$

Proof. We prove the first inequality by induction on $|xy|$ (the second inequality is symmetric). The cases where $|x| = 0$ or $|y| = 0$ are clear. Let us assume that $|x|, |y| > 0$ and $|y'|_{\mathrm{pal}} \leq |x'|_{\mathrm{pal}} + |x'y'|_{\mathrm{pal}}$ whenever $|x'y'| < |xy|$. Let $|x|_{\mathrm{pal}} = m$ and $|xy|_{\mathrm{pal}} = n$. Let $x = p_1 \cdots p_m$ and $xy = q_1 \cdots q_n$, where every p_i and every q_i is a nonempty palindrome. There are two (similar) cases: $|p_1| \leq |q_1|$ and $|p_1| > |q_1|$.

If $|p_1| \leq |q_1|$, then $q_1 = p_1 r$ for some word r, and $r = st$ for some palindromes s, t by Lemma 5. Let $x' = p_2 \cdots p_m$. Then $x'y = stq_2 \cdots q_n$. By the induction hypothesis,

$$|y|_{\mathrm{pal}} \leq |x'|_{\mathrm{pal}} + |x'y|_{\mathrm{pal}} \leq (m - 1) + (n + 1) = |x|_{\mathrm{pal}} + |xy|_{\mathrm{pal}}.$$

If $|p_1| > |q_1|$, then $p_1 = q_1 r$ for some word r, and $r = st$ for some palindromes s, t by Lemma 5. Let $x' = stp_2 \cdots p_m$. Then $x'y = q_2 \cdots q_n$. By the induction hypothesis,

$$|y|_{\mathrm{pal}} \leq |x'|_{\mathrm{pal}} + |x'y|_{\mathrm{pal}} \leq (m + 1) + (n - 1) = |x|_{\mathrm{pal}} + |xy|_{\mathrm{pal}}.$$

This completes the induction.   □

Now we are ready to prove the main result of this section and the equivalence of Conjectures 1 and 2.

Theorem 7. *Let w be a finite or infinite word. Then*

$$|\mathrm{Fact}(w)|_{\mathrm{pal}} \leq 2|\mathrm{Pref}(w)|_{\mathrm{pal}}.$$

Proof. Let y be any factor of w. There is a word x such that xy is a prefix of w. Then $|x|_{\mathrm{pal}}, |xy|_{\mathrm{pal}} \leq |\mathrm{Pref}(w)|_{\mathrm{pal}}$, and

$$|y|_{\mathrm{pal}} \leq |x|_{\mathrm{pal}} + |xy|_{\mathrm{pal}} \leq 2|\mathrm{Pref}(w)|_{\mathrm{pal}}$$

by Lemma 6.   □

Corollary 8. *Conjectures 1 and 2 are equivalent.*

Proof. For an aperiodic infinite word w, the condition $|\mathrm{Pref}(w)|_{\mathrm{pal}} = \infty$ implies $|\mathrm{Fact}(w)|_{\mathrm{pal}} = \infty$, because $\mathrm{Pref}(w) \subseteq \mathrm{Fact}(w)$, and the condition $|\mathrm{Fact}(w)|_{\mathrm{pal}} = \infty$ implies $|\mathrm{Pref}(w)|_{\mathrm{pal}} = \infty$ by Theorem 7. Therefore Conjectures 1 and 2 are equivalent.   □

The next example shows that the inequality $|\mathrm{Fact}(w)|_{\mathrm{pal}} \leq 2|\mathrm{Pref}(w)|_{\mathrm{pal}}$ in Theorem 7 is almost optimal. We do not know whether it could be replaced by $|\mathrm{Fact}(w)|_{\mathrm{pal}} \leq 2|\mathrm{Pref}(w)|_{\mathrm{pal}} - 1$.

Example 9. Let $\{a_1, \ldots, a_{n-1}, b_1, \ldots, b_{n-1}\}$ be an alphabet and let

$$A = a_1 \cdots a_{n-1} \qquad \text{and} \qquad B = b_1 \cdots b_{n-1}.$$

It is quite easy to see that all prefixes of the infinite word

$$w = (AA^R BB^R)^\omega = ((a_1 \cdots a_{n-1})(a_{n-1} \cdots a_1)(b_1 \cdots b_{n-1})(b_{n-1} \cdots b_1))^\omega$$

have palindromic length at most n. On the other hand, w has the factor

$$u = A^R BB^R AA^R B$$
$$= (a_{n-1} \cdots a_1)(b_1 \cdots b_{n-1})(b_{n-1} \cdots b_1)(a_1 \cdots a_{n-1})(a_{n-1} \cdots a_1)(b_1 \cdots b_{n-1}),$$

and we can show that u has palindromic length $2n - 1$. To see this, let $u = p_1 \cdots p_k$, where every p_i is a palindrome. We first note that u contains every letter exactly three times. Every letter appears an even number of times in every palindrome of even length, so every letter must appear in at least one p_i of odd length. But u does not contain a factor of the form aba for any letters a, b, so it does not have palindromic factors of odd length, except the letters. Therefore, for every letter a, there exists i such that $p_i = a$. This means that the sequence $p_1, \ldots, p_k$ contains at least $2n - 2$ letters. It follows that $k \geq 2n - 1$.

4   Binary Alphabet

In Example 9, we used an alphabet whose size depended on the parameter n. This raises the question of whether similar examples could be constructed using an alphabet of fixed sized, preferably a binary alphabet. It would be convenient if, for any alphabet $\{a_1, \ldots, a_n\}$, we could give a morphism $h : \{a_1, \ldots, a_n\}^* \to \{a, b\}^*$ that preserves palindromic lengths of words, and approximately preserves palindromic widths of sets of factors and prefixes. Then we could use this morphism also later to turn n-ary examples into binary ones. The first idea might be to define $h(a_i) = ab^i a$ for all i. This morphism preserves palindromicity, but it can significantly reduce the palindromic length of a word. A better morphism is given in the next lemma.

Lemma 10. *Let us define a morphism*

$$h : \{a_1, \ldots, a_n\}^* \to \{a, b\}^*, \quad h(a_i) = ab^i a^5 b^i a.$$

Let u be a finite word and w a finite or infinite word over $\{a_1, \ldots, a_n\}$. Then

$$|h(u)|_{\mathrm{pal}} = |u|_{\mathrm{pal}},$$
$$|\mathrm{Fact}(w)|_{\mathrm{pal}} \leq |\mathrm{Fact}(h(w))|_{\mathrm{pal}} \leq |\mathrm{Fact}(w)|_{\mathrm{pal}} + 6,$$
$$|\mathrm{Pref}(w)|_{\mathrm{pal}} \leq |\mathrm{Pref}(h(w))|_{\mathrm{pal}} \leq |\mathrm{Pref}(w)|_{\mathrm{pal}} + 3.$$

Proof. First, we prove that $|h(u)|_{\mathrm{pal}} \leq |u|_{\mathrm{pal}}$. If $u = p_1 \cdots p_k$, where every p_i is a palindrome, then $h(u) = h(p_1) \cdots h(p_k)$ and every $h(p_i)$ is a palindrome. The claim follows.

Second, we prove that $|u|_{\mathrm{pal}} \leq |h(u)|_{\mathrm{pal}}$. Let $h(u) = q_1 \cdots q_k$, where every q_i is a palindrome. We are going to define a factorization $u = p_1 \cdots p_k$ such that every p_i is a palindrome. The informal idea is to define the words p_i so that, for a letter c in u, if the centermost letter in the image $h(c)$ is inside q_j, then c will be inside p_j. This means that either $|p_j| \leq 1$ or $q_j = xh(p_j)y$, where x is either a suffix of $a^2 b^i a$ or the inverse of a prefix of $ab^i a^2$ for some i, and y is either a prefix of $ab^i a^2$ or the inverse of a suffix of $a^2 b^i a$ for some i. If $p_j = a_{j_0} \cdots a_{j_m}$, where $m \geq 1$ and $j_i \in \{1, \ldots, n\}$ for all i, then

$$q_j = x' a^3 b^{j_0} a \left(\prod_{i=1}^{m-1} ab^{j_i} a^5 b^{j_i} a \right) ab^{j_m} a^3 y',$$

where $x', y' \in \{a, b\}^*$ do not contain a^3 as a factor. Because q_j is a palindrome, it must be $j_i = j_{m-i}$ for all i, so also p_j is a palindrome. The claim follows.

Third, we prove that

$$|\mathrm{Fact}(w)|_{\mathrm{pal}} \leq |\mathrm{Fact}(h(w))|_{\mathrm{pal}}$$

If v is a factor of w of palindromic length k, then $h(v)$ is a factor of $h(w)$ of palindromic length k. The claim follows.

Finally, we prove that

$$|\mathrm{Fact}(h(w))|_{\mathrm{pal}} \leq |\mathrm{Fact}(w)|_{\mathrm{pal}} + 6.$$

Every factor of $h(w)$ is of the form $xh(v)y$, where v is a factor of w, x is a suffix of $ab^i a^5 b^i a$ for some i, and y is a prefix of $ab^i a^5 b^i a$ for some i. Then

$$|xh(v)y|_{\mathrm{pal}} \leq |x|_{\mathrm{pal}} + |h(v)|_{\mathrm{pal}} + |y|_{\mathrm{pal}} \leq |v|_{\mathrm{pal}} + 6.$$

The claim follows.

The inequalities about the sets of prefixes can be proved in a similar way. $\square$

Example 11. If w is the word of Example 9 and h is the morphism of Lemma 10, then the palindromic lengths of all prefixes of the binary infinite word $h(w)$ are at most $n + 3$, but $h(w)$ has a factor of palindromic length $2n - 1$.

5   Palindromic Jumps

In this section, we are going to prove a generalization of Lemma 6, which might be useful when studying palindromic lengths of factors. Let $w = a_0 a_1 a_2 \cdots$ (w could also be a finite word). In the following, it is convenient to think that the positions between the letters of w are labeled so that the position before a_0 is 0, the position between a_0 and a_1 is 1, and so on. We say that (i, j) is a *palindromic jump in w* if either $i \leq j$ and $a_i \cdots a_{j-1}$ is a palindrome or $j \leq i$ and $a_j \cdots a_{i-1}$ is a palindrome. If $i \leq j$, then (i, j) is a *forward palindromic jump*, and if $j \leq i$, then (i, j) is a *backward palindromic jump*.

If we can get from position i to position j with n forward palindromic jumps, then the factor between positions i and j is a product of n palindromes. The inequality $|y|_{\mathrm{pal}} \leq |x|_{\mathrm{pal}} + |xy|_{\mathrm{pal}}$ in Lemma 6 means that if we can get from position $|x|$ in the word xy to position 0 with m backward palindromic jumps, and we can get from position 0 to position $|xy|$ with n forward palindromic jumps, then we can get from position $|x|$ to position $|xy|$ with $m+n$ forward palindromic jumps. It follows that any sequence of m backward palindromic jumps followed by n forward palindromic jumps can be converted into a sequence of $m + n$ forward palindromic jumps. In the following theorem, we will generalize this by proving that any sequence of n palindromic jumps can be converted into a sequence of n forward palindromic jumps. So if we can get from position i to position j with n palindromic jumps, then the factor between positions i and j has palindromic length at most n.

Theorem 12. *Let $w = a_0 a_1 a_2 \cdots$. Let $k_0, \ldots, k_n \geq 0$ and $k_0 \leq k_n$. If (k_{i-1}, k_i) is a palindromic jump in w for all $i \in \{1, \ldots, n\}$, then $a_{k_0} \cdots a_{k_n-1}$ is a product of n palindromes.*

Proof. We can assume that $k_{i-1} \neq k_i$ for all $i \in \{1, \ldots, n\}$. The proof is by induction on $L = |k_0 - k_1| + \cdots + |k_{n-1} - k_n|$. If $L = k_n - k_0$, then the sequence $k_0, \ldots, k_n$ is increasing and the claim is clear. Let us assume that $L > k_n - k_0$ and that the claim is true for all values smaller than L. There is a number j such that either $k_j < k_{j-1}, k_{j+1}$ or $k_j > k_{j-1}, k_{j+1}$. By Lemma 5, there is a number k such that $(k_{j-1}, k), (k, k_{j+1})$ are palindromic jumps in w and either $k_{j-1} \leq k \leq k_{j+1}$ or $k_{j+1} \leq k \leq k_{j-1}$. Let $k'_j = k$ and $k'_i = k_i$ for all $i \neq j$. Then $|k'_0 - k'_1| + \cdots + |k'_{n-1} - k'_n| < L$ and every (k'_{i-1}, k'_i) is a palindromic jump in w, so $a_{k_0} \cdots a_{k_n}$ is a product of n palindromes by the induction hypothesis. $\square$

Example 13. Consider the word *abaca*. Then $(0, 3)$ is a forward palindromic jump, because *aba* is a palindrome, $(3, 2)$ is a backward palindromic jump, because *a* is a palindrome, and $(2, 5)$ is a forward palindromic jump, because *aca* is a palindrome. By Theorem 12, *abaca* is a product of three palindromes, which is of course very easy to see directly as well. The proof of Theorem 12 would convert the sequence $(0, 3), (3, 2), (2, 5)$ of palindromic jumps either into the sequence $(0, 1), (1, 2), (2, 5)$, which corresponds to the factorization $a \cdot b \cdot aca$, or to the sequence $(0, 3), (3, 4), (4, 5)$, which corresponds to the factorization $aba \cdot c \cdot a$.

6 Palindromes and Inverses of Palindromes

From now on, we view the word monoid Σ^* as a subset of the free group $(\Sigma^{\pm 1})^*$. If $x, y \in \Sigma^*$, then $y = x^{-1}xy$, so the inequality $|y|_{\mathrm{pal}} \leq |x|_{\mathrm{pal}} + |xy|_{\mathrm{pal}}$ of Lemma 6 can be formulated as follows: If $y = p_1 \cdots p_m q_1 \cdots q_n$, where every p_i is the inverse of a palindrome and every q_i is a palindrome, then y is a product of $m + n$ palindromes. This raises the following questions:

- If a word is a product of n elements of $(\Sigma^*)^{\pm 1}$ that are palindromes or inverses of palindromes, is the word necessarily a product of n palindromes?
- If a word is a product of n FG-palindromes, is the word necessarily a product of n palindromes?

The answer to both of these questions is negative, as is shown by the word

$$abca = aba \cdot a^{-2} \cdot aca,$$

which was already mentioned in Example 4. However, in Theorem 14 we prove a weaker result. This is essentially a reformulation of Theorem 12. We could also have proved Theorem 14 first and then Theorem 12 as a consequence.

Theorem 14. *Let $w = p_1 \cdots p_n$, where w is a word and every p_i is either a palindrome or the inverse of a palindrome. If $p_i \cdots p_j \in (\Sigma^*)^{\pm 1}$ whenever $1 \leq i \leq j \leq n$, then w is a product of n palindromes.*

Proof. For all $i \in \{0, \ldots, n\}$, let $q_i = p_1 \cdots p_i$. Let $R = \{q_i \mid q_i \in \Sigma^*\}$ and $S = \{q_i^{-1} \mid q_i^{-1} \in \Sigma^*\}$. Let r be a longest word in R and s be a longest word in S.

First we are going to show that every $q_i \in R$ is a prefix of r. If $q_i = xay$ and $q_j = xbz$, where x, y, z are words, a, b are different letters, and $i < j$, then

$$p_{i+1} \cdots p_j = q_i^{-1} q_j = y^{-1} a^{-1} bz \notin (\Sigma^*)^{\pm 1},$$

which is a contradiction. Therefore, one of q_i, q_j is a prefix of the other. This means that every $q_i \in R$ is a prefix of r.

Then we are going to show that every $q_i^{-1} \in S$ is a suffix of s. If $q_i^{-1} = xaz$ and $q_j^{-1} = ybz$, where x, y, z are words, a, b are different letters, and $i < j$, then

$$p_{i+1} \cdots p_j = q_i^{-1} q_j = xab^{-1} y^{-1} \notin (\Sigma^*)^{\pm 1},$$

which is a contradiction. Therefore, one of q_i^{-1}, q_j^{-1} is a suffix of the other. This means that every $q_i^{-1} \in S$ is a suffix of s.

Let $w = sr$. Then $(|sq_{i-1}|, |sq_i|)$ is a palindromic jump in w for all $i \in \{1, \ldots, n\}$. The claim follows from Theorem 12. □

7   Conjugates and Edit Distance

In this section, we will compare palindromic length and FG-palindromic length and show that they can be very different. We prove that FG-palindromic length has some nice properties that the ordinary palindromic length does not have: The FG-palindromic lengths of conjugates are almost the same, and if two elements are close to each other as measured by edit distance, then also their FG-palindromic lengths are close to each other.

Theorem 15. *For every conjugacy class of a free group, there is a number k such that the FG-palindromic lengths of all elements in the conjugacy class are in $\{2k-1, 2k\}$.*

Proof. Of all the members of a conjugacy class, let x be one with minimal FG-palindromic length. Let k be such that the FG-palindromic length of x is in $\{2k-1, 2k\}$. Then $x = p_1 \cdots p_{2k}$, where every p_i is an FG-palindrome (we can add the empty palindrome if necessary). For every conjugate yxy^{-1} of x we have

$$yxy^{-1} = y(p_1 \cdots p_{2k})y^{-1} = \prod_{i=1}^{k} yp_{2i-1}p_{2i}y^{-1} = \prod_{i=1}^{k} yp_{2i-1}y^R(y^R)^{-1}p_{2i}y^{-1}.$$

Here the two elements $yp_{2i-1}y^R$ and $(y^R)^{-1}p_{2i}y^{-1} = (y^{-1})^R p_{2i}y^{-1}$ are FG-palindromes for all $i \in \{1, \ldots, k\}$, so yxy^{-1} is a product of $2k$ FG-palindromes. This proves the claim. □

All conjugates of a product of two palindromes are also products of two palindromes, but a conjugate of a product of three palindromes can have arbitrarily high palindromic length, as is shown in the next example. This means that Theorem 15 does not hold for palindromic length.

Example 16. Let $\{a_1, \ldots, a_n, b, c\}$ be an alphabet and let $A = a_1 \cdots a_n$. The word

$$A^R Abc = (a_n \cdots a_1)(a_1 \cdots a_n)bc$$

has palindromic length three, but its conjugate

$$AbcA^R = (a_1 \cdots a_n)bc(a_n \cdots a_1)$$

has palindromic length $2n+2$. On the other hand, Theorem 15 guarantees that $AbcA^R$ is a product of four FG-palindromes. In fact, it is a product of three FG-palindromes:

$$AbcA^R = AbA^R \cdot (A^R)^{-1}A^{-1} \cdot AcA^R.$$

This also shows that the ratio of the palindromic length and the FG-palindromic length of a word can be arbitrarily large.

The edit distance (or Levenshtein distance) of two words can be defined as the smallest number of deletions, insertions and substitutions of letters that are required to transform the first word into the second word. A similar definition can be given for elements of a free group. Formally, we define the *FG-edit distance* of $x, y \in (\Sigma^{\pm 1})^*$ as follows:

- If $x = y$, the FG-edit distance is zero.
- If $x = uav \neq y = ubv$, where $u, v \in (\Sigma^{\pm 1})^*$ and $a, b \in \Sigma^{\pm 1} \cup \{\varepsilon\}$, the FG-edit distance is one.
- Otherwise, the FG-edit distance is the smallest number n for which there are $x_0, \ldots, x_n \in (\Sigma^{\pm 1})^*$ such that $x_0 = x$, $x_n = y$, and the edit distance of x_{i-1} and x_i is one for all $i \in \{1, \ldots, n\}$.

The FG-edit distance of two words can be smaller than their ordinary edit distance. For example, the edit distance of ε and ab is two, but the FG-edit distance of $\varepsilon = aa^{-1}$ and ab is one.

Next we will prove that if two elements are close to each other as measured by FG-edit distance, then also their FG-palindromic lengths are close to each other. The idea is that if we want to make a deletion, insertion or substitution in the middle of an element, we can first take a suitable conjugate, then make the deletion, insertion or substitution at the end of the element, and finally take another suitable conjugate. None of these operations can change the FG-palindromic length by much.

Theorem 17. *If the FG-edit distance of x and y is d, then the difference of their FG-palindromic lengths is at most $2d + 1$.*

Proof. First, consider the case $d = 1$. Let $x = uav$ and $y = ubv$, where $u, v \in (\Sigma^{\pm 1})^*$ and $a, b \in \Sigma^{\pm 1} \cup \{\varepsilon\}$. Let the FG-palindromic length of x be $2k - l$, where $l \in \{0, 1\}$. Then the FG-palindromic length of vua is at most $2k$ by Theorem 15, the FG-palindromic length of $vub = vua \cdot a^{-1} \cdot b$ is at most $2k + 2$, and the FG-palindromic length of ubv is at most $2k + 2$ by Theorem 15. This proves the claim for $d = 1$. The general case follows by iterating the above procedure. $\square$

The next example shows that Theorem 17 does not hold for palindromic length.

Example 18. Consider the word $AbcA^R$ that appeared in Example 16. It is within edit distance one of a palindrome, but its palindromic length is $2n + 2$. On the other hand, Theorem 17 guarantees that $AbcA^R$ is a product of four FG-palindromes. In fact, it is a product of three FG-palindromes, as we saw in Example 16.

8   Conclusion

In this article, we have studied palindromic length. In free monoids, we have compared the maximal palindromic lengths of factors and prefixes, proved the equivalence of two well-known conjectures, and given alternative equivalent ways to define palindromic length. In free groups, we have studied the relations between palindromic length, FG-palindromic length, conjugates, and edit distance. There are many open questions:

- Conjecture 1 remains open.
- The fact that the FG-palindromic length of a word can be much smaller than the palindromic length suggests the following question: Does there exist an aperiodic infinite word such that the FG-palindromic lengths of its factors are bounded by a constant?
- There are several small questions about the optimality of various results. For example, are there words such that all of their prefixes have palindromic length at most n but some of their factors have palindromic length $2n$? In the binary case, can we do better than using Lemma 10?
- We could also look at combinatorial and algorithmic questions related to FG-palindromic length. Finding an algorithm for determining the FG-palindromic length was mentioned as an open problem already in [3].

References

1. Allouche, J.P., Baake, M., Cassaigne, J., Damanik, D.: Palindrome complexity. Theoret. Comput. Sci. **291**(1), 9–31 (2003)
2. Bardakov, V.G., Gongopadhyay, K.: Palindromic width of finitely generated solvable groups. Comm. Algebra **43**(11), 4809–4824 (2015)
3. Bardakov, V.G., Shpilrain, V., Tolstykh, V.: On the palindromic and primitive widths of a free group. J. Algebra **285**(2), 574–585 (2005)
4. Blondin Massé, A., Brlek, S., Labbé, S.: Palindromic lacunas of the Thue-Morse word. In: Proceedings of GASCom, pp. 53–67 (2008)
5. Borchert, A., Rampersad, N.: Words with many palindrome pair factors. Electron. J. Comb. **22**(4), P4.23 (2015)
6. Brlek, S., Hamel, S., Nivat, M., Reutenauer, C.: On the palindromic complexity of infinite words. Int. J. Found. Comput. Sci. **15**(2), 293–306 (2004)
7. Bucci, M., Richomme, G.: Greedy palindromic lengths (Preprint). http://arxiv.org/abs/1606.05660
8. Fici, G., Gagie, T., Kärkkäinen, J., Kempa, D.: A subquadratic algorithm for minimum palindromic factorization. J. Discrete Algorithms **28**, 41–48 (2014)
9. Fink, E.: Palindromic width of wreath products. J. Algebra **471**, 1–12 (2017)
10. Frid, A.E., Puzynina, S., Zamboni, L.Q.: On palindromic factorization of words. Adv. Appl. Math. **50**(5), 737–748 (2013)
11. Glen, A., Justin, J., Widmer, S., Zamboni, L.Q.: Palindromic richness. Eur. J. Comb. **30**(2), 510–531 (2009)
12. Guo, C., Shallit, J., Shur, A.M.: On the combinatorics of palindromes and antipalindromes (Preprint). http://arxiv.org/abs/1503.09112
13. Holub, Š., Müller, M.: Fully bordered words. Theoret. Comput. Sci. **684**, 53–58 (2017)
14. I, T., Sugimoto, S., Inenaga, S., Bannai, H., Takeda, M.: Computing palindromic factorizations and palindromic covers on-line. In: Kulikov, A.S., Kuznetsov, S.O., Pevzner, P. (eds.) CPM 2014. LNCS, vol. 8486, pp. 150–161. Springer, Cham (2014). doi:10.1007/978-3-319-07566-2_16
15. Ravsky, O.: On the palindromic decomposition of binary words. J. Autom. Lang. Comb. **8**(1), 75–83 (2003)

Invariance: A Theoretical Approach for Coding Sets of Words Modulo Literal (Anti)Morphisms

Jean Néraud$^{(\boxtimes)}$ and Carla Selmi

Laboratoire d'Informatique, de Traitement de l'Information et des Systèmes, Université de Rouen, Avenue de l'Université, 76800 Saint-Étienne-du-Rouvray, France
neraud.jean@free.fr, carla.selmi@univ-rouen.fr

Abstract. Let A be a finite or countable alphabet and let θ be literal (anti)morphism onto A^* (by definition, such a correspondence is determinated by a permutation of the alphabet). This paper deals with sets which are invariant under θ (θ-invariant for short). We establish an extension of the famous defect theorem. Moreover, we prove that for the so-called thin θ-invariant codes, maximality and completeness are two equivalent notions. We prove that a similar property holds for some special families of θ-invariant codes such as prefix (bifix) codes, codes with a finite (two-way) deciphering delay, uniformly synchronous codes and circular codes. For a special class of involutive antimorphisms, we prove that any regular θ-invariant code may be embedded into a complete one.

Keywords: Antimorphism · Bifix · Circular · Code · Complete · Deciphering delay · Defect · Delay · Embedding · Equation · Literal · Maximal · Morphism · Prefix · Synchronizing delay · Variable length code · Verbal synchronizing delay · Word

1 Introduction

During the last decade, in the free monoid theory, due to their powerful applications, in particular in DNA-computing, one-to-one *morphic* or *antimorphic* correspondences play a particularly important part. Given a finite or countable *alphabet*, say A, any such mapping is a substitution which is fully determined by extending a unique permutation of A, to a mapping onto A^* (the *free monoid* that is generated by A). The resulting mapping is commonly referred to as *literal* (or *letter-to-letter*) moreover, in the case of a finite alphabet, it is well known that, with respect to the composition, some power of such a correspondence is the identity (classically, in the case where this power corresponds to the square, we say that the correspondence is *involutive*).

In that special case of involutive morphisms or antimorphisms -we write (anti)morphisms for short, lots of successful investigations have been done for extending the now classical combinatorial properties on words: we mention the study of the so-called pseudo-palindromes [3,5], or that of pseudo-repetitions [4,9,13]. The framework of some peculiar families of codes [12] and equations in

© Springer International Publishing AG 2017
S. Brlek et al. (Eds.): WORDS 2017, LNCS 10432, pp. 214–227, 2017.
DOI: 10.1007/978-3-319-66396-8_20

words [6,7] have been also concerned. Moreover, in the larger family of one-to-one (anti)morphisms, a nice generalization of the famous theorem of Fine and Wilf [14, Theorem 1.2.5] has been recently established in [8].

Equations in words are also the starting point of the study in the present paper, where we adopt the point of view from [14, Chap. 9]. Let A be a finite or countable alphabet; a one-to-one literal (anti)morphism onto A^*, namely θ, being fixed, consider a finite collection of unknown words, say Z. In view of making the present foreword more readable, in the first instance we take θ as an involutive literal substitution (that is $\theta^2 = id_{A^*}$). We assign that the words in Z and their images by θ to satisfy a given equation, and we are interested in the cardinality of any set T, whose elements allow by concatenation to compute all the words in Z. Actually, such a question might be more complex than in the classical configuration, where θ does not interfer: it is well known that in that classical case, according to the famous defect theorem [14, Theorem 1.2.5], the words in Z may be computed as the concatenation of at most $|Z| - 1$ words that don't satisfy any non-trivial equation. With the terminology of [10,14], T, the set of such words is a *code*, or equivalently T^*, the submonoid that it generates, is *free*: more precisely, with respect to the inclusion of sets it is the smallest free submonoid of A^* that contains Z.

Along the way, for solving our problem, applying the defect theorem to the set $X = Z \cup \theta(Z)$ might appear as natural. Such a methodology garantees the existence of a code T, with $|T| \leq |X| - 1$, and such that T^* is the smallest free submonoid of A^* that contains X. Unfortunately, since both the words in Z and $\theta(Z)$ are expressed as concatenations of words in T, among the elements of $T \cup \theta(T)$ non-trivial equations can remain; in other words, by applying that methodology, the initial problem would be transferred among the words in $T \cup \theta(T)$. This situation is particularly illustrated by [13, Proposition 3], where the authors prove that, given an involutive antimorphism θ, the solutions of the equation $xy = \theta(y)x$ are $x = (uv)^i u, y = vu$, where the elements u, v of T satisfy the non-trivial equation $vu = \theta(u)\theta(v)$.

In the general case where θ is a literal one-to-one (anti)morphism, we note that the union, say Y, of the sets $\theta^i(T)$, for all $i \in \mathbb{Z}$, is itself θ-invariant, therefore an alternative methodology will consist in asking for some code Y which is invariant under θ, and such that Y^* is the smallest free submonoid of A^* that contains $X = \bigcup_{i \in \mathbb{Z}} \theta^i(Z)$. By the way, it is straightforward to prove that the intersection of an arbitrary family of θ-invariant free submonoids is itself a θ-invariant free submonoid. In the present paper we prove the following result:

Theorem 1. *Let A be a finite or countable alphabet, let θ be a literal (anti)morphism onto A^*, and let X be a finite θ-invariant set. If X it is not a code, then the smallest θ-invariant free submonoid of A^* that contains X is generated by a θ-invariant code Y which satisfies $|Y| \leq |X| - 1$.*

For illustrating this result in term of equations, we refer to [6,7], where the authors considered generalizations of the famous equation in three unknowns of Lyndon-Shützenberger [14, Sect. 9.2]. They proved that, an involutive (anti)morphism θ being fixed, given such an equation with sufficiently long members, a word t exists such that any 3-uple of "solutions" can be expressed as a concatenation of words in $\{t\} \cup \{\theta(t)\}$. With the notation of Theorem 1, the elements of the θ-invariant set X are $x, y, z, \theta(x), \theta(y), \theta(z)$ and those of Y are t and $\theta(t)$: we verify that Y is a θ-invariant code with $|Y| \leq |X| - 1$.

In the sequel, we will continue our investigation by studying the properties of complete θ-invariant codes: a subset X of A^* is *complete* if any word of A^* is a factor of some words in X^*. From this point of view, a famous result from Schützenberger states that, for the wide family of the so-called *thin* codes (which contains regular codes) [10, Sect. 2.5], maximality and completeness are two equivalent notions. In the framework of invariant codes, we prove the following result:

Theorem 2. *Let A be a finite or countable alphabet. Given a thin θ-invariant code $X \subseteq A^*$, the three following conditions are equivalent:*

(i) X *is complete*
(ii) X *is a maximal code*
(iii) X *is maximal in the family of the θ-invariant codes.*

In the proof, the main feature consists in establishing that a non-complete θ-invariant code X cannot be maximal in the family of θ-invariant codes: actually, the most delicate step lays upon the construction of a convenient θ-invariant set $Z \subseteq A^*$, with $X \cap Z = \emptyset$ and such that $X \cup Z$ remains itself a θ-invariant code.

It is well known that the preceding result from Schützenberger has been successfully extended to some famous families of thin codes, such as *prefix* (*bifix, uniformly synchronous, circular*) codes (cf [10, Proposition 3.3.8], [10, Proposition 6.2.1], [10, Theorem 10.2.11], [15, Corollary 3.13] and [11, Theorem 3.5]) and codes with a *finite deciphering delay* (f.d.d. codes, for short) [10, Theorem 5.2.2]. From this point of view, we will examine the behavior of corresponding families of θ-invariant codes. Actually we establish a result similar to the preceding Theorem 2 in the framework of the family of prefix (bifix, f.d.d., two-way f.d.d, uniformly synchronized, circular codes). In the proof, a construction very similar to the previous one may be used in the case of prefix, bifix, f.d.d., two-way f.d.d codes. At the contrary, investigating the behavior of circular codes with regards to the question necessitates the computation of a more sofisticated set; moreover the family of uniformly synchronized codes itself impose to make use of a significantly different methodology.

In the last part of our study, we address to the problem of embedding a non-complete θ-invariant code into a complete one. For the first time, this question was stated in [2], where the author asked whether any finite code can be imbedded into a regular one. A positive answer was provided in [1], where was established a formula for embedding any regular code into a complete one. From the point of view of θ-invariant codes, we obtain a positive answer only in the case where θ is an involutive antimorphism which is different of the so-called miror image; actually the general question remains open.

We now describe the contents of the paper. Section 2 contains the preliminaries: the terminology of the free monoid is settled, and the definitions of some classical families of codes are recalled. Theorem 1 is established in Sect. 3, where an original example of equation is studied. The proof of Theorem 2 is done in Sect. 3, and extensions for special families of θ-invariant codes are studied in Sect. 4. The question of embedding a regular θ-invariant code into a complete one is examined in Sect. 5.

2   Preliminaries

We adopt the notation of the free monoid theory: given an alphabet A, we denote by A^* the free monoid that it generates. Given a word w, we denote by $|w|$ its length, the empty word, that we denote by ε, being the word with length 0. We denote by w_i the letter of position i in w: with this notation we have $w = w_1 \cdots w_{|w|}$. We set $A^+ = A^* \backslash \{\varepsilon\}$. Given $x \in A^*$ and $w \in A^+$, we say that x is a *prefix* (*suffix*) of w if a word u exists such that $w = xu$ ($w = ux$). Similarly, x is a *factor* of w if a pair of words u, v exist such that $w = uxv$. Given a non-empty set $X \subseteq A^*$, we denote by $P(X)$ ($S(X), F(X)$) the set of the words that are prefix (suffix, factor) of some word in X. Clearly, we have $X \subseteq P(X)$ ($S(X), F(X)$). A set $X \subseteq A^*$ is *complete* iff $F(X^*) = A^*$. Given a pair of words w, w', we say that it *overlaps* if words u, v exist such that $uw' = wv$ or $w'u = vw$, with $1 \leq |u| < |w|$ and $1 \leq |v| < |w'|$; otherwise, the pair is *overlapping-free* (in such a case, if $w = w'$, we simply say that w is overlapping-free).

It is assumed that the reader has a fundamental understanding with the main concepts of the theory of variable length codes: we only recall some of the main definitions and we suggest, if necessary, that he (she) report to [10]. A set X is a *variable length code* (a *code* for short) iff any equation among the words of X is trivial, that is, for any pair of sequences of words in X, namely $(x_i)_{1 \leq i \leq m}$, $(y_j)_{1 \leq i \leq n}$, the equation $x_1 \cdots x_m = y_1 \cdots y_n$ implies $m = n$ and $x_i = y_i$ for each integer $i \in [1, m]$. By definition X^*, the submonoid of A^* which is generated by X, is *free*. Equivalently, X^* satisfies the property of *equidivisibility*, that is $(X^*)^{-1}X^* \cap X^*(X^*)^{-1} = X^*$.

Some famous families of codes that have been studied in the literature: X is a *prefix* (*suffix*, *bifix*) *code* iff $X \neq \{\varepsilon\}$ and $X \cap XA^+ = \emptyset$ ($X \cap A^+X = \emptyset, X \cap XA^+ = X \cap A^+X = \emptyset$). X is a code with a *finite deciphering delay* (*f.d.d. code* for short) if it is a code and if a non-negative integer d exists such that $X^{-1}X^* \cap X^d A^+ \subseteq X^+$. With this condition, if another integer d' exists such that we have $X^*X^{-1} \cap A^+ X^{d'} \subseteq X^+$, we say that X is a *two-way f.d.d. code*. X is a *uniformly synchronized code* if it is a code and if a positive integer k exists such that, for all $x, y \in X^k, u, v \in A^+$: $uxyv \in X^* \implies ux, xv \in X^*$. X is a *circular code* if for any pair of sequences of words in X, namely $(x_i)_{1 \leq i \leq m}, (y_j)_{1 \leq j \leq n}$, and any pair of words s, p, with $s \neq \varepsilon$, the equation $x_1 \cdots x_m = sy_2 \cdots y_n p$, with $y_1 = ps$, implies $m = n$, $p = \varepsilon$ and $x_i = y_i$ for each $i \in [1, m]$.

In the whole paper, we consider a *finite* or *countable* alphabet A and a mapping θ which satisfies each of the three following conditions:

(a) θ is a one-to-one correspondence onto A^*
(b) θ is *literal*, that is $\theta(A) \subseteq A$
(c) either θ is a *morphism* or it is an *antimorphism* (it is an antimorphism if $\theta(\varepsilon) = \varepsilon$ and $\theta(xy) = \theta(y)\theta(x)$, for any pair of words x, y); for short in any case we write that θ is an *(anti)morphism*.

In the case where A is a finite set, it is well known that a positive integer n exists such that $\theta^n = id_{A^*}$. In the whole paper, we are interested in the family of sets $X \subseteq A^*$ that are invariant under the mapping θ (θ-*invariant* for short), that is $\theta(X) = X$.

3   A Defect Effect for Invariant Sets

Informally, the famous defect theorem says that if some words of a set X satisfy a non-trivial equation, then these words may be written upon an alphabet of smaller size. In this section, we examine whether a corresponding result may be stated in the frameword of θ-invariant sets. The following property comes from the definition:

Proposition 1. *Let M be a submonoid of A^* and let $S \subseteq A^*$ be such that $M = S^*$. Then M is θ-invariant if and only if S is θ-invariant.*

Clearly the intersection of a non-empty family of θ-invariant free submonoids of A^* is itself a θ-invariant free submonoid. Given a submonoid M of A^*, recall that its *minimal generating set* is $(M\backslash\{\varepsilon\})\backslash(M\backslash\{\varepsilon\})^2$.

Theorem 2. *Let A be a finite or countable alphabet, let $X \subseteq A^*$ be a θ-invariant set and let Y be the minimal generating set of the smallest θ-invariant free submonoid of A^* which contains X. If X is not a code, then we have $|Y| \leq |X| - 1$.*

Proof. With the notation of Theorem 2, since Y is a code, each word $x \in X$ has a unique factorization upon the words of Y, namely $x = y_1 \cdots y_n$, with $y_i \in Y$ ($1 \leq i \leq n$). In a classical way, we say that y_1 (y_n) is the *initial* (*terminal*) *factor* of x (with respect to such a factorization). At first, we shall establish the following lemma:

Lemma 3. *With the preceding notation, each word in Y is the initial (terminal) factor of a word in X.*

Proof. By contradiction, assume that a word $y \in Y$ that is never initial of any word in X exists. Set $Y_0 = (Y\backslash\{y\})\{y\}^*$ and $Y_i = \theta^i(Y_0)$, for each integer $i \in \mathbb{Z}$. In a classical way (cf e.g. [14, p. 7]), since Y is a code, Y_0 itself is a code. Since θ^i is a one-to-one correspondence, for each integer $i \in \mathbb{Z}$, Y_i is a code, that is Y_i^* is a free submonoid of A^*. Consequently, the intersection, namely M, of the family $(Y_i^*)_{i \in \mathbb{Z}}$ is itself a free submonoid of A^*. Moreover we have $\theta(M) \subseteq M$ (indeed, given a word $w \in M$, $\theta(w) \notin Y_i$ implies $w \notin Y_{i-1}$) therefore, since θ is onto, we obtain $\theta(M) = M$. Let x be an arbitrary word in X. Since $X \subseteq Y^*$, and according to the definition of y, we have $x = (y_1 y^{k_1})(y_2 y^{k_2}) \cdots (y_n y^{k_n})$, with $y_1, \cdots y_n \in Y \backslash \{y\}$ and $k_1, \cdots k_n \geq 0$. Consequently x belongs to Y_0^*, therefore we have $X \subseteq Y_0^*$. Since X is θ-invariant, this implies $X = \theta(X) \subseteq Y_i^*$ for each $i \in \mathbb{Z}$, thus $X \subseteq M$.

But the word y belongs to Y^* and doesn't belong to Y_0^* thus it doesn't belong to M. This implies $X \subseteq M \subsetneq Y^*$: a contradiction with the minimality of Y^*. ∎

Proof of Theorem 2. Let α be the mapping from X onto Y which, with every word $x \in X$, associates the initial factor of x in its (unique) factorization over Y^*. According to Lemma 3, α is onto. We will prove that it is not one-to-one. Classically, since X is not a code, a non-trivial equation may be written among its words, say: $x_1 \cdots x_n = x_1' \cdots x_m'$, with $x_i, x_j' \in X$ $x_1 \neq x_1'$ $(1 \leq i \leq n, 1 \leq j \leq m)$. Since Y is a code, a unique sequence of words in Y, namely $y_1, \cdots, y_p$ exists such that: $x_1 \cdots x_n = x_1' \cdots x_m' = y_1 \cdots y_p$. This implies $y_1 = \alpha(x_1) = \alpha(x_1')$ and completes the proof. ∎

In what follows we discuss some interpretation of Theorem 2 with regards to equations in words. For this purpose, we assume that A is finite, thus a positive integer n exists such that $\theta^n = id_{A^*}$. Consider a finite set of words, say Z, and denote by X the union of the sets $\theta^i(Z)$, for $i \in [1, n]$; assume that a non-trivial equation holds among the words of X, namely $x_1 \cdots x_m = y_1 \cdots y_p$. By construction X is θ-invariant therefore, according to Theorem 2, a θ-invariant code Y exists such that $X \subseteq Y^*$, with $|Y| \leq |X| - 1$. This means that each of the words in X can be expressed by making use of at most $|X| - 1$ words of type $\theta^i(u)$, with $u \in Y$ and $1 \leq i \leq n$. It will be easily verified that the examples from [6, 7, 13] corroborate this fact, moreover below we mention an original one:

Example 4. Let θ be a literal antimorphism such that $\theta^3 = id_{A^*}$. Consider two different words x, y, with $|x| > |y|$, which satisfy the following equation:

$$x\theta(y) = \theta^2(y)\theta(x).$$

With these conditions, a pair of words u, v exists such that $x = uv$, $\theta^2(y) = u$, thus $y = \theta(u)$, moreover we have $v = \theta(v)$ and $u = \theta(u) = \theta^2(u)$. With the preceding notation, we have $Z = \{x, y\}$, $X = Z \cup \theta(Z) \cup \theta^2(Z)$, $Y = \{u\} \cup \{v\} \cup \{\theta(u)\} \cup \{\theta(v)\} \cup \{\theta^2(u)\} \cup \{\theta^2(v)\}$. It follows from $y = \theta(y) = \theta^2(y)$ that $X = \{x\} \cup \{\theta(x)\} \cup \{\theta^2(x)\} \cup \{y\}$.

- At first, assume that no word t exists such that $u, v \in t^+$. In a classical way, we have $uv \neq vu$, thus $X = \{x, \theta(x), \theta^2(x), y\}$ and $Y = \{u, v\}$. We verify that $|Y| \leq |X| - 1$.
- Now, assume that we have $u, v \in t^+$. We obtain $X = Z = \{x, y\}$ and $Y = \{t\}$. Once more we have $|Y| \leq |X| - 1$.

4 Maximal θ-Invariant Codes

Given set $X \subseteq A^*$, we say that it is *thin* if $A^* \neq F(X)$. Regular codes are well known examples of thin codes. From the point of view of maximal codes, below we recall one of the famous result stated by Schützenberger:

Theorem 5. [10, Theorem 2.5.16] *Let X be an thin code. Then the following conditions are equivalent:*

(i) *X is complete*
(ii) *X is a maximal code.*

The aim of this section is to examine whether a corresponding result may be stated in the family of thin θ-invariant codes.

In the case where $|A| = 1$, we have $\theta = id_{A^*}$, moreover the codes are all the singletons in A^+. Therefore any code is θ-invariant, maximal and complete. In the rest of the paper, we assume that $|A| \geq 2$.

Some notations. Let X be a non-complete θ-invariant code, and let $y \notin F(X^*)$. Without loss of generality, we may assume that the initial and the terminal letters of y are different (otherwise, substitute to y the word $ay\bar{a}$, with $a, \bar{a} \in A$ and $a \neq \bar{a}$), we may also assume that $|y| \geq 2$. Set:

$$y = ax\bar{a}, \quad z = \bar{a}^{|y|}ya^{|y|} = \bar{a}^{|y|}ax\bar{a}a^{|y|}. \tag{1}$$

Since θ is a literal (anti)morphism, for each integer $i \in \mathbb{Z}$, a pair of different letters $b, \bar{b}$ and a word x' exist such that $|x'| = |x| = |y| - 2$, and:

$$\theta^i(z) = \bar{b}^{|y|}\theta^i(y)b^{|y|} = \bar{b}^{|y|}bx'\bar{b}b^{|y|}. \tag{2}$$

Given two (not necessarily different) integers $i, j \in \mathbb{Z}$, we will accurately study how the two words $\theta^i(z), \theta^j(z)$ may overlap.

Lemma 6. *With the notation in (2), let $u, v \in A^+$ and $i, j \in \mathbb{Z}$ such that $|u| \leq |z| - 1$ and $\theta^i(z)v = u\theta^j(z)$. Then we have $|u| = |v| \geq 2|y|$, moreover a letter b and a unique positive integer k (depending of $|u|$) exist such that we have $\theta^i(z) = ub^k$, $\theta^j(z) = b^k v$, with $k \leq |y|$.*

Proof. According to (2), we set $\theta^i(z) = \bar{b}^{|y|}bx'\bar{b}b^{|y|}$ and $\theta^j(z) = \bar{c}^{|y|}cx''\bar{c}c^{|y|}$, with $b, \bar{b}, c, \bar{c} \in A$ and $b \neq \bar{b}, c \neq \bar{c}$. Since θ is a literal (anti)morphism, we have $|\theta^i(z)| = |\theta^j(z)|$ thus $|u| = |v|$; since we have $1 \leq |u| \leq 3|y| - 1$, exactly one of the following cases occurs:

Case 1: $1 \leq |u| \leq |y| - 1$. With this condition, we have $(\theta^i(z))_{|u|+1} = \bar{b} = \bar{c} = (u\theta^j(z))_{|u|+1}$ and $(\theta^i(z))_{|y|+1} = b = \bar{c} = (u\theta^j(z))_{|y|+1}$, which contradicts $b \neq \bar{b}$.

Case 2: $|u| = |y|$. This condition implies $(\theta^i(z))_{|u|+1} = b = \bar{c} = (u\theta^j(z))_{|u|+1}$ and $(\theta^i(z))_{2|y|} = \bar{b} = \bar{c} = (u\theta^j(z))_{2|y|}$, which contradicts $b \neq \bar{b}$.

Case 3: $|y| + 1 \leq |u| \leq 2|y| - 1$. We obtain $(\theta^i(z))_{2|y|} = \bar{b} = \bar{c} = (u\theta^j(z))_{2|y|}$ and $(\theta^i(z))_{2|y|+1} = b = \bar{c} = (u\theta^j(z))_{2|y|+1}$ which contradicts $b \neq \bar{b}$.

Case 4: $2|y| \leq |u| \leq 3|y| - 1$. With this condition, necessarily we have $b = \bar{c}$, therefore an integer $k \in [1, |y|]$ exists such that $\theta^i(z) = ub^k$ and $\theta^j(z) = b^k v$.

∎

Set $Z = \{\theta^i(z) | i \in \mathbb{Z}\}$. Since $y \notin F(X^*)$ and since X is θ-invariant, for any integer $i \in \mathbb{Z}$ we have $\theta^i(z) \notin F(X^*)$, hence we obtain $Z \cap F(X^*) = \emptyset$. By construction, all the words in Z have length $|z|$ moreover, as a consequence of Lemma 6:

Lemma 7. *With the preceding notation, we have $A^+ Z A^+ \cap Z X^* Z = \emptyset$.*

Proof. By contradiction, assume that $z_1, z_2, z_3 \in Z$, $x \in X^*$ and $u, v \in A^+$ exist such that $u z_1 v = z_2 x z_3$. By comparing the lengths of the words u and v with $|z|$, exactly one of the three following cases occurs:

Case 1: $|z| \leq |u|$ and $|z| \leq |v|$. With this condition, we have $z_2 \in P(u)$ and $z_3 \in S(v)$, therefore the word z_1 is a factor of x: this contradicts $Z \cap F(X^*) = \emptyset$.

Case 2: $|u| < |z| \leq |v|$. We have in fact $u \in P(z_2)$ and $z_3 \in S(v)$. We are in the condition of Lemma 6: the words z_2, z_1 overlap. Consequently, $u \in A^+$ and $b \in A$ exist such that $z_2 = ub^k$ and $z_1 = b^k z_1'$, with $1 \leq k \leq |y|$. But by construction we have $|u z_1| = |z_2 x z_3| - |v|$: since we assume $|v| \geq |z|$, this implies $|u z_1| \leq |z_2 x z_3| - |z| = |z_2 x|$, therefore we obtain $u z_1 = ub^k z_1' \in P(z_2 x)$. It follows from $z_2 = ub^k$ that $z_1' \in P(x)$. Since $z_1 \in Z$ and according to (2), $i \in \mathbb{Z}$ and $\bar{b} \in A$ exist such that we have $z_1 = b^k z_1' = b^{|y|} \theta^i(y) \bar{b}^{|y|}$. Since by Lemma 6 we have $|z_1'| = |u| \geq 2|y|$, we obtain $\theta^i(y) \in F(z_1')$, which contradicts $y \notin F(X^*)$.

Case 3: $|v| < |z| \leq |u|$. Same arguments on the reversed words lead to a conclusion similar to that of Case 2.

Case 4: $|z| > |u|$ and $|z| > |v|$. With this condition, both the pairs of words z_2, z_1 and z_1, z_3 overlap. Once more we are in the condition of Lemma 6: letters c, d, words u, v, s, t, and integers h, k exist such that the two following properties hold:

$$z_2 = uc^h, \quad z_1 = c^h s, \quad |u| = |s| \geq 2|y|, \quad h \leq |y|, \tag{3}$$

$$z_1 = td^k, \quad z_3 = d^k v, \quad |v| = |t| \geq 2|y|, \quad k \leq |y|. \tag{4}$$

It follows from $u z_1 v = z_2 x z_3$ that $u z_1 v = (uc^h) x (d^k v)$, thus $z_1 = c^h x d^k$. Once more according to (2), $i \in \mathbb{Z}$ and $\bar{c} \in A$ exist such that we have $z_1 = c^{|y|} \theta^i(y) \bar{c}^{|y|}$. Since we have $h, k \leq |y|$, this implies $d = \bar{c}$ moreover $\theta^i(y)$ is a factor of x. Once more, this contradicts $y \notin F(X^*)$ (Fig. 1). ∎

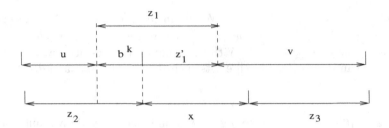

Fig. 1. Proof of Lemma 7: Case 2

Thanks to Lemma 7 we will prove some meaningful results in Sect. 5. Presently, we will apply it in a special context:

Corollary 8. *With the preceding notation, X^*Z is a prefix code.*

Proof. Let $z_1, z_2 \in Z$, $x_1, x_2 \in X^*, u \in A^+$, such that $x_1 z_1 u = x_2 z_2$. For any word $z_3 \in Z$, we have $(z_3 x_1) z_1 (u) = z_3 x_2 z_2$, a contradiction with Lemma 7. ∎

We are now ready to prove the main result of the section:

Theorem 9. *Let A be a finite or countable alphabet and let $X \subseteq A^*$ be a thin θ-invariant code. Then the following conditions are equivalent:*

(i) X is complete
(ii) X is a maximal code
(iii) X is maximal in the family θ-invariant codes.

Proof. Let X be a θ-invariant code. According to Theorem 5, if X is thin and complete, then it is a maximal code, therefore X is maximal in the family of θ-invariant codes. For proving the converse, we consider a set X which is maximal in the family of θ-invariant codes.

Assume that X is not complete and let $y \notin F(X^*)$. Define the word z as in (1) and consider the set $Z = \{\theta^i(z)|i \in \mathbb{Z}\}$. At first, we will prove that $X \cup Z$ remains a code. In view of that, we consider an arbitrary equation between the words in $X \cup Z$. Since X is a code, without loss of generality, we may assume that at least one element of Z has at least one occurrence in one of the two sides of this equation. As a matter of fact, with such a condition and since $Z \cap F(X^*) = \emptyset$, two sequences of words in X^*, namely $(x_i)_{1 \leq i \leq n}, (x'_j)_{1 \leq j \leq p}$ and two sequences of words in Z, namely $(z_i)_{1 \leq i \leq n-1}, (z'_j)_{1 \leq j \leq p-1}$ exist such that the equation takes the following form:

$$x_1 z_1 x_2 z_2 \cdots x_{n-1} z_{n-1} x_n = x'_1 z'_1 x'_2 z'_1 \cdots x'_{p-1} z'_{p-1} x'_p. \tag{5}$$

Without loss of generality, we assume $n \geq p$. At first, according to Corollary 8, necessarily, we have $x_1 = x'_1$, therefore Eq. (5) is equivalent to: $z_1 x_2 z_2 \cdots x_{n-1} z_{n-1} x_n = z'_1 x'_2 z'_2 \cdots x'_{p-1} z'_{p-1} x'_p$, however, since all the words in Z have a common length, we have $z_1 = z_1'$ hence our equation is equivalent to $x_2 z_2 \cdots x_{n-1} z_{n-1} x_n = x_2' z_2' \cdots x_{p-1}' z_{p-1}' x_p'$. Consequently, by applying iteratively the result of Corollary 8, we obtain: $x_2 = x_2', \cdots, x_p = x_p'$, which implies $x_{p+1} z_{p+1} \cdots z_{n-1} x_n = \varepsilon$, thus $n = p$. In other words Eq. (5) is trivial, thus $X \cup Z$ is a code.

Next, since θ is one-to-one and since we have $\theta(X \cup Z) \subseteq \theta(X) \cup \theta(Z) = X \cup Z$, the code $X \cup Z$ is θ-invariant. It follows from $z \in Z \setminus X$ that X is strictly included in $X \cup Z$: this contradicts the maximality of X in the whole family of θ-invariant codes, and completes the proof of Theorem 9. ■

Example 10. Let $A = \{a, b, c\}$. Consider the antimorphism θ which is generated by the permutation $\sigma(a) = b, \sigma(b) = c, \sigma(c) = a$ and let $X = \{ab, cb, ca, ba, bc, ac\}$; it can be easily verified that X is a θ-invariant code. Since we have $c^3 \notin F(X^*)$, by setting $y = c^3 b$ and $z = b^4 \cdot c^3 b \cdot a^4$ we are in Condition (1). The corresponding set Z is $\{\theta^i(z) | i \in \mathbb{Z}\} = \{b^4 c b^3 c^4, a^4 c^3 a c^4, a^4 b a^3 b^4, c^4 b^3 c b^4, c^4 a c^3 a^4, b^4 a^3 b a^4\}$. Since $X \cup Z$ is a prefix set, this guarantees that $X \cup Z$ remains a θ-invariant code.

5   Maximality in Some Families of θ-Invariant Codes

In the literature, statements similar to Theorem 5 were established in the framework of some special families of thin codes. In this section we will draw similar investigations with regards to θ-invariant codes. We will establish the following result:

Theorem 11. *Let A be a finite or countable alphabet and let $X \subseteq A^*$ be a thin θ-invariant prefix (resp. bifix, f.d.d., two-way f.d.d, uniformly synchronized, circular) code. Then the following conditions are equivalent:*

(i) X *is complete*
(ii) X *is a maximal code*
(iii) X *is maximal in the family of prefix (bifix, f.d.d., two-way f.d.d, uniformly synchronized, circular) codes*
(iv) X *is maximal in the family θ-invariant codes*
(v) X *is maximal in the family of θ invariant prefix (bifix, f.d.d., two-way f.d.d, uniformly synchronized, circular) codes.*

Sketch proof. According to Theorem 9, and thanks to [10, Proposition 3.3.8], [10, Proposition 6.2.1], [10, Theorem 5.2.2], [15, Corollary 3.13] and [11, Theorem 3.5], if X is complete then it is maximal in the family of θ-invariant codes and maximal in the family of θ-invariant prefix (bifix, f.d.d., two-way f.d.d, uniformly synchronized, circular) codes. Consequently, the proof of Theorem 11 comes down to establish that if X is not complete, then it cannot be maximal in the family of θ-invariant prefix (bifix, f.d.d., wo-way f.d.d, uniformly synchronized, circular) codes.

(1) We begin by θ-invariant prefix codes. At first, we assume that θ is an antimorphism. Since $X \cap XA^+ = \emptyset$, and since θ is injective, we have $\theta(X) \cap \theta(XA^+) = \emptyset$, thus $X \cap A^+X = \emptyset$, hence X is also a suffix code. Assume that X is not complete. According to [10, Proposition 3.3.8], it is non-maximal in both the families of prefix codes and suffix codes. Therefore a pair of words $y, y' \in A^+\backslash X$ exists such $X \cup \{y\}$ $(X \cup \{y'\})$ remains a prefix (suffix) code. By construction $X \cup \{yy'\}$ remains a code which is both prefix and suffix.

Set $Y = \{\theta^i(yy')|i \in \mathbb{Z}\}$: since all the words in Y have same positive length, Y is a prefix code. From the fact that θ is one-to-one, for any integer $i \in \mathbb{Z}$ we obtain $\theta^i(\{yy'\}) \cap \theta^i(P(X)) = \theta^i(X) \cap P(\theta^i(yy')) = \emptyset$, consequently $X \cup Y$ remains a prefix code. By construction, Y is θ-invariant and it is not included in X, thus X is not a maximal prefix code.

In the case where θ is a morphism, the preceding arguments may be simplified. Actually, a word $y \in A^+\backslash X$ exists such that $X \cup \{y\}$ remains a prefix code, therefore by setting $Y = \{\theta^i(y)|i \in \mathbb{Z}\}$, $X \cup Y$ remains a prefix code.

(2) (sketch) The preceding arguments may be applied for proving that in any case, if X is a non-complete bifix code, then it is maximal.

(3, 4) (sketch) In the case where X is a (two-way) f.d.d.-code, according to [10, Proposition 5.2.1], similar arguments leads to a similar conclusion.

(5) In the case where X is a θ-invariant uniformly synchronized code with *verbal delay* k ([10, Sect. 10.2]), we must make use of different arguments. Actually, according to [15, Theorem 3.10], a complete uniformly synchronized code X' exists, with synchronizing delay k, and such that $X \subsetneq X'$. More precisely, X' is the minimal generating set of the submonoid M of A^* which is defined by $M = (X^{2k}A^* \cap A^*X^{2k}) \cup X^*$. According to Proposition 1 in the present paper, X' is θ-invariant. Since X is strictly included in X', it cannot be maximal in the family of θ-invariant uniformly synchronized codes with delay k.

(6) It remains to study the case where X is a non-complete θ-invariant circular code. Let $y \notin F(X^*)$ and let z and Z be computed as in Sect. 3: this guarantees that $X \cup Z$ is a θ-invariant set. For proving that $X \cup Z$ is a circular code, by contradiction we assume that some words $y_1, \cdots y_n, y_1', \cdots, y_m' \in X \cup Z$ (with $m + n$ minimal), $p \in A^*$ and $s \in A^+$, exist such that the following equation holds:

$$y_1 y_2 \cdots y_n = sy_2'y_3' \cdots y_m'p \quad \text{and} \quad y_1' = ps. \tag{6}$$

Once more since X is a code, and since $Z \cap F(X^*) = \emptyset$, without loss of generality we assume that at least one integer $i \in \mathbb{Z}$ exists such that $y_i \in Z$; similarly, at least one integer $j \in [1, m]$ exists such that $y_j' \in Z$. By construction, we have $y_i \in F(y_j' \cdots y_m'y_1' \cdots y_j' \cdots y_m'y_1' \cdots y_j')$; consequently, since all the words in Z have the same length, a pair of integers $h, k \in [1, m]$ and a pair of words u, v exist such that $uy_iv \in y_h'X^*y_k'$. According to Lemma 7, necessarily we have either $u = \varepsilon$ or $v = \varepsilon$; this implies $y_i = y_h'$ or $y_i = y_k'$, which contradicts the minimality of $m + n$, therefore $X \cup Z$ is a circular code. ∎

6 Embedding a Regular Invariant Code into a Complete One

In this section, we consider a non-complete regular θ-invariant code X and we are interested in the problem of computing a complete one, namely Y, such that $X \subseteq Y$. Historically, such a question appears for the first time in [2], where the author asked for the possibility of embedding a finite code into a regular complete one. With regards to θ-invariant codes, it seems natural to generalize the formula from [1] by making use of the code Z that was introduced in Sect. 4. More precisely we would consider the set $X' = X \cup (ZU)^*Z$, with $U = A^* \backslash (X^* \cup A^*ZA^*)$. Unfortunately, with such a construction we observe that some pairs of words in Z may overlap, therefore a non-trivial equation could hold among the words of X'.

Nevertheless, we shall see that in the very special case where θ is an involutive antimorphism, convenient invariant overlapping-free words can be computed. Denote by θ_0 the antimorphism which is generated by the identity onto A; in other words, with every word $w = w_1 \cdots w_n \in A^*$ (with $w_i \in A$, for $1 \le i \le n$), it associates $\theta_0(w) = w_n \cdots w_1$.

Proposition 12. *Let A be a finite alphabet and let θ be an antimorphism onto A^*, with $\theta \ne \theta_0$. If θ is involutive, then any non-complete regular θ-invariant code can be embedded into a complete one.*

Proof. Let X be such that $\theta(X) = X$. Assume that X is not complete. We will construct an overlapping-free word $t \notin F(X^*)$ such that $\theta(t) = t$. At first, we consider a word x such that $x \notin F(X^*)$ and $|x| \ge 2$. Without loss of generality, we assume that x is overlapping-free (otherwise, as in [10, Proposition 1.3.6], a word s exists such that xs is overlapping-free). If $\theta(x) = x$, then we set $t = x$, otherwise let $y = cx$, where c stands for the initial letter of x. Once more, without loss of generality we assume that y is overlapping-free. By construction we have $y \in ccA^+$, thus $|y| \ge 3$ and $y_1 = y_2 = c$. If $\theta(y) = y$, then we set $t = y$. Now assume $\theta(y) \ne y$; according to the condition of Proposition 12, we have $\theta|_A \ne id_A$, therefore a pair of letters a, b exists such that the following property holds:

$$a \ne b, \quad b \ne c, \quad \theta(a) = b, \quad \theta(b) = a. \tag{7}$$

Set $t = a^{|y|}b\theta(y)yab^{|y|}$. By construction, we have $\theta(t) = t$, moreover the following property holds:

Claim. t is an overlapping-free word.

Proof. Let $u, v \in A^*$ such that $ut = tv$, with $1 \leq |u| \leq |t| - 1$. According to the length of u, exactly one of the following cases occurs:

Case 1: $1 \leq |u| \leq |y|$. With this condition, we obtain $t_{|y|+1} = b = (ut)_{|y|+1} = a$: a contradiction with $a \neq b$.

Case 2: $|y| + 1 \leq |u| \leq 2|y|$. This condition implies $\theta(y_1) = t_{2|y|+1} = a$, therefore we obtain $c = y_1 = \theta(a) = b$: a contradiction with (7).

Case 3: $|u| = 2|y|+1$. We have $y = a^{|y|}$: since we have $|y| \geq 3$, this contradicts the fact that y is overlaping-free.

Case 4: $|u| = 2|y| + 2$. We have $t_{2|y|+3} = y_2 = c = (ut)_{2|y|+3} = a$. It follows from $y_1 = y_2 = c$ that $y = a^{|y|}$: once more this contradicts the fact that y is overlapping-free.

Case 5: $2|y| + 3 \leq |u| \leq 3|y| + 2$. By construction, we have $t_{|u|+|y|} = b = (ut)_{|uy|} = a$, a contradiction with (7).

Case 6: $3|y|+3 \leq |u| \leq |t|-1 = 4|y|+1$. We obtain $t_{|u|+1} = b = (ut)_{|u|+1} = a$: once more this contradicts (7).

In any case we obtain a contradiction: this establishes the claim.

Since we have $t \notin F(X^*)$, and since t is overlapping-free, the classical method from [1] may be applied without any modification to ensure that X may be embedded into a complete code, say X'. Recall that it computes in fact a code X' as $X \cup V$, with $V = t(Ut)^*$ and $U = A^* \backslash (X^* \cup A^* tA^*)$. Moreover, since $\theta(t) = t$, it is straightforward to verify that $\theta(X') = X'$ (Fig. 2). ∎

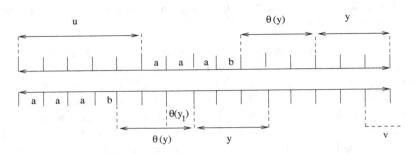

Fig. 2. Proof of Proposition 12: Case 2 with $|y| = 3$ and $|u| = 5$

With regards to the antimorphism θ_0, necessarily the words w, $\theta_0(w)$ overlap, therefore the preceding methodology seems to be unreliable in the most general case. We finish our paper by stating the following open problem:

Problem. Let A be a finite alphabet and let θ be an (anti)morphism onto A^*. Given a non-complete regular θ-invariant code $X \subset A^*$, can we compute a complete regular θ-invariant code Y such that $X \subseteq Y$?

References

1. Ehrenfeucht, A., Rozenberg, S.: Each regular code is included in a regular maximal one. Theor. Inform. Appl. **20**, 89–96 (1985)
2. Restivo, A.: On codes having no finite completion. Discrete Math. **17**, 309–316 (1977)
3. Bucci, M., de Luca, A., De Luca, A., Zamboni, L.Q.: On θ-episturmian words. Eur. J. Comb. **30**, 473–479 (2009)
4. Darshini, C.A.D.P., Rajkumar Dare, V., Venkat, I., Subramanian, K.G.: Factors of words under an involution. J. Math. Inf. **1**, 52–59 (2013–2014)
5. de Luca, A., De Luca, A.: Pseudopalindrome closure operators in free monoids. Theoret. Comput. Sci. **362**, 282–300 (2006)
6. Czeizler, El., Czeizler, Eu., Kari, L., Seki, S.: An extension of the Lyndon-Schützenberger result to pseudoperiodic words. Inf. Comput. **209**, 717–730 (2011)
7. Manea, F., Müller, M., Nowotka, D., Seki, S.: Generalised Lyndon-Schützenberger equations. In: Csuhaj-Varjú, E., Dietzfelbinger, M., Ésik, Z. (eds.) MFCS 2014. LNCS, vol. 8634, pp. 402–413. Springer, Heidelberg (2014). doi:10.1007/978-3-662-44522-8_34
8. Manea, F., Mercaş, R., Nowotka, D.: Fine and Wilf's theorem and pseudo-repetitions. In: Rovan, B., Sassone, V., Widmayer, P. (eds.) MFCS 2012. LNCS, vol. 7464, pp. 668–680. Springer, Heidelberg (2012). doi:10.1007/978-3-642-32589-2_58
9. Gawrychowski, P., Manea, F., Mercas, R., Nowotka, D., Tiseanu, C.: Finding pseudo-repetitions. In: Portier, N., Wilke, T. (eds) 30th International Symposium on Theoretical Aspects of Computer Science (STACS 2013). Leibniz International Proceedings in Informatics (LIPIcs), vol. 20, pp. 257–268. Dagstuhl, Germany (2013). Schloss Dagstuhl-Leibniz-Zentrum fuer Informatik
10. Berstel, J., Perrin, D., Reutenauer, C.: Codes and Automata. Cambridge University Press, New York (2010)
11. Néraud, J.: Completing circular codes in regular submonoids. Theoret. Comput. Sci. **391**, 90–98 (2008)
12. Kari, L., Mahalingam, K.: DNA codes and their properties. In: Mao, C., Yokomori, T. (eds.) DNA 2006. LNCS, vol. 4287, pp. 127–142. Springer, Heidelberg (2006). doi:10.1007/11925903_10
13. Kari, L., Mahalingam, K.: Watson-Crick conjugate and commutative words. In: Garzon, M.H., Yan, H. (eds.) DNA 2007. LNCS, vol. 4848, pp. 273–283. Springer, Heidelberg (2008). doi:10.1007/978-3-540-77962-9_29
14. Lothaire, M.: Combinatorics on Words. Cambridge University Press, Cambridge (1983). 2nd edn. in Cambridge University Press 1997
15. Bruyère, V.: On maximal codes with bounded synchronization delay. Theoret. Comput. Sci. **204**, 11–28 (1998)

Burrows-Wheeler Transform and Run-Length Enconding

Sabrina Mantaci[1]([✉]), Antonio Restivo[1]([✉]), Giovanna Rosone[2]([✉]),
and Marinella Sciortino[1]([✉])

[1] University of Palermo, Palermo, Italy
{sabrina.mantaci,antonio.restivo,marinella.sciortino}@unipa.it
[2] University of Pisa, Pisa, Italy
giovanna.rosone@unipi.it

Abstract. In this paper we study the clustering effect of the Burrows-Wheeler Transform (BWT) from a combinatorial viewpoint. In particular, given a word w we define the BWT-clustering ratio of w as the ratio between the number of clusters produced by BWT and the number of the clusters of w. The number of clusters of a word is measured by its Run-Length Encoding. We show that the BWT-clustering ratio ranges in $]0, 2]$. Moreover, given a rational number $r \in]0, 2]$, it is possible to find infinitely many words having BWT-clustering ratio equal to r. Finally, we show how the words can be classified according to their BWT-clustering ratio. The behavior of such a parameter is studied for very well-known families of binary words.

Keywords: Burrows-Wheeler transform · Run-length encoding · Clustering effect

1 Introduction

Burrows-Wheeler Transform is a popular method used for text compression (cf. [1,3]). It produces a permutation of the characters of an input word w in order to obtain a word easier to compress. Actually compression algorithms based on BWT take advantage of the fact that the word output of BWT shows a local similarity (occurrences of a given symbol tend to occur in clusters) and then turns out to be highly compressible. Several authors refer to such a property as the "clustering effect" of BWT. The aim of this paper is to study such a clustering effect of BWT from the point of view of combinatorics on words.

In order to measure the amount of local similarity, or clustering, in a word we consider its Run-Length Encoding (RLE). RLE is another fundamental string compression technique: it replaces in a word occurrences of repeated equal symbols with a single symbol and a non-negative integer (run length)

Partially supported by the project MIUR-SIR CMACBioSeq ("Combinatorial methods for analysis and compression of biological sequences") grant no. RBSI146R5 and by the Gruppo Nazionale per il Calcolo Scientifico (GNCS-INDAM).

S. Brlek et al. (Eds.): WORDS 2017, LNCS 10432, pp. 228–239, 2017.
DOI: 10.1007/978-3-319-66396-8_21

counting the number of times the symbol is repeated. *RLE* can be considered an efficient compression scheme when the input data is highly repetitive. In a more formal way, every word w over the alphabet Σ has a unique expression of the form $w = w_1^{l_1} w_2^{l_2} \cdots w_k^{l_k}$ with $l_i \in \mathbb{N}$ and $w_i \in \Sigma$ and $w_i \neq w_{i+1}$ for $i = 1, 2, \ldots, k$. The run-length encoding of w is the sequence $\mathtt{rle}(w) = (w_1, l_1)(w_2, l_2) \cdots (w_k, l_k)$. For instance if $w = aaabbbbbccbbbb$ the run-length encoding is $\mathtt{rle} = (a, 3)(b, 5)(c, 2)(b, 4)$. We set $\rho(w) = |\mathtt{rle}(w)|$, i.e., $\rho(w)$ is the number of maximal runs of equal letters in w. For instance, $\rho(aaabbbbbccbbbb) = 4$. It is straightforward that $1 \leq \rho(w) \leq |w|$. The quantity $|w|/\rho(w)$ provides a measure of the amount of local similarity of the word w, in the sense that the lower is the value $\rho(w)$ with respect to $|w|$, the greater is the length of the runs of individual symbols in w.

In this paper we are interested to investigate the "clustering effect" of *BWT*, extending some results presented in [8]. For this aim we introduce for any word its *BWT*-clustering ratio

$$\gamma(w) = \frac{\rho(\mathtt{bwt}(w))}{\rho(w)}$$

where $\mathtt{bwt}(w)$ denotes the output of *BWT* on the input word w. Our first result (Theorem 7) states that, for any word w, $0 < \gamma(w) \leq 2$. This means that, if the number of runs increases after the application of the *BWT* ("un-clustering effect"), in the worst case the number of runs in the output is at most twice the number of runs in the original word. In other words, whereas the "clustering effect" for some words w could be very high ($\gamma(w)$ close to 0), the "un-clustering effect" is in any case moderate. The fact that the worst case is not too bad provides an additional formal motivation of usefulness of *BWT* in Data Compression.

We further prove (Theorem 8) that, for any rational number r, with $0 < r \leq 2$, there exists a word w such that $\gamma(w) = r$.

Previous results suggest that the parameter $\gamma(w)$ could be an interesting tool for the study (or classification) of finite words. In particular, we derive a characterization of Christoffel words w in terms of $\gamma(w)$ and we determine the possible values of $\gamma(w)$ for a de Bruijn word w.

Finally in Sect. 5 we show the results of some statistical experiments that classify words in terms of their *BWT*-clustering ratio γ.

2   Burrows-Wheeler Transform

Let $\Sigma = \{a_1, a_2, \ldots, a_\sigma\}$ be a finite ordered alphabet with $a_1 < a_2 < \ldots < a_\sigma$, where $<$ denotes the standard lexicographic order. We denote by Σ^* the set of words over Σ. Given a finite word $w = w_1 w_2 \cdots w_n \in \Sigma^*$ with each $w_i \in \Sigma$, the length of w, denoted $|w|$, is equal to n. We denote by $alph(w)$ the subset of Σ containing all the letters that appear in w. Given a finite word $w = w_1 w_2 \cdots w_n$ with each $w_i \in \Sigma$, a *factor* of a word w is written as $w[i, j] = w_i \cdots w_j$ with $1 \leq i \leq j \leq n$. A factor of type $w[1, j]$ is called a *prefix*, while a factor of type $w[i, n]$ is called a *suffix*. We also denote by $w[i]$ the i-th letter in w for any $1 \leq i \leq n$.

We say that two words $x, y \in \Sigma^*$ are *conjugate*, if $x = uv$ and $y = vu$, where $u, v \in \Sigma^*$. Conjugacy between words is an equivalence relation over Σ^*. The *conjugacy class* (w) of $w \in \Sigma^n$ (or *necklace*) is the set of all words $w_i w_{i+1} \cdots w_n w_1 \cdots w_{i-1}$, for any $1 \leq i \leq n$. A necklace can be also thought as a cyclic word.

A nonempty word $w \in \Sigma^*$ is *primitive* if $w = u^h$ implies $w = u$ and $h = 1$.

A *Lyndon word* is a primitive word which is the minimum in its conjugacy class, with respect to the lexicographic order relation.

The *Burrows-Wheeler Transform* (*BWT*) can be described as follows: given a word $w \in \Sigma^*$, the output of *BWT* is the pair $(\text{bwt}(w), I)$, where:

- $\text{bwt}(w)$ is the permutation of the letters in the input word w obtained by considering the matrix M containing the lexicographically sorted list of the conjugates of w, and by concatenating the letters of the last column L of matrix M.
- I is the index of the row of M containing the original word w.

Note that if two words v and w are conjugate then $\text{bwt}(v) = \text{bwt}(w)$, i.e. the output of *BWT* is the same up to the second component of the pair. Note also that the first column F of the matrix M is the sequence of lexicographically sorted symbols of w.

The Burrows-Wheeler transform is reversible by using the properties (cf. [3]) described in the following proposition.

Proposition 1. *Let (L, I) be a pair produced by the BWT applied to a word w. Let F be the sequence of the sorted letters of $L = \text{bwt}(w)$. The following properties hold:*

1. *for all $i = 1, \ldots, n$, $i \neq I$, the letter $F[i]$ follows $L[i]$ in the original string w;*
2. *for each letter c, the r-th occurrence of c in F corresponds to the r-th occurrence of c in L;*
3. *the first letter of w is $F[I]$.*

From the above properties it follows that the *BWT* is reversible in the sense that, given L and I, it is possible to reconstruct the original string w. Note that when $I = 1$, one can build the Lyndon conjugate of the original word.

Actually, according to Property 2 of Proposition 1, we can define a permutation $\tau : \{1, \ldots, n\} \to \{1, \ldots, n\}$ where τ gives the correspondence between the positions of letters of F and L. The permutation τ is also called *FL-mapping*.

The permutation τ also represents the order in which we have to rearrange the elements of F to reconstruct the original word w. Hence, starting from I, we can recover the word w as follows:

$$w[i] = F[\tau^{i-1}(I)] , \text{ where } \tau^0(x) = x, \text{ and } \tau^i(x) = \tau(\tau^{i-1}(x)), \text{with } 1 \leq i \leq n.$$

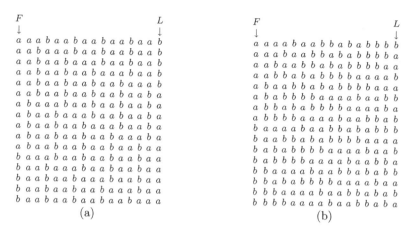

Fig. 1. On the left (a) the matrix of all lexicographic sorted conjugates of the Lyndon word $aaabaabaabaabaab$. In this case the output of BWT is the pair $(bbbbbaaaaaaaaaaa, 1)$. On the right (b) the matrix M of the word $aaaabaabbababbbb$. For such a word BWT outputs the pair $(baabababbababababa, 1)$

Example 2. Let us consider the words examined in Fig. 1.

Given the pair $(bbbbbaaaaaaaaaaa, 1)$ the permutation τ between the positions of $F = aaaaaaaaaaabbbbb$ and $L = bbbbbaaaaaaaaaaa$ is the following:

$$\tau = \begin{pmatrix} 1\ 2\ 3\ 4\ 5\ \ 6\ \ 7\ \ 8\ \ 9\ \ 10\ 11\ 12\ 13\ 14\ 15\ 16 \\ 6\ 7\ 8\ 9\ 10\ 11\ 12\ 13\ 14\ 15\ 16\ \ 1\ \ 2\ \ 3\ \ \ 4\ \ \ 5 \end{pmatrix}$$

So, we can reconstruct the word $w = aaabaabaabaabaab$.

If we consider the pair $(baabababbababababa, 1)$ the permutation τ between $F = aaaaaaaaaaaabbbb$ and $L = baabababbababababa$ is:

$$\tau = \begin{pmatrix} 1\ 2\ 3\ 4\ 5\ \ 6\ \ 7\ \ 8\ 9\ 10\ 12\ 14\ 16\ 1\ 4\ \ 6\ \ 8\ \ 9\ 11\ 13\ 15 \\ 2\ 3\ 5\ 7\ 10\ 12\ 14\ 16\ 1\ 4\ \ 6\ \ 8\ \ 9\ 11\ 13\ 15 \end{pmatrix}$$

So, the recovered word is $w = aaaabaabbababbbb$.

3 *BWT*-Clustering Ratio of a Word

The Run-Length Encoding is a fundamental string compression technique that replaces in a word occurrences of repeated equal symbols with a single symbol and a non negative integer (run length) counting the number of times the symbol is repeated. Formally, every word w over the alphabet Σ has a unique expression of the form $w = w_1^{l_1} w_2^{l_2} \cdots w_k^{l_k}$ with $l_i \in \mathbb{N}$ and $w_i \in \Sigma$ and $w_i \neq w_{i+1}$ for $i = 1, 2, \ldots, k$. The *run-length encoding* of a word w, denoted by $\mathrm{rle}(w)$, is a sequence of pairs (w_i, l_i) such that $w_i w_{i+1} \cdots w_{i+l_i-1}$ is a maximal run of a letter w_i (i.e., $w_i = w_{i+1} = \cdots = w_{i+l_i-1}$, $w_{i-1} \neq w_i$ and $w_{i+l_i} \neq w_i$), and all such maximal runs are listed in $\mathrm{rle}(w)$ in the order they appear in w. We denote by

$\rho(w) = |\text{rle}(w)|$ i.e., is the number of pairs in w, or equivalently the number of equal-letter runs (also called *clusters*) in w.

Moreover we denote by $\rho(w)_{a_i}$ the number of pairs (w_j, l_j) in $\text{rle}(w)$ where $w_j = a_i$.

It is clear that for all $w \in \Sigma^*$ one has that $|alph(w)| \leq \rho(w) \leq |w|$. Notice also that if $w = uv$ then $\rho(w) \leq \rho(u) + \rho(v)$, that is, ρ is sub-additive.

In this section we introduce a parameter that gives a measure on how much the application of the BWT to a given word modifies the number of its clusters.

Definition 3. The *BWT-clustering ratio* of a word w is

$$\gamma(w) = \frac{\rho(\text{bwt}(w))}{\rho(w)}$$

Example 4. Let us compute the BWT-clustering ratio for the words considered in Fig. 1. If $w = aaabaabaabaabaab$ we have that $\rho(w) = 10$ and $\rho(\text{bwt}(w)) = \rho(bbbbbaaaaaaaaaaaa) = 2$. So, $\gamma(w) = 1/5$.

Let us consider $w = aaaabaabbababbbb$. In this case we have that $\rho(w) = 8$. Since $\text{bwt}(w) = baababababbababababa$ then $\rho(\text{bwt}(w)) = 14$. So, $\gamma(w) = 7/4$.

Remark 5. We note that if w is not a primitive word (i.e., $w = v^k$ for some $k > 1$) one can prove that $\rho(v^k) \leq k\rho(v)$. Moreover, in [9] it has been proved that if $\text{bwt}(v) = v_1 v_2 \cdots v_n$, where $v_i \in \Sigma$, then $\text{bwt}(v^k) = v_1^k v_2^k \cdots v_n^k$. So, $\rho(\text{bwt}(v^k)) = \rho(\text{bwt}(v))$. This implies that $\gamma(v^k) \geq \frac{1}{k}\gamma(v)$. In particular, it was proved (cf. [8]) that if v is a Lyndon word (different from a single letter), then $\gamma(v^k) = \frac{1}{k}\gamma(v)$.

Remark 6. We recall that if u and v are conjugate words, then $\text{bwt}(u) = \text{bwt}(v)$. On the other hand, one has that $|\rho(u) - \rho(v)| \leq 1$ and, within the conjugacy class, a power of a Lyndon word is one of the conjugates having least number of clusters. Since we are interested in evaluating how the number of clusters produced by BWT can grow compared to the number of clusters in the input necklace, we can consider words that are power of a Lyndon word as input of the parameter γ. Moreover, due to the property described in Remark 5 we can limit our attention to Lyndon words.

The following theorem, also reported in [8], shows that the number of clusters can at most be doubled by the BWT.

Theorem 7. *Given a Lyndon word w, we have that $0 < \gamma(w) \leq 2$.*

Proof. Let $\Sigma = \{a_1, a_2, \ldots, a_\sigma\}$ with $a_1 < a_2 < \cdots < a_\sigma$ and let $\text{rle}(w) = (b_1, l_1), (b_2, l_2), \ldots, (b_k, l_k)$, where $b_1, b_2, \ldots b_k \in \Sigma$.

Recall that when computing $\text{bwt}(w)$, the column F of the matrix of sorted conjugates of w has the form $a_1^{|w|_{a_1}} a_2^{|w|_{a_2}} \cdots a_\sigma^{|w|_{a_\sigma}}$. It is then naturally defined a parsing of the column L according to the runs $(a_1, |w|_{a_1})(a_2, |w|_{a_2}) \cdots (a_\sigma, |w|_{a_\sigma})$ of F. We denote by u_{a_i} the factor in $L = \text{bwt}(w)$ associated to the run $(a_i, |w|_{a_i})$

of F, i.e. all the letters that in the original words precede an occurrence of the letter a_i. Then we can write $\texttt{bwt}(w) = u_{a_1} u_{a_2} \cdots u_{a_\sigma}$.

Consider any block u_{a_j}. In this block there are at most as many letters different from a_j as the number of different runs of a_j in w. In fact, in w, a_j is preceded by a letter different from a_j itself only in the beginning of each of its runs. So the greatest possible number of runs contained in u_{a_j} is achieved when all the letters different from a_j are spread in the block, never one next to another, producing on u_{a_j} a number of runs $\texttt{rle}(u_{a_j})$ at most equal to $2\rho(w)_{a_j}$. This happens for each block, then

$$\rho(\texttt{bwt}(w)) \le \sum_{i=1}^{\sigma} \rho(u_{a_i}) \le \sum_{i=1}^{\sigma} 2 \cdot \rho(w)_{a_i} = 2 \sum_{i=1}^{\sigma} \rho(w)_{a_i} = 2\,\rho(w).$$

□

In the following theorem we show that for any positive rational number r smaller than or equal to 2, it is possible to construct a binary word such that its BWT-clustering ratio is equal to r.

Theorem 8. *For any $r \in Q \cap (0, 2]$, there exists a Lyndon word $w \in \{a, b\}^*$ having $\gamma(w) = r$.*

Proof. Let p and q two coprime positive integers such that $r = \frac{p}{q}$. Let k be an integer such that $k \ge 2$.

Let us define $f_i = a^{2i-1}b^{2i-1}$ for $i = 2, 3, \ldots, k$ and $f_1 = abb$. Let h be an integer such that $h \ge 1$.

We can define a family of words

$$v_{h,k} = (f_k)^h f_{k-1} \cdots f_1 = (a^{2k-1}b^{2k-1})^h a^{2k-3}b^{2k-3} \cdots a^3 b^3 ab^2.$$

Since each f_i has two clusters, the first an a-cluster and the second a b-cluster, $\rho(v_{h,k}) = 2h + 2k - 2$.

We now compute $\rho(\texttt{bwt}(v_{h,k}))$.

First of all we consider the case $h = 1$. In fact, $\texttt{bwt}(v)$ can be factored in two parts: the first one corresponding to all the conjugates starting with a, and the second one corresponding to all the conjugates starting with b.

The first part starts with the only conjugate that has $a^{2k-1}b$ as prefix, then the one with $a^{2k-2}b$, then the two conjugates that start with $a^{2k-3}b$ (from the rightmost to the leftmost), and so on. From this we can see that the first part of $\texttt{bwt}(v_{1,k})$ is $baba^3 \cdots ba^{2(k-1)-1}ba^{k-1}$ that has $2k$ clusters.

For the second part, we have exactly all the conjugates starting in the second part of each f_i. In particular, there are k conjugates starting with ba. All these conjugates are cyclicly preceded by b. Then we have all the conjugates starting with bba. The lexicographically smallest in this group is the one corresponding to the block f_1, then we have the conjugates corresponding to the other f_i from the leftmost to the rightmost. Such conjugates are lexicographically followed by the conjugates starting with $bbba$ that correspond to the blocks from f_k to f_2 and so on. This means that the second part of $\mathtt{bwt}(v_{1,k})$ is $b^k ab^{2k-3} ab^{2k-5} a \cdots ba$ that has $2k$ clusters. It follows that

$$\mathtt{bwt}(v_{1,k}) = baba^3 \cdots ba^{2(k-1)-1} ba^{k-1} b^k ab^{2k-3} ab^{2k-5} a \cdots ba,$$

that has $2k + 2k$ clusters. So $\rho(\mathtt{bwt}(v_{1,k})) = 4k$.

Finally, one can prove that for any $h \geq 1$, $\rho(\mathtt{bwt}(v_{h,k})) = \rho(\mathtt{bwt}(v_{1,k}))$. In fact, $\mathtt{bwt}(v_{h,k})$ is obtained by concatenating

$$b^h a^h ba^{2h+1} ba^{2h+3} \cdots ba^{2h+2k-5} ba^{k-2+h}$$

and

$$b^{k-1+h} ab^{2k-5+2h} ab^{2k-7+2h} a \cdots b^{1+2h} ab^h a^h.$$

So, the thesis follows since

$$\gamma(v_{h,k}) = \frac{4k}{2h + 2k - 2} = \frac{2k}{h + k - 1} = \frac{p}{q}.$$

It is then sufficient to find suitable integer solutions to unknown h and k to the above equation.   □

Example 9. Let us consider the rational number $6/5$ (>1). In this case a solution to the equation

$$\frac{2k}{h + k - 1} = \frac{6}{5}$$

is $k = 3$ and $h = 3$. In fact one can verify that if $w = (a^5 b^5)^3 a^3 b^3 abb$ then $\mathtt{bwt}(w) = b^3 a^3 ba^7 ba^4 b^5 ab^7 ab^3 a^3$, so $\rho(w) = 10$ and $\rho(\mathtt{bwt}(w)) = 12$.

On the other hand if we consider the rational number $4/5$ (< 1) a solution to

$$\frac{2k}{h + k - 1} = \frac{4}{5}$$

is $k = 2$ and $h = 4$. One can verify that $w = (a^3 b^3)^4 abb$ and $\mathtt{bwt}(w) = b^4 a^4 ba^4 b^5 ab^4 a^4$, so $\rho(w) = 10$ and $\rho(\mathtt{bwt}(w)) = 8$.

Corollary 10. *For any rational number $0 < r \leq 2$, there are infinitely many words w with $\gamma(w) = r$.*

Proof. The solutions of the equation

$$\frac{2k}{h + k - 1} = \frac{p}{q}$$

corresponds to the integer solutions to all of the following systems:

$$\begin{cases} 2k = lp \\ h + k - 1 = lq \end{cases}$$

for any choice of l that gives integer solutions to h and k. In particular, if p is even, any integer value of l is allowed, if p is odd, only even values of l are allowed.

4   Special Cases on Binary Alphabet

In this section we give some characterization and properties of families of words over two letters alphabets well known in combinatorics on words, according to their BWT-clustering ratio γ.

4.1   Clusters in Christoffel Words

In this subsection we take into account the BWT-clustering ratio of a class of words over a binary alphabet known in literature as Christoffel words (cf. [2,6]). We start by giving the definition of a class of words strictly related to them, i.e. the Standard words. There exist many equivalent definitions of Standard words. Here we use the one that makes evident their relationships with the notion of characteristic Sturmian word.

Let $d_1, d_2, \ldots, d_n, \ldots, n \geq 1$ be a sequence of natural integers, with $d_1 \geq 0$ and $d_i > 0$ for $i = 2, \ldots, n, \ldots$. Consider the sequence of words $\{s_n\}_{n \geq 0}$ recursively defined by:

$$s_0 = b, \quad s_1 = a, \quad \text{and} \quad s_{n+1} = s_n{}^{d_n} s_{n-1} \text{ for } n \geq 1.$$

Each finite word s_n is called a *standard word*. It is univocally determined by the (finite) directive sequence $(d_1, d_2, \ldots, d_n)$. Such sequences are very important, since their limit, for $n \to \infty$ converges to infinite words called characteristic Sturmian words, well known in literature for its numerous and interesting combinatorial properties.

For any standard word w, the Lyndon word in its class is also called *Christoffel word*. We are now considering Christoffel words since, as usual, we take the Lyndon word for each class. For instance the word $aaabaabaabaabaab$ considered in Fig. 1(a) is a Christoffel word.

The following proposition gives a new characterization of Christoffel words in terms of the γ ratio.

Proposition 11. *A word w is a Christoffel word $\Longleftrightarrow \gamma(w) = \frac{1}{\min\{|w|_a, |w|_b\}}$.*

Proof. Let w be a Christoffel word and suppose that $|w|_b = h$ and $|w|_a = k$ with $h < k$ (the other case has an analogous proof). Then no b in w appears next to another b, therefore w has $2h$ clusters, i.e. $\rho(w) = 2|w|_b$. On the other side in [9] it has been proved that any conjugate of a standard word (in particular any Christoffel word) has a totally clustered bwt; in particular $\mathtt{bwt}(w) = b^h a^k$, i.e. $\rho(\mathtt{bwt}(w)) = 2$. Therefore

$$\gamma(w) = \frac{2}{2h} = \frac{1}{|w|_b}.$$

Suppose now that $\gamma(w) = 1/|w|_b$, that is

$$\frac{\rho(\mathtt{bwt}(w))}{\rho(w)} = \frac{1}{|w|_b}.$$

Then $\rho(\mathtt{bwt}(w)) \cdot |w|_b = \rho(w) \leq 2|w|_b$, i.e. $\rho(\mathtt{bwt}(w)) \leq 2$.

But on binary words $\rho(\mathtt{bwt}(w)) \geq 2$, then $\rho(\mathtt{bwt}(w)) = 2$ and this is true if and only if w is a Christoffel word (cf [9]).   □

Remark 12. For any $\epsilon > 0$ there exists a Christoffel word w such that $\gamma(w) < \epsilon$. In fact let us consider a Christoffel word where $|w| = 2n + 1$, $|w|_b = n$ and $|w|_a = n + 1$. By Proposition 11 $\gamma(w) = \frac{1}{n}$. For n sufficiently large, $1/n < \epsilon$.

4.2 Clusters in Binary de Bruijn Words

In this section we consider another famous class of words called de Bruijn words. In particular here we consider de Bruijn words over a binary alphabet.

A de Bruijn word of order n on an alphabet Σ of size k is a cyclic word in which every word of length n on Σ occurs exactly once as a factor. By Remark 6, in the following when we refer to a de Bruijn word we mean the corresponding Lyndon word in its necklace. Such a word is denoted by $B(k,n)$ and has length k^n, which is also the number of distinct factors of length n on Σ. There are $\frac{(k!)^{k^{n-1}}}{k^n}$ many distinct de Bruijn words $B(k,n)$. In particular for two letters alphabets, all de Bruijn words $B(2,n)$ have length 2^n, and there are $\frac{2^{2^{n-1}}}{2^n}$ many distinct de Bruijn words $B(2,n)$.

One can verify that the word $aaaabaabbababbbb$ considered in Fig. 1(b) is a de Bruijn word of order 4 over the alphabet $\{a,b\}$, since every word in $\{a,b\}^4$ appears once in the corresponding cyclic word.

In the following proposition, also reported in [8], we find the number of runs of a de Bruijn word of order n on a binary alphabet. Note that this result can be inferred by using some combinatorial properties analyzed in [4].

Proposition 13. *Let $B(2,n)$ be any de Bruijn word of order n over a binary alphabet. Then $\rho(B(2,n)) = 2^{n-1}$.*

Proof. We first consider the runs of a's. First of all there is no run a^i with $i > n$ otherwise a^n would be a word of length n that appears more than once in $B(2,n)$. The run a^n is a particular word of length n, then, by definition, it appears exactly once as a factor in $B(2,n)$ (in particular as factor of $ba^n b$).

The words ba^{n-1} and $a^{n-1}b$ also appear once, but since they are factors of $ba^n b$, we have no runs of a's of length $n-1$.

For any $1 \leq i \leq n-2$ consider the runs of the form a^i. They appear as factors of all the words of the form $ba^i bw$ where w is any word of length $n-i-2$. Each of the words $ba^i bw$ appear exactly once. There are 2^{n-i-2} of such words, therefore there are 2^{n-i-2} runs a^i. We have overall:

$$1 + \sum_{i=1}^{n-2} 2^{n-i-2} = 1 + \sum_{i=0}^{n-3} 2^i = 1 + 2^{n-2} - 1 = 2^{n-2}$$

So there are 2^{n-2} runs of a's. For the same reason there are 2^{n-2} runs of b's, then overall $2 \cdot 2^{n-2} = 2^{n-1}$ runs.   □

Remark 14. As a byproduct of the theorem proved by Higgins in [5] (cf. also [10]) we have that if $B(n, k)$ is a de Bruijn word of order n, then $\mathtt{bwt}(B(k, n)) \in G^{k^{n-1}}$, where G is the set of all sequences of Σ of length $|\Sigma|$ obtained by permuting all the letters in Σ. In particular, if $k = 2$, $\mathtt{bwt}(B(2, n)) \in \{ab, ba\}^{2^{n-1}}$.

The following theorem is a consequence of Proposition 13 and of the above remark.

Theorem 15. *If w is a binary de Bruijn word then:*

$$1 + \frac{4}{|w|} \leq \gamma(w) \leq 2 - \frac{4}{|w|}.$$

Proof. Recall that any binary de Bruijn word of order n has length 2^n, with $n \geq 2$.

As remarked above, by Higgins's Theorem, $\mathtt{bwt}(w) \in \{ab, ba\}^{2^{n-1}}$. Moreover, one can note that ba must be the prefix and the suffix of $\mathtt{bwt}(B(2, n))$, since for any word its $\mathtt{bwt}$ cannot start with the smallest symbol and cannot end with the biggest symbol. Since, as proved in [8], $\rho(\mathtt{bwt}(B(2, n))) < |B(2, n)| = 2^n$, then $\mathtt{bwt}(w) \neq (ba)^{2^{n-1}}$ then both aa and bb must be factors of $\mathtt{bwt}(B(2, n))$. So, the upper bound follows because $\rho(\mathtt{bwt}(w)) \leq 2^n - 2$. The lower bound on the number of runs is reached when $\mathtt{bwt}(w) = b(aabb)^{2^{n-2}-1}aba$. In this case this value is $2^{n-1} + 2$. Then, the thesis follows.

5   Experimental Results

It is commonly said that the application of BWT as a preprocessing to the application of a statistical compressor is useful since BWT tends to cluster together equal letters that appear in equal contexts, generating a so called "clustering effect". In this paper we highlight that this is not always the case, that is, there are words that are "un-clustered" by the BWT, that is, the application of BWT generate on such words a greater number of shorter clusters.

The BWT-clustering ratio γ allows to classify words into BWT-*good words*, if $0 < \gamma(w) < 1$, and BWT-*bad words*, if $1 < \gamma(w) \leq 2$. The qualities *good* and *bad* reflects a good or bad behavior of BWT with respect to clustering, that is a good requirement for compression. For instance, since for any Christoffel word w, $\gamma(w) < 1$, then Christoffel words are BWT-good. On the other hand, any binary Bruijn word is BWT-bad.

Of course, the other special case is when this ratio is 1, i.e. the words, called BWT-*neutral*, where the BWT has no effect in terms of clustering. Among these words we can find fixed points (i.e. words w such that $\mathtt{bwt}(w) = w$), that are studied in [7].

In this section, we show some experiments that highlight the distribution of the BWT-neutral, BWT-good and BWT-bad binary words when the length is fixed. In particular, table in Fig. 2 shows such a distribution for all Lyndon words w of length 16 and 24.

length	number of words	$\gamma(w) = 1$	$\gamma(w) < 1$	$\gamma(w) > 1$	$\gamma(w) = 2$
16	4.080	1.160	1.247	1.673	142
24	698.870	156.652	237.636	304.582	4.362

Fig. 2. Distribution of Lyndon words of length 16 and 24

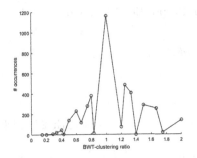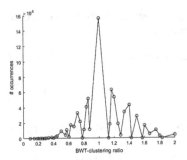

Fig. 3. Lyndon words of length 16 and 24

On the other hand, table in Fig. 4 shows such a distribution for all Lyndon words w of length $16, 20, 24$ and 28. With the same number of occurrences of letter a and letter b.

For completeness, the graphs in Figs. 3 and 5 show the number of Lyndon words of length 16 and 24 as a function of the BWT-clustering ratio. It is interesting to point out that the graphs show that the trend does not change substantially

length	number of words	$\gamma(w) = 1$	$\gamma(w) < 1$	$\gamma(w) > 1$	$\gamma(w) = 2$
16	800	224	239	337	26
20	9.225	2.183	3.042	4.000	129
24	112.632	23.866	38.884	49.882	666
28	1.432.613	288.485	504.505	639.623	3.556

Fig. 4. Distribution of Lyndon words of length $16, 20, 24$ and 28. With the same number of letters a and b

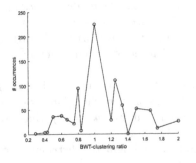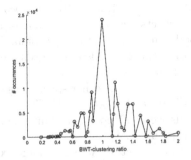

Fig. 5. Lyndon words of length 16 and 24 with the same number of letters a and b

when words having the same number of a and b are considered. A possible further work could be to develop an analytic study of this behavior.

Acknowledgements. We thank the anonymous reviewers for providing us with many helpful comments and suggestions.

References

1. Adjeroh, D., Bell, T., Mukherjee, A.: The Burrows-Wheeler Transform: Data Compression, Suffix Arrays, and Pattern Matching. Springer, New York (2008)
2. Berstel, J., Lauve, A., Reutenauer, C., Saliola, F.: Combinatorics on Words: Christoffel Words and Repetitions in Words. CRM Monograph Series, vol. 27. American Mathematical Soc., Providence (2008)
3. Burrows, M., Wheeler, D.J.: A block sorting data compression algorithm. Tech. report, DIGITAL System Research Center (1994)
4. Fredricksen, H.: A survey of full length nonlinear shift register cycle algorithms. SIAM Rev. **24**(2), 195–221 (1982)
5. Higgins, P.M.: Burrows-Wheeler transformations and de Bruijn words. Theoret. Comput. Sci. **457**, 128–136 (2012)
6. Lothaire, M.: Applied Combinatorics on Words. Encyclopedia of Mathematics and its Applications. Cambridge University Press, New York (2005)
7. Mantaci, S., Restivo, A., Rosone, G., Russo, F., Sciortino, M.: On fixed points of the Burrows-Wheeler Transform. Fundamenta Informaticae **154**, 277–288 (2017)
8. Mantaci, S., Restivo, A., Rosone, G., Sciortino, M., Versari, L.: Measuring the clustering effect of BWT via RLE. Theor. Comput. Sci. (in press)
9. Mantaci, S., Restivo, A., Sciortino, M.: Burrows-Wheeler transform and Sturmian words. Inf. Process. Lett. **86**, 241–246 (2003)
10. Perrin, D., Restivo, A.: Words. In: Bona, M. (ed.) Handbook of Enumerative Combinatorics. CRC Press (2015)

A Permutation on Words in a Two Letter Alphabet

Niccolò Castronuovo[1], Robert Cori[2(✉)], and Sébastien Labbé[3]

[1] Dipartimento di Matematica, Università di Ferrara, Ferrara, Italy
niccol.castronuovo@unife.it
[2] LaBRI, Université de Bordeaux, Bordeaux, France
robert.cori@labri.fr
[3] CNRS, LaBRI, UMR 5800, 33400 Talence, France
sebastien.labbe@labri.fr

Abstract. We define a permutation Γ_n on the set of words with n occurrences of the letter a and $n+1$ occurrences of the letter b. The definition of this permutation is based on a factorization of these words that allows to associate a non crossing partition to them. We prove that all the cycles of this permutation are of odd lengths. We will prove also other properties of this permutation Γ_n, one of them allows to build a family of strips of stamps.

Keywords: Dyck words · Permutations · Strips of stamps

1 Introduction

In this section we define a transformation Γ_n on the family of words on the alphabet $A = \{a, b\}$ having one more occurrence of the letter b than that of the letter a. This transformation will be generalized to all words in A^* in Sect. 6, however the more interesting properties are obtained mainly for the restricted family considered in this section. The terminology and notation for words follow that of Lothaire's books [10,11].

Words in a 2-Letter Alphabet

We consider the mapping δ from the set A^* of words to the ring $\mathcal{Z}$ of integers such that for any word w, $\delta(w)$ is equal to $|w|_a - |w|_b$, where $|w|_x$ denotes the number of occurrences of the letter x in the word w.

A *Dyck word* is a word containing the same number of occurrences of letters a and b and such that no prefix of it contains more occurrences of b than that of a. Hence a Dyck word f is such that $\delta(f) = 0$ and $\delta(u) \geq 0$ for any prefix u of f. We use the convention that the empty word is a Dyck word.

The following very simple Lemma introduces a decomposition of words in A^*. It was called *Catalan decomposition* in [4] since it allows to prove bijectively that the number c_n of Dyck words of length $2n$ satisfies $(n+1)c_n = \binom{2n}{n}$. This shows that c_n is the Catalan number.

© Springer International Publishing AG 2017
S. Brlek et al. (Eds.): WORDS 2017, LNCS 10432, pp. 240–251, 2017.
DOI: 10.1007/978-3-319-66396-8_22

Lemma 1. *For any word f on the two letter alphabet A, there exists a unique decomposition:*

$$f = u_1 \, b \, u_2 \, b \cdots u_p \, b \, w \, a \, v_q \, a \, v_{q-1} \cdots a \, v_1 \tag{1}$$

such that $p, q \geq 0$ and $u_1, u_2, \cdots, u_p, v_1, v_2, \cdots, v_q, w$ are Dyck words.

Proof. It is easy to check that u_i is such that $u_1 \, b \, u_2 \, b \cdots b \, u_i \, b$ is the shortest prefix of f whose image by δ is equal to $-i$. Similarly v_j is such that $a \, v_j \, a \, v_{j-1} \cdots a \, v_1$ is the shortest suffix of f whose image by δ is equal to j. If f is a Dyck word then $p = q = 0$ and $w = f$.

Example 2. This Lemma may be illustrated by the decomposition of the word $f = b \, a \, a \, b \, b \, b \, a \, b \, a$ as:

$$f = b \, (a \, a \, b \, b) \, b \, (a \, b) \, a \,,$$

Here $p = 2, q = 1$, the words u_1, v_1 are both empty, $u_2 = a \, a \, b \, b$, and $w = a \, b$.

Denote A_n the set of words of length $m = 2n + 1$ on the alphabet $A = \{a, b\}$ having n occurrences of the letter a. If f is a word in A_n, the decomposition given by Lemma 1 is such that $p = q + 1$. We denote $\mathcal{D}_n$ the set of words w in A_n such that $\delta(w') \geq 0$ for any prefix $w' \neq w$ of w. These words are Dyck words to which is added an occurrence of the letter b at their ends. Note that a word $f \in A_n$ is in $\mathcal{D}_n$ if and only if in the decomposition given in Lemma 1 we have $p = 1$ and $q = 0$.

The above decomposition also allows to prove:

Lemma 3 (Cyclic Lemma [7]). *Any word w of A_n has exactly one conjugate in $\mathcal{D}_n$ that is a decomposition into two factors $f = u \, v$ such that $v \, u \in \mathcal{D}_n$.*

Proof. Indeed given the decomposition of f in Eq. (1) one has:

$$u = u_1 \, b \, u_2 \, b \cdots u_p \, b \text{ and } v = w \, a \, v_q \, a \, v_{q-1} \cdots a \, v_1$$

Definition of Γ_n

Definition 4. *Let $f = f_1 \, f_2 \cdots f_m$ be a word of A_n, where $f_i \in A$. The pivot of f is the positive integer j equal to the length of $u_1 \, b \, u_2 \, b \cdots b \, u_p \, b$ in the decomposition of f given by Lemma 1. The map Γ_n is the function from A_n into itself such that:*

$$\Gamma_n(f) := \overline{f}_1 \, \overline{f}_2 \cdots \overline{f}_{j-1} \, b \, \overline{f}_{j+1} \cdots \overline{f}_m \tag{2}$$

where the letter $\overline{f}_i$ is b when $f_i = a$ and is a when $f_i = b$.

For our example we have that the pivot of f is 6 and:

$$\Gamma_n(b \, a \, a \, b \, b \, b \, a \, b \, a) = a \, b \, b \, a \, a \, b \, b \, a \, b$$

If f is a word in $\mathcal{D}_n$ then $f = u_1 b$ where u_1 is a Dyck word. Hence we have that the pivot of f is equal to m, the length of f.

Properties of Γ_n

The two main results of this paper are:

Theorem 5. *The mapping Γ_n is a permutation of the words in A_n. The cycles of this permutation have odd lengths.*

Theorem 6. *For each cycle C of the permutation Γ_n let $\pi_i(C)$ be the number of words in C such that i is the pivot of w. Then all the $\pi_i(C)$ are equal.*

The paper is organized as follows. In the next section we consider some combinatorial objects that correspond bijectively to words in A_n. We then prove Theorems 1 and 2, in the next two sections. After that we show how to build a strip of stamps from the pair $w, \Gamma_n(w)$ in Sect. 5. We generalize the transformation Γ such that it acts on all words of $A*$ in Sect. 6 and suggest how the two main results of the paper could be generalized for all words.

2   Combinatorial Objects Corresponding to A_n

In this section we propose three combinatorial objects on which the transformation Γ_n may be applied. The third one will be useful in the proofs given in Sects. 3 and 4.

2-Colored Chord Diagrams with a Pivot

A simple way to represent graphically a word f in A_n is to draw a circle and put $m = 2n + 1$ points numbered in increasing order clockwise. Then color the point i in white when $f_i = a$ and in black when $f_i = b$.

We now show that given a 2-coloring of $2n + 1$ points on a circle such that n are colored white and $n + 1$ are colored black we can draw n chords in the cycle such that each chord joins a white point to a black one and no two chords cross (in particular no two chords have a common vertex). The set of chords P is obtained by drawing iteratively a chord between two points using the following algorithm:

Algorithm: Building the chord diagram from the 2-coloring

1. For any white point i immediately followed by a black one j while turning clockwise (hence $j = i + 1$ or $j = 1$ if $i = 2n + 1$), draw the chord $\{i, j\}$.
2. Then repeat the following action: find a white point i and a black one j, both not adjacent to a chord, such that going clockwise from i to j one meets only points adjacent to a chord then join i to j by a chord.
3. When step 2 cannot be repeated any more one ends with a unique black point, the pivot of f (Fig. 1).

The fact that this algorithm ends with a unique chord diagram, independently of the choices made for the points i, j in step **2** of the algorithm is obtained by considering two facts. The first is the decomposition described in Eq. (1). The second is that in a Dyck word each occurrence of the letter a is matched with an occurrence of a letter b such that the word between them is a Dyck word (which may be empty).

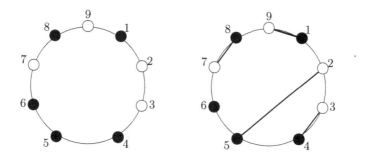

Fig. 1. The diagram of the word $b\,a\,a\,b\,b\,b\,a\,b\,a$ in A_4

Non Crossing Partitions

A chord diagram may be considered as a non crossing partition where all the blocks contain two elements (the points adjacent to a chord), except a block which contains only one element (the pivot). This may also be seen as an involution α in $\mathcal{S}_m$ such that $\alpha(p) = p$ for the pivot p and $\alpha(i) = j$ if $\{i, j\}$ is a chord of the diagram (Fig. 2).

Fig. 2. The partition built from the word f in A_n

The partition associated to the word $b\,a\,a\,b\,b\,b\,a\,b\,a$ is denoted P_0. It consists of 4 blocks with 2 elements and one singleton block: $\{1,9\}, \{2,5\}, \{3,4\}, \{7,8\}, \{6\}$.

Pairs Consisting of a Dyck Word and an Integer

In the rest of the paper, we will often associate to each word f in A_n a pair consisting of a word g in $\mathcal{D}_n$ and an integer k. This is done by considering the decomposition of f given by Lemma 1:

$$f = u_1\,b\,u_2\,b\cdots u_p\,b\,w\,a\,v_q\,a\,v_{q-1}\cdots a\,v_1$$

then using Lemma 3 to obtain the word g in $\mathcal{D}_n$ by:

$$g = w\,a\,v_q\,a\,v_{q-1}\cdots a\,v_1\,u_1\,b\,u_2\,b\cdots u_p\,b.$$

The pivot of f is equal to the length of the word $u_1\,b\,u_2\,b\cdots u_p\,b$.

For our example $f = b\,a\,a\,b\,b\,b\,a\,b\,a$ we have $g = a\,b\,a\,b\,a\,a\,b\,b\,b$ and $k = 6$.

3   Length of the Cycles of Γ_n

In this section we prove Theorem 5. We first obtain the following:

Lemma 7. *Let $w \in A_n$ and $w' = \Gamma_n(w)$ then the pivot p of w is the length of the longest prefix u' of w' that ends with an occurrence of the letter b and for which $\delta(u')$ is the maximum value of δ on those prefixes of w' which end with b.*

Proof. Since w and w' have the same length, to each prefix u of w corresponds a prefix u' of w' of the same length. If the length of u is less than p then $\delta(u') = -\delta(u)$, if this length is greater than or equal to p then $\delta(u') = -\delta(u) - 2$. When $|u| = p$, the value of $\delta(u)$ is minimal. Denote $-k$ this minimal value, then $k \geq 1$ since $\delta(w) = -1$. Since p is the length of the shortest prefix of w attaining the minimal value of $\delta(u)$ it is clear that the maximal value of $\delta(x)$ for a prefix x of w' is $k-1$. This maximal value is attained by the prefix u' of w' of length $p-1$. For the prefixes y of w' ending with an occurrence of the letter b, the maximal value of $\delta(y)$ is $k - 2$. It is attained by the prefix $u'b$ and may be attained by a prefix v' of w' ending with b, but this prefix would be shorter than $u'b$.

Corollary 8. *The transformation Γ_n is a permutation on A_n.*

Proof. From the above Lemma one can easily show that for any word w' in A_n there exists a unique word w such that $w' = \Gamma_n(w)$. Indeed one can find from w' the pivot p of the word w using the characterization given by this Lemma and then obtain w replacing each occurrence of a in w' by an occurrence of b and each occurrence of b which is not in the position of the pivot by an occurrence of a.

We now prove that the cycles of Γ_n have odd lengths. Our main tool is the use of the map γ defined in [2] where it is proved that it is a permutation on words in $\mathcal{D}_n$ which has cycles of odd lengths.

This map γ was defined in [2] as the composition of two involutions. We give here a direct definition, using the Cyclic Lemma in the following way.

Recall that for a word w, the word $\overline{w}$ is obtained from w replacing each occurrence of a by an occurrence of b and each occurrence of b by an occurrence of a.

Definition 9. *For $f = f'b \in \mathcal{D}_n$, the word $\gamma(f)$ is the unique conjugate of the word $\overline{f}'b$ belonging to $\mathcal{D}_n$.*

It is also possible to give a description of γ in terms of what we call the *principal prefix* of a word f in $\mathcal{D}_n$; this principal prefix is the prefix of f for which the map δ attains its maximum value for the first time. If the word f has decomposition uvb where u is its principal prefix, then we have (see Proposition 4 in [2]) $\gamma(f) = \overline{v}b\overline{u}$.

Now we wish to describe the corresponding action of the map Γ_n over the set of pairs (f, i) where f is an element of $\mathcal{D}_n$ and i is an integer such that

$1 \le i \le 2n + 1$. In what follows we will denote $\oplus_m$ the operation on the set $E_n = \{1, 2, \cdots, 2n + 1\}$, where $m = 2n + 1$ defined for any pair i, j by:

$$i \oplus_m j = \begin{cases} i + j & \text{if } i + j \le m \\ i + j - m & \text{otherwise} \end{cases}$$

Notice that this operation is the sum *mod m* where the set $\{0, \ldots, m-1\}$ of the representatives of the classes in $\mathcal{Z}/m\mathcal{Z}$ is replaced by $\{1, \ldots, m\}$.

Proposition 10. *Let* $(f, i) \in \mathcal{D}_n \times \{1, 2, \ldots, m\}$. *Then*

$$\Gamma_n(f, i) = (\gamma(f), i \oplus_m p(f))$$

where $p(f)$ *is the length of the principal prefix of* f.

Proof. In order to compute $\Gamma_n(f, i)$ it is convenient to write $f = u v b$ where the length $|v b|$ of the word $v b$ is equal to i. Then consider $w = v b u$, the word of A_n represented by (f, i). Let $w' = \Gamma_n(w) = \overline{v} b \overline{u}$. Then $\Gamma_n(f, i)$, may be represented by (f', i') where f' is the conjugate of w' belonging to $\mathcal{D}_n$. Since w and $\overline{u} \overline{v} b$ are conjugate, f' is also the conjugate of $\overline{u} \overline{v} b$ belonging to $\mathcal{D}_n$. Hence using the definition of $\gamma(f)$ and the Cyclic Lemma, we have that $f' = \gamma(f)$.

We now determine i'. For that we consider three sequences of integers δ_j, δ'_j and ε_j.

-   The sequence $\delta_1, \delta_2, \cdots, \delta_m$ is such that $\delta_j = \delta(w^{(j)})$, where $w^{(j)}$ is the prefix of w of length j.
-   The sequence of integers $\delta'_1, \delta'_2, \cdots, \delta'_m$ is such that $\delta'_j = \delta(w'^{(j)})$, where $w'^{(j)}$ is the prefix of w' of length j. Since $w = v b u$ and $w' = \overline{v} b \overline{u}$ we have:

$$\delta'_j = \begin{cases} -\delta_j & \text{if } j < i \\ -\delta_j - 2 & \text{if } j \ge i \end{cases} \tag{3}$$

-   The sequence of integers $\varepsilon_1, \varepsilon_2, \cdots, \varepsilon_m$ is such that $\varepsilon_j = \delta(f^{(j)})$, where $f^{(j)}$ is the prefix of f of length j. Since $f = u v b$ and $w = v b u$, we have:

$$\varepsilon_j = \begin{cases} -\delta_i + \delta_{i+j} & \text{if } j \le m - i \\ -\delta_i - 1 + \delta_{j-(m-i)} & \text{if } m - i < j \end{cases} \tag{4}$$

To end the proof Proposition 10 we notice that i is the smallest index where the sequence $\delta_1, \delta_2, \cdots, \delta_m$ attains its minimal value, i' is the smallest index where the sequence $\delta'_1, \delta'_2, \cdots, \delta'_m$ attains its minimal value, $p(f)$ is the smallest index where the sequence $\varepsilon_1, \varepsilon_2, \cdots, \varepsilon_m$ attains its maximal value. Hence we have to play with the Eqs. (3) and (4) to obtain $i' = i \oplus_m p(f)$. This will be done in an extended version of this paper.

We now prove the second part of Theorem 1: As a consequence of the previous theorem we have the following result.

Corollary 11. *The cycles of Γ_n have odd lengths.*

Proof. Consider a word $w \in A_n$ or, equivalently, the corresponding pair $(f, i) \in \mathcal{D}_n \times [1, 2, ..., m]$. Consider the orbit of w under the action of Γ_n. Let ℓ be the length of the cycle containing f under the action of the map γ. Then by Proposition 10 the length of the cycle of Γ_n containing (f, i) is a multiple of ℓ. The main result of [2] states that all the cycles of γ have odd cardinality, hence ℓ is odd. Let $p^*(w)$ be given by:

$$p^*(w) = p(w) \oplus_m p(\gamma(w)) \oplus_m p(\gamma^2(w)) \oplus_m \cdots \oplus_m p(\gamma^{\ell-1}(w)).$$

Then $\Gamma_n^\ell(w, i)$ is equal to $(w, i \oplus_m p^*(w))$. And for any integer k:

$$\Gamma_n^{k\ell}(w, i) = (w, i \oplus_m kp^*(w)).$$

Hence the cycle of Γ_n containing u has length $k\ell$ such that k is the smallest positive integer satisfying $kp^*(w) \equiv 0 \pmod{m}$. This gives for the length of the cycle of Γ_n:

$$\frac{\ell m}{\gcd(m, p^*(w))}$$

This number is odd since ℓ and $m = 2n + 1$ are both odd numbers.

4   Pivots in the Orbits of Γ_n

In this section we prove Theorem 6, this follows directly from:

Proposition 12. *Let w in A_n and C the cycle of Γ_n containing w then the number of times 1 appears as the pivot of an element of C is equal to the number of times 2 appears as the element of a pivot of C.*

Proof. To do that we consider the following automata with outputs. In this automata the transitions translate the action of Γ_n on the first two letters of the word w on which Γ_n acts. The output is equal to the pivot if it is 1 or 2; it is the empty word when the pivot is not equal to one of these values. The states represent the first two letters of the words on which Γ_n acts. If the pivot is not

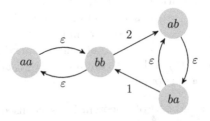

Fig. 3. The automata used in the proof of Proposition 12

1 or 2 then the transition goes from u to $\bar{u}$. The pivot could be 1 only if the first two letters u of the word w are ba it could be 2 only if these first two letters are bb. The transitions in these two cases go to bb and ab. The whole action of a cycle of Γ_n corresponds to a circuit in this automata, clearly each circuit contains the same number of outputs 2 than that of outputs 1.

Example 13. Let $n = 7$ and $w = aaababbbaababbb$, then w may be represented by $(f, 15)$, where $f = w$ since $w \in \mathcal{D}_7$.

The length ℓ of the cycle of γ containing f is 21, and that of Γ_7 containing w is 105. The value of $p^*(w)$ is 3, since $m = 15$ we have $gcd(m, p^*(w)) = 3$. However the values of the first 21 pivots are

$$15, 3, 8, 2, 7, 10, 4, 9, 14, 8, 11, 1, 10, 15, 3, 12, 2, 5, 1, 4, 9$$

the next ones are obtained are obtained adding 3 (mod 15) to these numbers, then adding 6, 9 and 12. The first 13 transitions on the automata are:

$$aa \rightarrow bb \rightarrow aa \rightarrow bb \rightarrow ab \rightarrow ba \rightarrow ab \rightarrow ba \rightarrow ab \rightarrow ba \rightarrow ab \rightarrow ba \rightarrow bb \rightarrow aa$$

And the corresponding output is $2, 1$, this will be repeated 7 times for the whole cycle of Γ_7.

From Proposition 12 we prove Theorem 2:

Corollary 14. *In a cycle of Γ_n all the integers i such that $1 \leq i \leq 2n+1$ appear the same number of times as a pivot.*

Proof. Indeed what was done for $1, 2$ in Proposition 12 may be done for any pair $i, i+1$ showing i and $i+1$ appear the same number of times as a pivot.

Remark 15. Using Γ_n on a word $w \in A_n$ we may define a sequence of integers p_i such that $1 \leq p_i \leq 2n + 1$ each p_i being the pivot of $\Gamma_n^i(w)$. The length of this sequence is the length of the cycle of Γ_n. Notice that in this sequence each j between 1 and $2n+1$ appears the same number of times. It reminds the so called Difference Sets which are and important object in the chapters of Combinatorics concerning Configurations and Block Designs (see [3, 12]).

5   Folding of a Strip of Stamps Associated to an Element of $\mathcal{F}_n$

In this section we show how to build a folding of a strip of stamps using the transformation Γ. The folding problem is a classical in enumerative combinatorics since the contribution of J. Touchard [13], where a presentation of it is given. We denote $E_n = \{1, 2, \cdots, 2n+1\}$. As shown in Sect. 2, to any word f in A_n we associate an element of the set $\mathcal{F}_n$ consisting of non crossing partitions with n blocks with 2 elements and a singleton block. We will build a folding of strip of stamps by using the partitions P and $\Gamma_n(P)$.

We first prove the following result:

Theorem 16. *Let f in A_n and let $P, P' \in \mathcal{F}_n$ be the partitions associated to f and $f' = \Gamma_n(f)$, let $G(P)$ be the graph with vertex set E_n and whose edges are the pairs of vertices (i, j) such that $\{i, j\}$ is a block of P or of P'. Then $G(P)$ is a path joining the pivot of P to the pivot of P'.*

This Theorem may be illustrated on the example of the partition P_0 considered above, which gives the graph drawn in Fig. 4, where the edges coming from P_0 are represented by thick segments while those coming from $\Gamma_n(P_0)$ are represented by dashed ones.

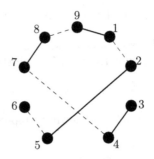

Fig. 4. The graph $G(P_0)$

The proof of this Theorem will be given in an extended version of this paper. It proceeds by induction on n considering a word $w = w' \, a \, b \, w"$ of length $2n + 3$ and the word $w' \, w"$ of length $2n + 1$. The main point in the proof is to describe how to get $G(w)$ from $G(w' \, w")$ and showing that if $G(w' \, w")$ is a path then so is $G(w)$.

Corollary 17. *For any P in $\mathcal{F}_n$, the pair $(P, \Gamma_n(P))$ defines a folding of a strip of (numbered) stamps of length $m = 2n + 1$ where the first stamp is on top.*

Proof. In order to obtain a folding of a strip of stamps like it is usually represented (see [9] page 277) one has to draw a representation of the partitions P and $P' = \Gamma_n(P)$ as follows:

- Consider $m = 2n + 1$ rows numbered from 1 to m from top to bottom of the figure.
- For each block $\{i, j\}$ in P draw a vertical segment on the left of the figure from row i to row j and two horizontal segments in these rows, going in the east direction starting from the end of the vertical segment.
- Draw a horizontal segment in the row p corresponding to the pivot of P Since P is non crossing all these $n + 1$ segments do not intersect.
- Draw a similar set of segments in order to represent P' on the right of the figure with horizontal segments going in the west direction.
- Glue together the horizontal segments on the same row.

This construction is illustrated in Fig. 5.

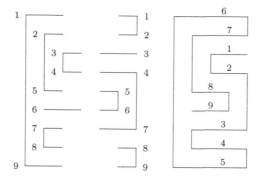

Fig. 5. The folding of the strip of stamps corresponding to P_0

By Theorem 16 the Figure obtained in this way is a path from the row corresponding to the pivot of P to that corresponding to the pivot of P'. It is possible to recover the graph $G(P)$ considering the labels of the rows met along this path.

Notice that given a folding of a strip of stamps of length $2n + 1$ one can build two partitions P and Q reversing the procedure described in the proof. It seems to be difficult to give a characterization of the foldings such that $Q = \Gamma_n(P)$.

From Partitions to Permutations

It is tempting to represent a partition P in $\mathcal{F}_n$ by a permutation α on $m = 2n+1$ elements having n cycles of length 2 corresponding to the blocks of P and a fixed point corresponding to the pivot of P. For instance the partition P_0 gives the permutation α_0 which representation by cycles is:

$$\alpha_0 = (1, 9)\,(2, 5)\,(3, 4)\,(6)\,(7, 8).$$

All the permutations obtained in this way are involutions with a unique fixed point. The fact that P is non crossing can be translated into the fact that the genus of α is 0 (see [5] for the definition of the genus of a permutation). A recent using of non crossing partitions in relation to the Cyclic Lemma is done in [1].

The genus of a permutation α is the non-negative integer $g(\alpha)$ given by:

$$m + 1 - 2g(\alpha) = z(\alpha) + z(\alpha^{-1}\zeta_m),$$

where ζ_m is the the circular permutation such that for all i, $\zeta_m(i) = i+1$ (where $m + 1$ means 1) and where z is the function that associates each permutation with its number of cycles.

Since the permutation α associated to P is an involution with $n+1$ cycles, the previous equation reduces to:

$$2g(\alpha) = n + 1 - z(\alpha\zeta_m).$$

Using the construction in Corollary 17 we have

Proposition 18. *There is a bijection between the foldings of a strip of stamps of length m where the stamp 1 is on top and the pairs of involutions (α, β) on m elements satisfying the following conditions:*

- *if m is odd α and β have a unique fixed point, if m is even α has 2 fixed points and β has no fixed point.*
- *$\alpha(1) = 1$ and $g(\alpha) = g(\beta) = 0$.*
- *The permutation $\alpha\beta$ obtained by the composition of α and β has a unique cycle.*

Proof. We can associate to the two involutions satisfying the first two conditions above two partitions P and Q in $\mathcal{F}_n$. The construction given above builds a figure which represents a folding of a strip of stamps. If m is odd the graph which edges are the blocks of P and Q is a path between the pivot of P and that of Q. this corresponds to the fact that $\alpha\beta$ has a unique cycle. If m is even then the graph is a path between the two fixed points of α.

Meanders studied in [6,8] have strong relations to strip of stamps foldings and can also be described in terms of involutions of genus 0.

6   Generalizing Γ_n to All Words in A^*

We may generalize Γ_n to build a transformation Γ on all words in A^*. We use the decomposition given by Eq. (1) as follows:

Definition 19. *Let f be a word in A^* and let $k = |\delta(w)|$, by Eq. (1) we have: $f = u_1\, b\, u_2\, b \cdots u_p\, b\, w\, a\, v_q\, a\, v_{q-1} \cdots a\, v_1$ Then the pivots of f are given by:*

- *if $p > q$ then $k = p - q$ and the pivots are the k occurrences of b appearing just after the words $u_{p-q+1}, u_{p-q+2}, \dots, u_p$.*
- *if $p < q$ then $k = q - p$ and the pivots are the k occurrences of a appearing just before $v_{q-p+1}, v_{q-p+2}, \dots, v_q$.*
- *if $p = q$ there are no pivots.*

$\Gamma(f)$ is obtained from f by transforming all the occurrences of a into occurrences of b and all the occurrences of b into occurrences of a except for those who are pivots.

Remark 20. It is clear that Γ_n is the restriction of Γ on A_n. The words w such that $\delta(w) = 0$ have no pivot and are in cycles of Γ of lengths 2. The words in a^* and b^* are fixed points. It is not difficult to prove that Γ is a permutation.

Theorem 2 can be generalized to Γ. Its proof uses an automata obtained from that of Fig. 3 by adding for $k > 1$ a loop on state bb with output $1, 2$ when 1 and 2 are both pivots. Theorem 1 is no more true for Γ. However we have the following:

Conjecture. The cycles of Γ containing words of odd lengths are also of odd lengths. Those containing words of even lengths with an odd number of occurrences of a are also of even lengths. Those containing words of even lengths with an even number of occurrences of a may have either odd or even length.

Acknowledgement. We are very grateful for the many valuable comments of the reviewers which improved the article presentation.

References

1. Armstrong, C., Mingo, J.A., Speicher, R., Wilson, J.C.H.: The non-commutative cycle lemma. J. Combin. Theory Ser. A **117**(8), 1158–1166 (2010)
2. Barnabei, M., Bonetti, F., Castronuovo, N., Cori, R.: Some permutations on Dyck words. Theoret. Comput. Sci. **635**, 51–63 (2016)
3. Baumert, L.D.: Cyclic Difference Sets. LNM, vol. 182. Springer, Heidelberg (1971). doi:10.1007/BFb0061260
4. Chottin, L., Cori, R.: Une preuve combinatoire de la rationalité d'une série génératrice associée aux arbres. RAIRO Inform. Théor. **16**(2), 113–128 (1982)
5. Cori, R., Hetyei, G.: Counting genus one partitions and permutations. Sém. Lothar. Combin. **70**, 29 (2013). Art. B70e
6. Di Francesco, P., Golinelli, O., Guitter, E.: Meanders: a direct enumeration approach. Nucl. Phys. B **482**(3), 497–535 (1996)
7. Dvoretzky, A., Motzkin, T.: A problem of arrangements. Duke Math. J. **14**, 305–313 (1947)
8. Lando, S.K., Zvonkin, A.K.: Plane and projective meanders. Theoret. Comput. Sci. **117**(1–2), 227–241 (1993)
9. Legendre, S.: Foldings and meanders. Australas. J. Combin. **58**, 275–291 (2014)
10. Lothaire, M.: Combinatorics on words. In: Encyclopedia of Mathematics and its Applications, vol. 17. Addison-Wesley Publishing Co., Reading (1983)
11. Lothaire, M.: Algebraic combinatorics on words. In: Encyclopedia of Mathematics and its Applications, vol. 90. Cambridge University Press, Cambridge (2002)
12. Storer, T.: Cyclotomy and difference sets. In: Lectures in Advanced Mathematics, no. 2. Markham Publishing Co., Chicago (1967)
13. Touchard, J.: Contribution à l'étude du problème des timbres poste. Can. J. Math. **2**, 385–398 (1950)

Symmetric Dyck Paths and Hooley's Δ-Function

José Manuel Rodríguez Caballero[(✉)]

Université du Québec à Montréal, Montréal, QC, Canada
rodriguez_caballero.jose_manuel@uqam.ca

Abstract. Hooley [6] introduced the function

$$\Delta(n) := \max_{u \in \mathbb{R}} \# \{d|n : \quad u < \log d \leqslant u + 1\},$$

where log is the natural logarithm. Changing the base of the logarithm from e to an arbitrary real number $\lambda > 1$, we define

$$\Delta_\lambda(n) := \max_{u \in \mathbb{R}} \# \{d|n : \quad u < \log_\lambda d \leqslant u + 1\}.$$

The aim of this paper is to express $\Delta_\lambda(n)$ as the height of a symmetric Dyck path defined in terms of the distribution of the divisors of n.

Keywords: Dyck path · Palindrome · Hooley's Δ-function

1 Introduction

Hooley's Δ-function,

$$\Delta(n) := \max_{u \in \mathbb{R}} \# \{d|n : \quad u < \log d \leqslant u + 1\},$$

was introduced in [6], where log is the natural logarithm. The applications of Hooley's Δ-function are widely spread in number theory, from Erdös's statistical theory of the distribution of divisors of a normal integer (see [4]) to Waring's problem (see [6]). This function corresponds to the integer sequence **A226898** in [10]. It is natural to extend Hooley's Δ-function to an arbitrary real number $\lambda > 1$ by means of the formula

$$\Delta_\lambda(n) := \max_{u \in \mathbb{R}} \# \{d|n : \quad u < \log_\lambda d \leqslant u + 1\}, \tag{1}$$

where $\log_\lambda d := \frac{\log d}{\log \lambda}$.

Höft [5] used symmetric Dyck paths in number theory in order to study the sum of divisors $\sigma(n)$ by means of what he[1] calls the "symmetric representation

---

[1] Höft described his own research as follows (personal communication, March 24, 2017): "My work in this context has been to find formulas, develop Mathematica code to compute the sequences and their associated irregular triangles, and use those to computationally verify conjectures for initial segments of some sequences. In addition, I tried to find 'elementary' arguments for conjectures stated in OEIS about the 'symmetric representation of sigma' and to prove in special cases that the area defined by two adjacent Dyck paths actually equals sigma (thus justifying the phrase used in OEIS in those cases)".

© Springer International Publishing AG 2017
S. Brlek et al. (Eds.): WORDS 2017, LNCS 10432, pp. 252–261, 2017.
DOI: 10.1007/978-3-319-66396-8_23

of sigma". In the present paper, we will introduce a symmetric Dyck path $\langle\!\langle n \rangle\!\rangle_\lambda$ associated to an integer $n \geqslant 1$ and a real number $\lambda > 1$ by means of the following definition.

Definition 1. *Consider a real number $\lambda > 1$ and a 2-letter alphabet $\Sigma = \{a, b\}$.*

(i) Given a finite set of positive real numbers S, the λ-class of S is the word

$$\langle\!\langle S \rangle\!\rangle_\lambda := w_0 \, w_1 \, w_2 \dots w_{k-1} \in \Sigma^*, \tag{2}$$

such that each letter is given by

$$w_i := \begin{cases} a & \text{if } \mu_i \in S, \\ b & \text{if } \mu_i \in \lambda S, \end{cases} \tag{3}$$

for all $0 \leqslant i \leqslant k - 1$, where μ_0, μ_1, ..., μ_{k-1} are the elements of the symmetric difference $S \triangle \lambda S$ written in increasing order, i.e.

$$\lambda S := \{\lambda s : \quad s \in S\},$$
$$S \triangle \lambda S = \{\mu_0 < \mu_1 < \dots < \mu_{k-1}\}. \tag{4}$$

(ii) If S is the set of divisors of n, then we will write $\langle\!\langle n \rangle\!\rangle_\lambda := \langle\!\langle S \rangle\!\rangle_\lambda$. The word $\langle\!\langle n \rangle\!\rangle_\lambda$ will be called the λ-class of n.

The aim of this paper is to prove that $\langle\!\langle n \rangle\!\rangle_\lambda$ and $\Delta_\lambda(n)$ are related by the following theorem.

Theorem 2. *Let $\lambda > 1$ be a real number. For any integer $n \geqslant 1$ the following statements hold.*

(i) The word $\langle\!\langle n \rangle\!\rangle_\lambda$ is a symmetric Dyck path.

(ii) The height of the Dyck path $\langle\!\langle n \rangle\!\rangle_\lambda$ is height $(\langle\!\langle n \rangle\!\rangle_\lambda) = \Delta_\lambda(n)$.

(iii) If $\frac{n}{\lambda} \notin \mathbb{Z}$ then the cardinality of the set of solutions of the inequalities

$$d_1 < d_2 < \dots < d_h < \lambda \, d_1, \tag{5}$$

where $d_1, d_2, ..., d_h$ are divisors of n and $h = \Delta_\lambda(n)$, coincides with the number of maximum-height-peaks in the Dyck path $\langle\!\langle n \rangle\!\rangle_\lambda$.

2   Preliminaries

Throughout this paper we will mainly work with the 2-letter alphabet $\Sigma = \{a, b\}$, although we will consider another letter in the proof of Proposition 12. The definitions that we will introduce in this section are well-known in the theory of formal languages and Dyck paths.

A word $w \in \Sigma^*$ is said to be a *Dyck path* (or a *Dyck word*) if

(i) $|w|_a = |w|_b$,
(ii) $|u|_a \geqslant |u|_b$, for each prefix u of w,

where $|w|_x$ is the number of occurrences of the letter $x \in \Sigma$ in the word w.

Dyck paths are just *well-formed parentheses*, but we preferred to use the symbols a and b in place of the parentheses in order to avoid ambiguities in the notation. Nevertheless, if there is no risk of confusion, we will use "(" and ")" in place of a and b, respectively.

The (graphical) representation of a Dyck path $w \in \Sigma^*$ is the lattice path in the complex plane, starting at $z = 0$ and ending at $z = |w|$, consisting of up steps $1 + \sqrt{-1}$ and down steps $1 - \sqrt{-1}$ associated to the letters a and b respectively, reading the word w from left to right. An example of this graphical representation is shown in Fig. 1. We prefer to use the term "Dyck path" rather than "Dyck word" in order to follow Höft's terminology from [5]. Also, it is more natural to talk about the "height of a lattice path" than the "height of a word".

The *height* of a Dyck path w is

$$\text{height}(w) := \max\{|u|_a - |u|_b : \quad u \text{ prefix of } w\}.$$

A *maximum-height-peak* in a Dyck path w is a prefix u of w satisfying $|u|_a - |u|_b = \text{height}(w)$.

3   Auxiliary Results About Arbitrary Dyck Paths

The word $\langle\!\langle S \rangle\!\rangle_\lambda$ given in Definition 1(i) will be our main tool to study the inequalities

$$s_1 < s_2 < \ldots < s_h < \lambda s_1, \tag{6}$$

with $s_1, s_2, \ldots, s_h$ belonging to a finite set S of positive real numbers. We will distinguish two cases, corresponding to the values of λ.

Definition 3. *Let S be a finite set of positive real numbers. A real number $\lambda > 1$ is said to be* regular *(with respect to S) if S and λS are disjoint. Otherwise, λ is said to be* singular *(with respect to S).*

Remark 4. Notice that for any finite set of real numbers S, there are only finitely many singular values of λ.

It is natural to extend the definition of the generalization of Hooley's Δ-function given in (1) as follows.

Definition 5. *Let $\lambda > 1$ be a real number. Define the* Hooley Δ_λ-function *of a finite set S of positive real numbers by the expression*

$$\Delta_\lambda(S) := \max\{h : \quad \exists s_1, s_2, \ldots, s_h \in S; \quad s_1 < s_2 < \ldots < s_h < \lambda s_1\}.$$

Lemma 6. *For each integer $n \geqslant 1$,*

$$\Delta_\lambda(n) = \Delta_\lambda(S),$$

where S is the set of divisors of n and $\Delta_\lambda(n)$ is given by (1).

Proof. In virtue of (1), there exist $d_1, d_2, ..., d_h \in S$, with $h = \Delta_\lambda(n)$, satisfying

$$\lambda^u < d_1 < d_2 < d_3 < ... < d_h \leqslant \lambda^{u+1}, \tag{7}$$

for some $u \in \mathbb{R}$. The inequality $\lambda^u < d_1$ implies $d_h \leqslant \lambda^{u+1} < \lambda d_1$. So, the inequalities (5) hold. Using Definition 5 we conclude that $\Delta_\lambda(n) \leqslant \Delta_\lambda(S)$.

In virtue of Definition 5, there exist $d_1, d_2, ..., d_h \in S$, with $h = \Delta_\lambda(S)$, satisfying (5). For all $\epsilon > 0$ small enough (see Remark 13), the inequalities (7) hold for $u = \log_\lambda(d_1 - \epsilon)$. Hence, $\Delta_\lambda(S) \leqslant \Delta_\lambda(n)$.

Therefore, $\Delta_\lambda(n) = \Delta_\lambda(S)$. $\square$

Lemma 7. *Let S be a finite set of positive real numbers. Consider an arbitrary real number $\lambda > 1$ and suppose that λ is regular. Consider μ_i, with $0 \leqslant i \leqslant k - 1$, from (4). Let $u = w_0 w_1 ... w_j$ be a prefix of $\langle\!\langle S \rangle\!\rangle_\lambda = w_0 w_1 ... w_{k-1}$, where each w_i, with $0 \leqslant i \leqslant k - 1$, is a letter. Define the sets*

$$A_u := \{i \in \mathbb{Z} : \quad 0 \leqslant i \leqslant j \text{ and } w_i = a\},$$
$$B_u := \{i \in \mathbb{Z} : \quad 0 \leqslant i \leqslant j \text{ and } w_i = b\}.$$

(i) *The map $\omega : B_u \longrightarrow A_u$ given by $i \mapsto i'$, where $\mu_i = \lambda \mu_{i'}$, is well-defined.*
(ii) *The map ω is injective.*
(iii) *There are exactly $|u|_a - |u|_b$ integer values of i satisfying $0 \leqslant i \leqslant j$, $\mu_i \in S$ and $\mu_j < \lambda \mu_i$.*

Proof. The sets S and λS are disjoint, because λ is regular.
Proof of (i). Take $i \in B_u$. By definition of B_u, $\mu_i \in \lambda S$, i.e. $\mu_i = \lambda s$ for some $s \in S$. Using that $S \triangle \lambda S = S \cup \lambda S$, there is some integer i' satisfying $0 \leqslant i' < i$ and $\mu_{i'} = s$ (this is no necessarily true if λ is singular). In particular, $i' \in A_u$.

Now, suppose that there is another integer i'' satisfying $\mu_i = \lambda \mu_{i''}$. Then $\lambda \mu_{i''} = \lambda \mu_{i'}$. So, $i'' = i'$. Therefore, the function ω is well-defined.
Proof of (ii). Suppose that $\omega(i_1) = \omega(i_2)$. Then $\lambda \mu_{i_1} = \lambda \mu_{i_2}$. So, $i_1 = i_2$. Hence, ω is injective.
Proof of (iii). By definition of A_u, there are exactly $|u|_a$ integers i such that $0 \leqslant i \leqslant j$, $\mu_i \in S$. The equality $\#B_u = \#\omega(B_u)$ implies that there are exactly $|u|_b$ integers i such that $0 \leqslant i \leqslant j$, $\mu_i \in S$ and $\lambda \mu_i \leqslant \mu_j$.

Therefore, there are exactly $|u|_a - |u|_b$ integers i such that $0 \leqslant i \leqslant j$, $\mu_i \in S$ and $\mu_j < \lambda \mu_i$. $\square$

The next result provides a link between the height of the Dyck path $\langle\!\langle S \rangle\!\rangle_\lambda$ and the inequalities (6), with $s_1, s_2, ..., s_h \in S$, provided that λ is regular.

Lemma 8. *Let S be a finite set of positive real numbers. Consider an arbitrary real number $\lambda > 1$. Suppose that λ is regular.*

(i) The word $\langle\!\langle S \rangle\!\rangle_\lambda$ is a Dyck path.

(ii) For any prefix u of $\langle\!\langle S \rangle\!\rangle_\lambda$ there is a solution $s_1, s_2, ..., s_h \in S$ of the inequalities (6) with $|u|_a - |u|_b = h$.

(iii) The height of the Dyck path $\langle\!\langle S \rangle\!\rangle_\lambda$ is height $(\langle\!\langle S \rangle\!\rangle_\lambda) = \Delta_\lambda(S)$.

(iv) The number of maximum-height-peaks in $\langle\!\langle S \rangle\!\rangle_\lambda$ coincides with the cardinal of the set of solutions of the inequalities (6) with $s_1, s_2, ..., s_h \in S$ and $h = \Delta_\lambda(S)$.

Proof. We have $S \cup \lambda S = S \triangle \lambda S$, because λ is regular. Consider μ_i, with $0 \leqslant i \leqslant k - 1$ from (4).

Proof of (i). Define the function $\alpha : S \cup \lambda S \longrightarrow S \cup \lambda S$ by $s \mapsto \lambda s, \lambda s \mapsto s$, where s is an arbitrary element from S. Notice that α is well-defined because S and λS are disjoint. It is clear that α is an involution such that $\alpha(S) = \lambda S$ and $\alpha(\lambda S) = S$. Hence, the number of $0 \leqslant i \leqslant k - 1$ satisfying $\mu_i \in \lambda S$ coincides with the number of $0 \leqslant i \leqslant k - 1$ satisfying $\mu_i \in S$. Applying (3) we conclude that $|\langle\!\langle S \rangle\!\rangle_\lambda|_a = |\langle\!\langle S \rangle\!\rangle_\lambda|_b$.

By Lemma 7, $|u|_a \geqslant |u|_b$ for each nonempty prefix u of $\langle\!\langle S \rangle\!\rangle_\lambda$. Therefore, $\langle\!\langle S \rangle\!\rangle_\lambda$ is a Dyck path.

Proof of (ii). Take a nonempty prefix $u = w_0 w_1 ... w_j$ of $\langle\!\langle S \rangle\!\rangle_\lambda = w_0 w_1 ... w_{k-1}$, where each w_i is a letter. In virtue of Lemma 7, there are precisely $h := |u|_a - |u|_b$ elements $s_1, s_2, ..., s_h \in S$ such that

$$s_1 < s_2 < ... < s_h \leqslant \mu_j < \lambda s_1 < \lambda s_2 < ... < \lambda s_h.$$

In particular, the inequalities (6) hold.

Proof of (iii). Combining the statement (ii) of this lemma and Definition 5, we obtain that height $(\langle\!\langle S \rangle\!\rangle_\lambda) \leqslant \Delta_\lambda(S)$.

By Definition 5, there are $s_1, s_2, ..., s_h \in S$, with $h = \Delta_\lambda(S)$, satisfying the inequalities (6). Let j be the unique integer such that $\mu_j = s_h$. Define the nonempty prefix $u := w_0 w_1 ... w_j$ of $\langle\!\langle S \rangle\!\rangle_\lambda = w_0 w_1 ... w_{k-1}$, where each w_i is a letter. The inequality $|u|_a - |u|_b \geqslant h$ follows by Lemma 7. Hence, height $(\langle\!\langle S \rangle\!\rangle_\lambda) \geqslant \Delta_\lambda(S)$.

Therefore, height $(\langle\!\langle S \rangle\!\rangle_\lambda) = \Delta_\lambda(S)$.

Proof of (iv). Combining the parts (ii) and (iii) of this lemma, it follows that the cardinality of the set of solutions of the inequalities (6) with $s_1, s_2, ..., s_h \in S$ and $h = \Delta_\lambda(S)$, is at least the number of maximum-height-peaks in $\langle\!\langle S \rangle\!\rangle_\lambda$.

Each solution of the inequalities (6) with $s_1, s_2, ..., s_h \in S$ and $h = \Delta_\lambda(S)$ corresponds to a unique value of s_h (this is no longer true if $h < \Delta_\lambda(S)$). Consider one of these solutions and let j be the unique integer such that $\mu_j = s_h$. Define the nonempty prefix $u := w_0 w_1 ... w_j$ of $\langle\!\langle S \rangle\!\rangle_\lambda = w_0 w_1 ... w_{k-1}$, where each w_i is a letter. In virtue of Lemma 7, u is a maximum-height-peak in $\langle\!\langle S \rangle\!\rangle$. Hence, the cardinality of the set of solutions of the inequalities (6) with $s_1, s_2, ..., s_h \in S$ and $h = \Delta_\lambda(S)$ is at most the number of maximum-height-peaks in $\langle\!\langle S \rangle\!\rangle_\lambda$.

Therefore, the cardinality of the set of solutions of the inequalities (6) with $s_1, s_2, ..., s_h \in S$ and $h = \Delta_\lambda(S)$ coincides with the number of maximum-height-peaks in $\langle\!\langle S \rangle\!\rangle_\lambda$.   □

Example 9. For $S = \{1, 2, 3, 6, 7\}$ and $\lambda = e$ (Euler's number), the Dyck path $\langle\!\langle S \rangle\!\rangle_\lambda$ is represented in Fig. 1.

Fig. 1. Representation of $\langle\!\langle 1, 2, 3, 6, 7 \rangle\!\rangle_e = (\,(\,)\,(\,)\,(\,(\,)\,)\,)$

This Dyck word can be computed using the following inequalities.

$$
\begin{array}{llllllllll}
S \,\triangle\, e\,S & = \{1 < & 2 < & e\,1 < & 3 < & e\,2 < & 6 < & 7 < & e\,3 < & e\,6 < e\,7\} \\
\langle\!\langle 1, 2, 3, 6, 7 \rangle\!\rangle_e = & a & a & b & a & b & a & a & b & b & b
\end{array}
$$

Lemma 10. *The step function* $]1, +\infty[\longrightarrow \mathbb{Z}$ *given by* $\lambda \mapsto \Delta_\lambda(S)$, *is continuous from the left, i.e. for any* $\lambda > 1$,

$$
\lim_{\lambda' \to \lambda^-} \Delta_{\lambda'}(S) = \Delta_\lambda(S). \tag{8}
$$

Proof. Let $\lambda > 1$ be a fixed real number. We recall that the expression (8) means that for any $\epsilon > 0$ there is $\delta > 0$ such that for any $\lambda' \in]1, +\infty[$, if $\lambda - \delta < \lambda' < \lambda$ then $|\Delta_\lambda(S) - \Delta_{\lambda'}(S)| < \epsilon$.

Take an arbitrary $\lambda' \in]1, \lambda[$. By Definition 5, there are $s_1, s_2, ..., s_h \in S$, with $h = \Delta_{\lambda'}(S)$, satisfying

$$
s_1 < s_2 < ... < s_h < \lambda' s_1. \tag{9}
$$

Using the fact that $\lambda' < \lambda$, we obtain (6). Hence, $\Delta_\lambda(S) \geqslant \Delta_{\lambda'}(S)$.

By Definition 5, there are $s_1, s_2, ..., s_h \in S$, with $h = \Delta_\lambda(S)$, satisfying (6). For all real numbers λ' near enough (see Remark 13) to λ and constrained by the inequality $\lambda' < \lambda$, we guarantee that $s_h < \lambda' s_1 < \lambda s_1$. So, (9) follows. Hence, $\Delta_\lambda(S) \leqslant \Delta_{\lambda'}(S)$.

We conclude that $\Delta_\lambda(S) = \Delta_{\lambda'}(S)$ holds for all real numbers λ' near enough to λ and constrained by the inequality $\lambda' < \lambda$. Therefore, the function $\lambda \mapsto \Delta_\lambda(S)$ is continuous from the left.   □

Lemma 11. *If* $w := u\,b\,a\,v$ *is a Dyck path, for two word* $u, v \in (\Sigma)^*$, *then* $w' := u\,v$ *is also an Dyck path and* height$(w) =$ height(w').

Proof. Suppose that w is a Dyck path. We have that $|w|_a - |w|_b = 0$, because w is a Dyck path. Using that $|w'|_a = |w|_a - 1$ and $|w'|_b = |w|_b - 1$, the equality $|w'|_a - |w'|_b = |w|_a - |w|_b = 0$ follows.

Let p' be a nonempty prefix of w'. Suppose that p' is a prefix of u. Then, the word $p := p'$ is a prefix of w. So $|p|_a - |p|_b \geqslant 0$, because $|w|$ is a Dyck word. Hence, $|p'|_a - |p'|_b = |p|_a - |p|_b \geqslant 0$.

Now, suppose that p' is not a prefix of u. Then $p' = u\,\hat{v}$ for some nonempty prefix $\hat{v}$ of v. So, the word $p := u\,b\,a\,\hat{v}$ is a prefix of w. Using that w is a Dyck path, we obtain $|p|_a - |p|_b \geqslant 0$. The equalities $|p'|_a = |p|_a - 1$ and $|p'|_b = |p|_b - 1$ imply that $|p'|_a - |p'|_b = |p|_a - |p|_b \geqslant 0$.

Therefore, w' is a Dyck path.

The only prefixed of w that we did not used in the above argument are $p = u\,b$ and $p = u\,b\,a$, but in both cases we have $\mathrm{height}(p) \leqslant \mathrm{height}(w)$. Hence, $\mathrm{height}(w) = \mathrm{height}(w')$. $\qquad\square$

Proposition 12. *Consider a finite set of positive real numbers S. For any real number $\lambda > 1$, the following statements hold.*

(i) The word $\langle\!\langle S \rangle\!\rangle_\lambda$ is a Dyck path.
(ii) The height of the Dyck path $\langle\!\langle S \rangle\!\rangle_\lambda$ is $\mathrm{height}\left(\langle\!\langle S \rangle\!\rangle\right) = \Delta_\lambda(S)$.

Proof. The proposition holds if λ is regular, because of Lemma 8. From now on, assume that λ is singular.

Consider the 3-letter alphabet $\Gamma = \{a, b, c\}$. Define the word

$$[\![S]\!]_\lambda := u_0\, u_1\, u_2 \ldots u_{r-1} \in \Gamma^*,$$

whose letters are given by

$$u_i := \begin{cases} a \text{ if } \nu_i \in S \setminus (\lambda\, S), \\ b \text{ if } \nu_i \in (\lambda\, S) \setminus S, \\ c \text{ if } \nu_i \in S \cap \lambda\, S, \end{cases}$$

for all $0 \leqslant i \leqslant r - 1$, where $\nu_0, \nu_1, \ldots, \nu_{r-1}$ are the elements of the union $S \cup \lambda\, S$ written in increasing order, i.e.

$$S \cup \lambda\, S = \{\nu_0 < \nu_1 < \ldots < \nu_{r-1}\}.$$

By Definition 1, the word $\langle\!\langle S \rangle\!\rangle_\lambda$ can be obtained from $[\![S]\!]_\lambda$ just deleting all the occurrences of the letter c. In virtue of the continuity and the monotony of the function $]1, +\infty[\longrightarrow]0, +\infty[$ given by $\lambda \mapsto \lambda\, s$, with $s \in S$ arbitrary, it follows that, for all $\lambda' \in]1, \lambda[$, the word $\langle\!\langle S \rangle\!\rangle_{\lambda'}$ can be obtained from $[\![S]\!]_\lambda$ just substituting all the occurrences of the letter c by the word $b\,a$, provided that λ' is near enough to λ (see Remark 13). So, the word $\langle\!\langle S \rangle\!\rangle_\lambda$ can be obtained from the word $\langle\!\langle S \rangle\!\rangle_{\lambda'}$ just deleting some factors of the form $b\,a$, for any $\lambda' \in]1, \lambda[$ near enough to λ. By Lemma 11, $\langle\!\langle S \rangle\!\rangle_\lambda$ is a Dyck path and $\mathrm{height}\left(\langle\!\langle S \rangle\!\rangle_\lambda\right) = \mathrm{height}\left(\langle\!\langle S \rangle\!\rangle_{\lambda'}\right)$, for each $\lambda' \in]1, \lambda[$ near enough to λ.

By Lemma 8, $\mathrm{height}\left(\langle\!\langle S \rangle\!\rangle_{\lambda'}\right) = \Delta_{\lambda'}(S)$ for each regular value $\lambda' \in]1, +\infty[$. By Lemma 10, $\Delta_{\lambda'}(S) = \Delta_\lambda(S)$ for all $\lambda' \in]1, \lambda[$ near enough to λ. Therefore, $\mathrm{height}\left(\langle\!\langle S \rangle\!\rangle_\lambda\right) = \Delta_\lambda(S)$. $\qquad\square$

Remark 13. Formalization is only required in analysis when we are dealing with pathological objects, e.g. a conditional convergent series which is not absolutely convergent, a nonuniformly convergent sequence of functions which is pointwise convergent, etc. When there is no risk of confusion, as in all analytical arguments throughout this paper, we can use informal statements like "$\epsilon > 0$ small enough" and "λ' is near enough to λ". These meta-mathematical statements can be easily formalized using $\epsilon - \delta$ language, but this formalization makes the argument unnecessarily harder to read. A way to formalize these statements preserving their simplicity is to use nonstandard analysis.

4   Symmetric Dyck Paths

Definition 14. *A Dyck path $w \in \Sigma^*$ is symmetric if for each $0 \leqslant i \leqslant k - 1$, either*

(i) $w_i = a$ and $w_{k-1-i} = b$, or
(ii) $w_i = b$ and $w_{k-1-i} = a$,

where $w = w_0 \, w_1 \ldots w_{k-1}$ and each w_i is a letter for all $0 \leqslant i \leqslant k - 1$.

Using the theory of f-palindromes (see [1]), we can rephrase Definition 14 as follows: a Dyck path w is symmetric if and only if w is an f-palindrome (i.e. $\tilde{w} = f(w)$), where $\tilde{w}$ is the mirror image of w and f is the morphism $a \mapsto b$ and $b \mapsto a$.

We now proceed to prove our main result.

Proof (of Theorem 2). Let S be the set of divisors of n.
Proof of (i). By Proposition 12, $\langle\!\langle n \rangle\!\rangle$ is a Dyck path.
    The function $\beta :]0, +\infty[\longrightarrow]0, +\infty[$, given by $x \mapsto \lambda \frac{n}{x}$, is strictly decreasing (in particular β is injective). It is straightforward to check that $\beta(S) = \lambda S$, $\beta(\lambda S) = S$ and $\beta(S \cap \lambda S) = S \cap \lambda S$.
    Consider μ_i, with $0 \leqslant i \leqslant k - 1$, from (4). We have that $\beta(\mu_i) = \mu_{k-1-i}$ for all $0 \leqslant i \leqslant k - 1$, because β restricted to $S \triangle \lambda S$ is a strictly decreasing bijection $S \triangle \lambda S \longrightarrow S \triangle \lambda S$. Furthermore, for any i, satisfying $0 \leqslant i \leqslant k - 1$, either

(i) for some $d|n$ we have $\mu_i = d \in S$ and $\mu_{k-1-i} = \beta(d) \in \lambda S$, or
(ii) for some $d|n$ we have $\mu_i = \lambda d \in \lambda S$ and $\mu_{k-1-i} = \beta(\lambda d) \in S$,

because $\beta(S \cap (S \triangle \lambda S)) = (\lambda S) \cap (S \triangle \lambda S)$ and $\beta((\lambda S) \cap (S \triangle \lambda S)) = S \cap (S \triangle \lambda S)$.
    In virtue of (3), for all i, satisfying $0 \leqslant i \leqslant k - 1$, either

(i) $w_i = a$ and $w_{k-1-i} = b$, or
(ii) $w_i = b$ and $w_{k-1-i} = a$,

where $\langle\!\langle s \rangle\!\rangle = w_0\, w_1 \ldots w_{k-1}$ and each w_i is a letter for all $0 \leqslant i \leqslant k-1$. Using Definition 14, we conclude that the Dyck path $\langle\!\langle n \rangle\!\rangle_\lambda$ is symmetric.

Proof of (ii). By Proposition 12, the height of $\langle\!\langle S \rangle\!\rangle_\lambda$ is $\Delta_\lambda(S)$. By Lemma 6, $\overline{\Delta_\lambda(S)} = \Delta_\lambda(n)$. Hence, the height of $\langle\!\langle n \rangle\!\rangle$ is precisely $\Delta_\lambda(n)$.

Proof of (iii). Let $\lambda > 1$ be a real number such that $\frac{n}{\lambda} \notin \mathbb{Z}$. Suppose that λ is singular. There are two divisors of n, denoted d and d', satisfying $d = \lambda d'$. Then $\frac{n}{\lambda} = d' \frac{n}{d} \in \mathbb{Z}$. By reductio ad absurdum, λ is regular.

In virtue of Lemma 8, the number of maximum-height-peaks in $\langle\!\langle n \rangle\!\rangle_\lambda$ coincides with the cardinality of the set of solutions of the inequalities (5) with $h = \Delta_\lambda(n)$, where $d_1, d_2, \ldots, d_h$ are divisors of n.   □

Example 15. Using Proposition 2 we can derive that $\Delta(126) = 4$, because the height of the symmetric Dyck path $\langle\!\langle 126 \rangle\!\rangle_e$ is 4. This Dyck path is represented in Fig. 2.

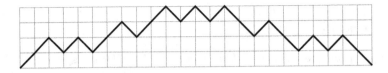

Fig. 2. Representation of $\langle\!\langle 126 \rangle\!\rangle_e = (\,(\,)\,(\,)\,(\,(\,)\,(\,(\,)\,(\,)\,(\,)\,)\,(\,)\,)\,(\,)\,(\,)\,)$.

The three maximum-height-peaks in $\langle\!\langle 126 \rangle\!\rangle_e$ correspond to the following solutions of the inequalities (5), with $h = \Delta(n)$,

$$6 < 7 < 9 < 14 < e\,6,$$
$$7 < 9 < 14 < 18 < e\,7,$$
$$9 < 14 < 18 < 21 < e\,9.$$

5   Final Remarks

Kassel and Reutenauer [7] proved that, for any integer $n \geqslant 1$, there is a unique polynomial $P_n(q)$ such that for any prime power $q = p^\ell$, the number of ideals of codimension[2] n of the group algebra $\mathbb{F}_q[\mathbb{Z} \oplus \mathbb{Z}]$ is precisely $(q-1)^2 P_n(q)$, where $\mathbb{Z} \oplus \mathbb{Z}$ is the free abelian group of rank 2 and $\mathbb{F}_q$ is the finite field with q elements.

The polynomials $P_n(q)$ are related to identity (9.2) in [3] (see [8]),

$$\frac{(q)_\infty^2}{(t^{-2}q)_\infty^2\,(t^2 q)_\infty^2} = 1 + (t - t^{-1}) \sum_{N \geqslant 1} q^N \sum_{\omega \mid N} \left(t^{2N/\omega - \omega} - t^{-2N/\omega + \omega} \right),$$

---

[2] Let k be a field and $\mathcal{R}$ be a k-algebra. The *codimension* of an ideal I of $\mathcal{R}$ is the dimension of the quotient $\mathcal{R}/I$ as a vector space over k.

where ω runs by the *odd* divisors of N and $(a)_\infty = \prod_{n \geqslant 0} (1 - a\,q^n)$ is the q-Pochhammer symbol. Notice that we can write

$$\sum_{\omega \mid N} \left(t^{2N/\omega - \omega} - t^{-2N/\omega + \omega} \right) = \sum_{i=0}^{k-1} (-1)^{\mu_i - 1} \frac{t^{\mu_{k-1-i}}}{t^{\mu_i}},$$

where each μ_i, with $0 \leqslant i \leqslant k - 1$, is from (4) taking S as the set of divisors of N and $\lambda = 2$.

Combining this observation with Theorem 2 and the generating function of $P_n(q)$ due to Kassel and Reutenauer [8], it follows that the largest coefficient of $P_n(q)$ is precisely[3] $\Delta_2(n)$. A complete proof of this result, without explicitly involving symmetric Dyck paths, can be found in [9].

Acknowledgement. The author thanks S. Brlek, C. Kassel and C. Reutenauer for they valuable comments and suggestions concerning this research. Also, the author want to express his gratitude to H. F. W. Höft for the useful exchanges of information.

References

1. Blondin-Massé, A., Brlek, S., Garon, A., Labbé, S.: Combinatorial properties of f-palindromes in the Thue-Morse sequence. Pure Math. Appl. **19**(2–3), 39–52 (2008)
2. Erdös, P., Nicolas, J.L.: Méthodes probabilistes et combinatoires en théorie des nombres. Bull. Sci. Math. **2**, 301–320 (1976)
3. Fine, N.J.: Basic Hypergeometric Series and Applications, vol. 27. American Mathematical Soc., Providence (1988)
4. Hall, R.R., Tenenbaum, G.: Divisors. Cambridge Tracts in Mathematics, vol. 90. Cambridge University Press, Cambridge (1988).
5. Höft, H.F.W.: On the symmetric spectrum of odd divisors of a number. https://oeis.org/A241561/a241561.pdf
6. Hooley, C.: On a new technique and its applications to the theory of numbers. Proc. London Math. Soc. **3**(1), 115–151 (1979)
7. Kassel, C., Reutenauer, C.: Counting the ideals of given codimension of the algebra of Laurent polynomials in two variables. arXiv preprint arXiv:1505.07229 (2015)
8. Kassel, C., Reutenauer, C.: Complete determination of the zeta function of the Hilbert scheme of n points on a two-dimensional torus. arXiv preprint arXiv:1610.07793 (2016)
9. Rodríguez Caballero, J.M.: On a function introduced by Erdös and Nicolas (To appear)
10. Sloane, N.J.A., et al.: The on-line encyclopedia of integer sequences (2012)

---

[3] The function $\Delta_2(n)$ was introduced by Erdös and Nicolas [2], using the notation $F(n)$, before Hooley's paper [6].

Author Index

Printed in the United States
By Bookmasters